ALSO BY SHARON WEINBERGER

A Nuclear Family Vacation

Imaginary Weapons

THE IMAGINEERS OF WAR

THE IMAGINEERS OF WAR

THE UNTOLD HISTORY OF **DARPA**,

THE PENTAGON AGENCY THAT CHANGED THE WORLD

Sharon Weinberger

ALFRED A. KNOPF | NEW YORK | 2017

THIS IS A BORZOI BOOK
PUBLISHED BY ALFRED A. KNOPF

Published in the United States by Alfred A. Knopf,
a division of Penguin Random House LLC, New York,
and distributed in Canada by Random House of Canada,
a division of Penguin Random House Canada Limited, Toronto.
www.aaknopf.com

Knopf, Borzoi Books, and the colophon are registered trademarks
of Penguin Random House, LLC.

Page 477 constitutes an extension of this copyright page.

Names: Weinberger, Sharon, author.
Title: The imagineers of war : the untold history of DARPA,
the Pentagon agency that changed the world / by Sharon Weinberger.
Other titles: Untold history of DARPA, the Pentagon agency that changed the world
Description: New York : Alfred A. Knopf, [2017]
Identifiers: LCCN 2016017245 (print) | LCCN 2016017718 (ebook) | ISBN 9780385351799
(hardcover) | ISBN 9780385351805 (ebook)
Subjects: LCSH: United States. Defense Advanced Research Projects Agency—History. |
Military research—United States. | Military art and science—Technological
innovations—United States. | Science and state—United States. | National security—
United States—History. | United States—Defenses—History.
Classification: LCC U394.A75 W45 2016 (print) | LCC U394.A75 (ebook) |
DDC 355/.040973—dc23
LC record available at https://lccn.loc.gov/2016017245

Jacket design by John Vorhees

Manufactured in the United States of America

First Edition

For my father, Miles Weinberger

If there are to be yet unimagined weapons affecting the balance of military power tomorrow, we want to have the men and the means to imagine them first.

—JAMES KILLIAN,
science adviser to Dwight D. Eisenhower, 1956

Science as science should no longer be served; indeed scientists ought to be made to serve.

—WILLIAM H. GODEL,
former deputy director of the Advanced Research Projects Agency, 1975

Contents

THE IMAGINEERS OF WAR

Prologue

Guns and Money

In June 1961, William Godel set off on a secret mission to Vietnam carrying a briefcase stuffed with cash. At a stopover in Hawaii, he converted some of the cash to traveler's checks to make space for a small bottle of liquor that he carried with him on business trips. Even that did not quite leave enough room, so he moved some of his secret Pentagon papers to another case to make space for the bottle. The money, $18,000, was for a classified project that would play a critical role in President John F. Kennedy's plan to battle communism in Southeast Asia.

At thirty-nine years old, Godel still wore the short buzz cut of his Marine Corps days, but his reputation had been forged in the world of intelligence. A drinker, a practical joker, and a master bureaucratic negotiator, Godel was the type of man who could one day offer to detonate a nuclear bomb in the Indian Ocean to make a crater for the National Security Agency's new radio telescope and the next day persuade the president to launch the world's first communications satellite to broadcast a Christmas greeting. Colleagues described him as someone you could drop in a foreign country, and a few months later he would emerge with signed agreements in hand, whether it was for secret radar tracking stations—something he did indeed set up in Turkey and Australia—or, in this case, winning the support of South Vietnam's president for a new American proposal. Bill Bundy, a former CIA official and White House adviser, called Godel an "operator" with a "rather legendary reputation for effectiveness" working overseas.

At five feet ten inches tall, Godel was not a physically imposing figure, but he had a way of impressing both admirers and enemies with his presence. "He was one of the more glamorous people to stride the halls of the Pentagon," recalled Lee Huff, who was recruited by Godel to the Defense Department. Godel was never the most famous man in the Pentagon, but for several years he was one of its most influential. And by the early 1960s, that influence was focused on Southeast Asia.

Godel arrived to the summer heat of Saigon, a congested city of semi-controlled chaos where cycle rickshaws, bicycles, mopeds, cars, and other motorized contraptions wove through the packed streets like schools of fish in a sea. The city was booming economically and culturally, even as it attracted an increasing number of American military advisers, spooks, and diplomats, who were looking to advise South Vietnam's president on how best to run his newly independent country.

Parisian-style sidewalk cafés still dotted the main city streets, and the city's French colonial heritage was reflected in everything from the fresh baguettes in the local bakeries to the city's grand villas. Vietnamese women dressed in the *áo dài,* the formfitting silk dress worn over pantaloons, mixed easily with teenage girls clad in miniskirts. It was still several years before the influx of American troops would provide a boon to the city's brothels, or frequent Vietcong terrorist attacks in Saigon would drive patrons away from sidewalk cafés, but signs of that unrest were on the horizon. In December of the previous year, the Vietcong bombed the kitchen of the Saigon Golf Club, marking the start of a series of terrorist attacks in the capital. In neighboring Laos, a civil war fueled by Soviet and American involvement was spilling over into Vietnam. More disquieting was that the Vietcong, the communist insurgents in South Vietnam, were getting weapons from North Vietnam, using the Ho Chi Minh Trail, the illicit supply route that snaked through Vietnam's mountains and jungle, and parts of Laos.

Godel had been traveling frequently to Vietnam for more than a decade. What made this trip unusual was that he was now working for the Advanced Research Projects Agency, known by its acronym, ARPA. Founded in 1958 to get America into space after the Soviets launched the world's first artificial satellite, ARPA had lost its space mission after less than two years. Now the young organization, hated by the military and distrusted by the intelligence community, was struggling to find a new role for itself. Godel figured if ARPA could not battle the communists in space, perhaps it could beat them in the jungles.

President Kennedy had taken office just five months prior and was still in the process of formulating a new policy for Southeast Asia. He had already decided to support South Vietnam's anticommunist president, Ngo Dinh Diem, a Catholic who hailed from a family of Mandarins, the bureaucrats who ran Vietnam under Chinese rule. The month before Godel's trip, Vice President Lyndon B. Johnson visited South Vietnam's president, calling Diem the "Winston Churchill of Asia," and in April, Kennedy sent four hundred Green Berets to South Vietnam to serve as special advisers, helping to train the South Vietnamese military and the Montagnards, the indigenous tribes who lived in the country's central plains. Diem was a deeply religious man, a lifelong bachelor who chose politics over the priesthood. Some in Western circles regarded him as an out-of-touch crackpot; others, like Godel, saw him as a flawed but promising leader.

In the early 1960s, South Vietnam was already battling a communist insurgency, but it was a war being fought in the shadows; that summer, astronauts and celebrities still dominated the covers of *Life* and *Time* magazines. Yet there were hints that this new conflict was beginning to occupy America's leaders in Washington. The October 27, 1961, cover of *Life* magazine featured a soldier peering out from jungle underbrush with the caption "GI trains for guerilla warfare." The cover lines read, "Vietnam: Our Next Showdown." Guerrilla warfare was precisely why Godel was in Vietnam. The money he carried with him to Saigon was a down payment on an initial $20 million that the American government expected to allocate for a combat center to develop technology suited for fighting insurgents in Vietnam's jungles. Located in Saigon and run by ARPA, the combat center would be used to help American military advisers and South Vietnam's military. Godel, however, was not just focused on Vietnam; ARPA's Combat Development and Test Center was the starting point for a global solution to counterinsurgency, relying on science and technology to guide the way.

The cash in Godel's bag, and his list of proposals for Diem, would alter the course of events in Vietnam and more broadly lay the groundwork for modern warfare. From stealthy helicopters that would slip over the border of Pakistan on a hunt for Osama bin Laden to a worldwide campaign using drones to conduct targeted killings, Godel's wartime experiments would later become military technologies that changed the way America wages war. His programs in Vietnam, many of which arose from that meeting with Diem, would be credited with some of the best and worst military innovations of the century. Within just a few months of that trip, Godel would bring over to Vietnam a new gun better suited for jungle warfare, the Armalite AR-15. He would also send social scientists to Vietnam, hoping that a better understanding of the people and culture would stem the insurgency. Some of Godel's work became infamous, like a plan to relocate Vietnamese peasants to new fortified villages, known as strategic hamlets. That plan became one of the more resounding failures of the war. Similarly, ARPA's introduction to Vietnam of chemical defoliants, including Agent Orange, is now held responsible for countless deaths and illnesses among Vietnamese and Americans.

At its height, the ARPA program he established employed hundreds of people spread across Southeast Asia—more than five hundred in Thailand alone—and then expanded later to the Middle East. The program sought to understand the roots of insurgency and develop methods to prevent it so that American forces would not have to get involved in regional wars they were unprepared to fight. ARPA developed new technologies, sponsored social science research, and published books on counterinsurgency warfare that would later influence a new generation of military leaders fighting in Iraq and Afghanistan. More than any single technology, Godel's single-minded promotion of the need to understand the nature of guerrilla warfare would have an impact decades later, when the army general David Petraeus, and his advisers known as the "strategic whizzes," found themselves studying the writing of David Galula, whose seminal work, *Pacification in Algeria,* was published in 1963, paid for by ARPA. Four decades before Petraeus made "counterinsurgency" a household phrase, Godel created a worldwide research program dedicated to insurgent warfare that dwarfed anything done in the years after 9/11.

The nascent counterinsurgency program Godel started inadvertently played a critical role in shaping the future agency whose name would become synonymous with innovation. The Vietnam counter-insurgency work eventually became the backbone of ARPA's Tactical Technology Office, the seminal division that would produce stealth aircraft, precision weaponry, and drones—the fundamentals of the modern battlefield. The space age might have given birth to ARPA, but Vietnam thrust the agency into the center of Cold War strategic debates, and it was Godel, more than any other ARPA official, who shaped the agency's future.

Yet it was not all counterinsurgency. In the early 1960s, the esoteric agency Godel helped build was planting the seeds for work that would bear fruit many years later. In the first two years, Godel helped create the agency's space program, providing cover to the world's first reconnaissance satellite, a top secret project. He also persuaded the president to launch the world's first communications satellite and helped build a worldwide network for nuclear test monitoring. By the end of the decade, a descendant of one of ARPA's first projects, the Saturn rocket, would launch Neil Armstrong and the other Apollo 11 astronauts on their journey to the moon. And just a month before Godel traveled to Vietnam, ARPA was handed a new assignment in command and control, which would in less than a decade grow into the ARPANET, the predecessor to the modern Internet. The following year, Godel personally signed off on the first computer-networking study, giving it money from his Vietnam budget.

Godel's seminal role was largely expunged from the record in later years, and his name rarely mentioned in official materials, forgotten except by a few loyal friends and dedicated enemies. The AR-15, the weapon that Godel personally carried over to Vietnam, eventually became the M16, the standard-issue infantry weapon for the entire U.S. military. The rest of Godel's Vietnam-era work would be dismissed as a onetime diversion for an agency now more closely associated with high technology than strategic thinking. His story did not fit an agency touted as a model for innovation. Yet the real key to the ARPA legacy lies in understanding how all these varied projects—satellites, drones, and computers—could come to exist in a single agency.

The Central Intelligence Agency sits on a compound in Langley, Virginia, made famous by countless movies and television shows. The NSA's massive headquarters is ringed by barbed wire and located on a military base in Maryland. Yet the agency responsible for some of the most important military and civil technologies of the past hundred years resides in relative obscurity behind a generic glass facade at 675 North Randolph Street in Arlington, Virginia. The unremarkable office tower stands across from a dying four-level brown-brick shopping mall that houses a mix of fast-food restaurants and discount stores.

Behind the nondescript exterior of the office building, just beyond the guards, is a panoramic wall display that covers more than fifty years of the agency's history. It begins in the fall of 1957, when the Soviet Union launched the first man-made satellite into orbit. Sputnik, as the satellite was called in the West, did little more than emit a simple beep. But that beach-ball-size sphere orbiting harmlessly around the earth touched off a storm of news reports that shook the American people's feeling of invulnerability by demonstrating that the Soviet Union might soon be able to launch a nuclear-armed missile that could reach the continental United States.

As the story goes, Sputnik sparked a national hysteria, and the American public demanded that the government take action. In response, President Dwight Eisenhower in early 1958 authorized the establishment of a central research agency independent from the military services, whose bickering had contributed to the Soviet Union's lead in space. This new agency, called the Advanced Research Projects Agency, was the nation's first space agency—established eight months before the National Aeronautics and Space Administration, or NASA. The organization today known as DARPA—the *D* for "Defense" was added in 1972 (and then dropped, and added again in later years)—has grown into an approximately $3-billion-a-year research agency, with projects that have ranged from space planes to cyborg insects. The display in the lobby is a monument to more than fifty years of this unusual government agency, which has produced marvelous and sometimes terrifying technological achievements: precision weapons, drones, robots, and networked computing, to name a few.

By thinking about fundamental problems of national security, DARPA created solutions that did far more than give the military a few novel weapons. In some cases, the agency changed the nature of warfare; in others, it helped prevent the nation from going to war. By thinking about how to deal with Soviet conventional military superiority without resorting to nuclear weapons, it introduced the era of precision weaponry. By looking for ways to detect underground nuclear explosions, it revolutionized the field of seismology and enabled the negotiation of critical arms control treaties. And by exploring ways to improve nuclear command and control, it created the ARPANET, the precursor to the modern Internet.

Not all solutions are so tidy, however. In trying to tackle the problem of communist insurgency, DARPA embarked on a decade-long worldwide experiment that ended in failure. It is tempting to carve out unsuccessful work, like the counterinsurgency programs, by claiming this was an aberration in the agency's history. This book argues, however, that DARPA's Vietnam War work and the ARPANET were not two distinct threads but rather pieces of a larger tapestry that held the agency together. What made DARPA successful was its ability to tackle some of the most critical national security problems facing the United States, unencumbered by the typical bureaucratic oversight and uninhibited by the restraints of scientific peer review. DARPA's history of innovation is more closely tied to this turbulent period in the 1960s and early 1970s, when it delved into questions of nuclear warfare and counterinsurgency, than to its brief life as a "space agency." Those two crucial decades represent a time when senior Pentagon officials believed the agency should play a critical role in shaping world events, rather than just develop technological novelties.

The Internet and the agency's Vietnam War work were proposed solutions to critical problems: one was a world-changing success, and the other a catastrophic failure. That muddied history of Vietnam and counterinsurgency might not fit well with DARPA's creation story, but it is the key to understanding its legacy. It is also the history that is often the most challenging to get many former agency officials to address. DARPA may brag about its willingness to fail, but that does not mean that it is eager to have those failures examined.

DARPA is now more than fifty-five years old, and much of its history has never been recorded in any systematic way. One effort was made, in 1973, when DARPA approached its fifteenth anniversary. Stephen Lukasik, then the director, commissioned an independent history of the agency to better understand its origins and purpose. The final document was regarded as so sensitive that the authors were only authorized to make six copies, all of which had to be handed over to the government. Although it was supposed to be an unclassified history, the new director was aghast at what he felt was an overly personal account; he stamped the final product as classified and locked it away. It took more than a decade for it to be released.

Agencies, like people, make sense of themselves through stories. And like people, they are selective about the facts that go into their stories, and as time passes, the stories are increasingly suspect and often apocryphal. No other research organization has a history as rich, complex, important, and at times strange as DARPA. Whether it was a mechanical elephant to trudge through the jungles of Vietnam or a jet pack for Special Forces, DARPA's projects have been ambitious, sometimes to the point of absurdity. Some of these fanciful ideas, like the concept of an invisible aircraft named after a fictional, eight-foot-tall rabbit, actually succeeded, but many more failed.

At some point, the successes, and the failures, began to get smaller, because the problems assigned to the agency grew narrower. The key to DARPA's success in the past was not just its flexibility but also its focus on solving high-level national security problems. DARPA today runs the risk of irrelevancy, creating marvelous innovations that have, unlike previous years, little impact on either the way the military fights or the way we live our lives. The price of success is failure, and the price of an important success is a significant failure, and the consequences of both should be weighed in assessing any institution's legacy. Conversely, if the stakes are not high, then neither the successes nor the failures matter, and that is where the agency is in danger of heading today, investing in technological novelties that are unlikely to have a significant impact on national security.

Current DARPA officials may disagree with this pessimistic assessment of the agency's current role or argue about which failures, and successes, should be highlighted. Yet the research for this book is based on thousands of pages of documents, many recently declassified, held in archives around the country, and hundreds of hours of interviews with former DARPA officials. Most past directors share a very similar sentiment: DARPA continues to produce good solutions to problems, but the problems it is assigned, or assigns itself, are no longer critical to national security. To understand why this narrowing of scope happened, it is important to examine the real history of DARPA. The agency's origins may begin with the space race, but DARPA's legacy lies elsewhere.

Godel and his trip to Vietnam were seminal to the agency's history—both its high and its low points. That trip helped create the modern agency and its greatest and worst legacies. Yet Godel's story is one that DARPA officials today do not talk about, or even know about. It is a story buried in long-forgotten court records and has been nearly written out of the agency's history, because it no longer fits the narrative of DARPA as an agency dedicated to technological surprise. Yet it is a story that illustrates the true tensions within DARPA, an agency that enlists science—and scientists—in the service of national security.

PART I

AN AGENCY FOR UNIMAGINED WEAPONS

CHAPTER 1

Scientia Potentia Est

Michiaki Ikeda was a chubby-faced six-year-old when the nuclear age smacked him in the face with a blinding flash of light. Just as he was stepping out of an elevator at Nagasaki Medical University's hospital, a nuclear weapon code-named Fat Man detonated seven hundred meters away from him. The bomb had the explosive equivalent in force of more than twenty kilotons of TNT and flattened almost everything within a kilometer radius. The concrete hospital building was mostly left standing, but the majority of the people inside were killed. The steel elevator shaft likely saved his life.

When he came to, it was pitch-dark, and the first sensation he recalled was the sound of something burning. Then the smell of smoke reached his nostrils, bringing him to his feet. As he stumbled out into what had been the hospital's corridor, his eyes adjusted to the darkness, and he realized he was standing on dirt. The wood floors had been blown away. In the corner, he saw a nurse on the ground surrounded by shattered glass, and her face covered in blood. To Michiaki, it was as if someone had poured a bucket of blood over her head. Yet her eyes were open, and she was staring at him.

"Call the ambulance service," she ordered, her expression a mix of shock and rage.

He looked around, but all he could see were shards of glass and wood panels blown from the ground. He crawled out a window frame and stepped down into what had been, just a little while before, a tranquil garden with water. Now, as he looked up, he could see some trees were toppled and the ones that still stood were in flames. When his eyes moved from the burning treetops down to the ground, the scene was pure horror. The hospital's garden was strewn with corpses with hair burned into frizzy clumps. Some had eyeballs hanging down on their cheeks, and faces with their lips and flesh burned away, leaving the teeth and jaw exposed. There were some bodies with stomachs bloated to twice their normal size, and others with internal organs spilling out.

He fled the burning hospital grounds and instinctively started walking toward the city, thinking he would find help. Instead, he found more horror. The main boulevards of Nagasaki were cluttered with debris of blown-out buildings. The living were walking, their arms dripping with scorched flesh outstretched in front of them to avoid the pain of having burned skin touch their bodies. Dazed, they walked down the street, calling for water and looking for help that was not there.

Three days earlier, the United States had dropped an atomic bomb called Little Boy, which used highly enriched uranium, on Hiroshima, instantly killing some seventy thousand people. Many more would die later from burns and radiation sickness. Nagasaki had not been the primary target of Fat Boy, a plutonium implosion bomb. A B-29 Superfortress, *Bockscar,* was planning to drop Fat Boy on the city of Kokura, but cloud cover forced the pilot to divert to Nagasaki, a secondary target. Nagasaki's natural geography of mountains and valleys protected part of the population, preventing many of the immediate deaths that took place in Hiroshima, but the city center was devastated.

Along with a bomb, a second airplane flying over Nagasaki dropped canisters containing scientific instrumentation. The canisters also contained copies of a personal letter several Manhattan Project scientists addressed to a prominent Japanese scientist. "You have known for several years that an atomic bomb could be built if a nation were willing to pay the enormous cost of preparing the necessary material," the letter, written by the nuclear physicist Luis Alvarez, read. "Now that you have seen that we have constructed the production plants, there can be no doubt in your mind that all the output of these factories, working 24 hours a day, will be exploded on your homeland."

In Japan, the bomb had now decimated two cities. Six-year-old Michiaki was fortunate: miraculously uninjured, he was found by a nurse and taken to a bomb shelter in the mountains, where he was eventually reunited with his family. Michiaki did not know anything about what had happened that day. He only knew that this was not like the other bombings the city endured during the war, a routine so common that residents often ignored the sirens warning of enemy aircraft. "I had no clue what a nuclear or atomic bomb was—that something like that existed," he recalled. "I just thought it was many, many big bombs that had fallen."

The bomb dropped on Nagasaki was the third atomic device ever detonated. The first atomic explosion, called the Trinity Test, was conducted in secrecy on July 16, 1945, at Alamogordo, New Mexico. Americans learned about this new weapon after Little Boy was dropped on Hiroshima on August 6 of that year. The *New York Times* announced the nuclear age to the world with the headline "First Atomic Bomb Dropped on Japan; Missile Is Equal to 20,000 Tons of TNT; Truman Warns Foe of a 'Rain of Ruin.'" In Japan, however, what little news was reported about Hiroshima was only that incendiary bombs were used.

Speaking the day the bomb on Hiroshima was dropped, President Harry Truman revealed not just the existence of this terrifying new weapon but a massive project conducted in secrecy to build it. Across the country, over two and a half years, as many as 125,000 people had been involved in this secret project, Truman announced. Many workers did not even know exactly what they were working on, only that it was an important war project. "We have spent two billion dollars on the greatest scientific gamble in history," he said, "and won."

Truman was right: Less than a week after Nagasaki was bombed, the Japanese emperor announced the country's unconditional surrender, telling the nation in a broadcast speech that despite great sacrifice "the war situation has developed not necessarily to Japan's advantage." More directly, he acknowledged the devastation wrought in Hiroshima and Nagasaki, saying "the enemy has begun to employ a new and most cruel bomb, the power of which to do damage is, indeed, incalculable, taking the toll of many innocent lives. Should we continue to fight, it would not only result in an ultimate collapse and obliteration of the Japanese nation, but also it would lead to the total extinction of human civilization."

A few weeks after the Japanese surrender, Herbert F. York, a young physicist who had been one of the thousands of workers on the secret project Truman had referred to, brought his father to Oak Ridge, Tennessee, where uranium had been enriched for the Little Boy bomb dropped on Hiroshima. The work inside the plant itself was still secret, but its existence no longer was. Standing at the top of a hill, York pointed down proudly to the facility hidden in the valley below, where he had labored in secret for two years of the war. "We have made war obsolete," he triumphantly told his father. It did not take York long to realize he was completely wrong.

In Japan, the power of the atomic bomb left people feeling helpless. In America, for that brief moment, it made people feel invincible. The idea that this same powerful weapon could soon threaten the United States had not yet sunk in. It would soon. The United States might have beaten the rest of the world in building an atomic bomb, but the Germans during the war had achieved something that the Americans, British, and Soviets had not: a guided ballistic missile. The V-2, a liquid-propelled rocket developed by Wernher von Braun and his team of scientists, could travel more than two hundred miles, with an engine thrust eighteen times greater than anything the Allies had achieved. The Nazis used it to terrorize England during the war.

The bombing of Hiroshima and Nagasaki hastened the end of World War II, and it also marked the beginning of a new war for scientific talent and engineering. The atomic bomb had proved that knowledge was power, and whatever nation had the most knowledge would have an edge in the next war. The Soviet Union might have been allies with the United States in its victory over Germany, but the two countries' interests diverged even before Japan surrendered. In Germany, the Soviets and the Americans were already engaged in a race to capture knowledge.

Standing in Frankfurt's Hauptbahnhof in 1949, twenty-eight-year-old William Godel paused to admire the grand arches and curved glass above the train terminal. Outside, most of the city was still many feet deep in rubble—the aftermath of bombing during the war. It was not just the station's neo-Renaissance design Godel was admiring but also the fact that it had survived the war with only superficial damage. The strategic bombing of Germany had been highly effective at causing civilian casualties but not at stopping the industrial war machine.

"Hey, you," an American woman snapped. "Come put this baggage aboard and I'll give you a cigarette."

"Jawohl, gnädige Frau," Godel answered, picking up her bag. As he carried it to the train, he walked with a slight limp—a war injury, something not uncommon to see in a German man his age in Frankfurt; Germany was flooded with crippled veterans. The train station was also filled with Americans, mostly military service members and their families stationed in Germany. The Americans who walked through the station were smartly dressed, whether in military uniform or civilian clothing. The Germans, on the other hand, shambled about the train station in threadbare suits. Germany was still under Allied occupation. The Americans controlled Frankfurt, and many still harbored a deep resentment of the Germans. Sometimes the Americans would tell him a compartment was for "Americans only."

Godel was accustomed to being given orders by Americans in the train station, and the woman's request to carry her bag was a relief; it meant that he was passing for what he was meant to pass for: Hermann Buhl, a former member of Germany's Wehrmacht, and not an American covert operative.

The young American was posing as a German veteran so he could slip across Soviet-occupied areas in Germany and Austria, and even into the Soviet Union, recruiting Russian and German scientists, engineers, and military officers to work for the United States. His German was fluent, but not native, good enough to pass with the Americans and Russians, and even Germans, in many cases. German veterans could quickly figure out he was not really ex-Wehrmacht, but that did not so much matter; they had other things to worry about in the late 1940s. "It was a high-risk undertaking, replete with forged documents, black-market funds, bribery, loose women, and all manner of illegalities and immoralities," he later wrote. He was also on his own when it came to the Russians. "Don't get caught," one army general told him, "because I cannot help you worth a damn over there."

Godel's work was under the larger rubric of Operation Paperclip, the military intelligence program that was scooping up German scientists and engineers to bring to the United States. The project, so named for the paper clip attached to each scientist's dossier, had already garnered the biggest bounty: von Braun and his team of rocket scientists. At the end of the war, von Braun had actively sought out the American military, knowing that he and his team would likely fare better with the United States than with the Soviet Union. In the spring of 1945, the Soviets dispatched specialized military intelligence teams to Germany to gather anything that could be found in the way of military technology, including missiles, radar, and nuclear research. The Soviets took Peenemünde, where von Braun and his rocket team had been based, but they had already fled, taking much of their design work with them. "This is absolutely intolerable," Joseph Stalin said. "We defeated Nazi armies; we occupied Berlin and Peenemünde; but the Americans got the rocket engineers. What could be more revolting and more inexcusable?"

The Soviets eventually took whatever they could, sending hundreds of German personnel back to the Soviet Union, not to mention trainloads of equipment. The Soviets' hunt for technical expertise was broad, but it also lacked focus. As von Braun put it, "The Americans looked for brains, the Russians for hands."

In Germany before the war, von Braun had been part of a visionary group that dreamed of building rockets for space travel but agreed to work for the military, and eventually the Nazis, on weapons. In going with the Americans, he hoped again to work on space travel. Instead, von Braun and more than a hundred other rocket scientists were taken to the United States, initially to Fort Bliss, Texas, and relegated to showing the Americans how to build and operate the V-2. Unsure of what to do with the Germans, and unwilling to give them money to design new rockets, let alone fulfill von Braun's ambitions of space travel, the Americans allowed his team to languish in the South.

The Soviets did not suffer from indecision, however. Using captured German know-how, the Soviets moved forward swiftly with designing rockets that could travel even greater distances than the V-2. "Do you realize the tremendous strategic importance of machines of this sort?" Stalin told a senior Russian rocket scientist after the war. "It could be an effective straightjacket for that noisy shopkeeper, Harry Truman.

We must go ahead with it, comrades." In the Soviet Union, the goal was clear. "What we really need," said Pavel Zhigarev, the commander in chief of the Soviet air forces, "are long-range, reliable rockets that are capable of hitting the American continent."

As the Soviets moved forward with their ballistic missile program, William Godel, disguised as Hermann Buhl, was on a parallel mission: trying to collect intelligence on Soviet military capabilities. He was growing increasingly convinced that the American military was pursuing weapons based on its own bureaucratic interests and not based on what intelligence was telling it was needed.

William Hermann Godel was born as Hermann Adolph Herbert Buhl Jr. on June 29, 1921, in Denver, Colorado, to Hermann Buhl Sr. and Lumena Buhl, German immigrants. Hermann Buhl Sr. died of pneumonia in 1931, and Lumena soon married another German immigrant, named William Frederick Godel, who ran his own insurance business and prior to World War II served as the German consul in Denver. The next year, Lumena's new husband legally adopted his stepson and, at the suggestion of the judge, officially changed the boy's name to William H. Godel. Relations between the two were icy at best. At one point, the younger Godel built a shack in the backyard to avoid living in the same house as his adoptive father.

After high school, Godel attended the New Mexico Military Institute in Roswell and then, later, Georgetown's School of Foreign Service. He initially went to work for the War Department's military intelligence division, but when the Japanese attacked Pearl Harbor, Godel was commissioned as an officer in the Marine Corps and participated in the initial landings in the Pacific. He was wounded twice, including at Guadalcanal in January 1943, where he was hit by a hand grenade. The fragments shattered the bone in his left leg and destroyed a good portion of its muscle. He was awarded the Purple Heart and sent back home to recuperate. For the rest of his life, he would need a leg brace and walk with a limp.

Godel desperately wanted to stay in the Marine Corps and insisted he was fit to serve, but by 1947, after a series of medical reviews, he lost the battle. The wound in his left leg was still not completely healed, and Godel was forcibly retired from the Marine Corps, declared medically unfit for service. He made enough of a name for himself that after the war General William "Wild Bill" Donovan, the wartime head of the Office of Strategic Services, recruited Godel to Washington to work as an intelligence research specialist for the army focusing on the Soviet Union.

It was a chaotic but exciting time to be involved in intelligence. Before the war, intelligence was regarded as something of a dirty business. "Gentlemen don't read other people's mail," Secretary of State Henry Stimson declared in 1929, when explaining why the United States should halt its cryptanalysis work. Pearl Harbor and World War II might have discredited that view, but there was still nothing approaching a robust intelligence machine even after the war ended. There were, however, powerful personalities lobbying for power, particularly those who formed part of a close-knit community of military and intelligence operatives who had served together in World War II. Men like the air force brigadier general Edward Lansdale, a legendary spy, and William Colby, the future director of the CIA, emerged during this period. So, too, did William Godel.

In 1947, Harry Truman signed the National Security Act, which attempted to impose order on the bureaucratic chaos that emerged after World War II. The war had created a multitude of people and organizations vying for power, and the legal reorganization was supposed to bring some clarity with the establishment of the National Security Council and the Central Intelligence Agency while also streamlining the Department of Defense and creating the Department of the Air Force, splitting it off from the army. The National Security Act, in reality, simply spawned an array of new organizations all competing for resources. The army, the navy, and the newly created air force all claimed ownership of rocket and missile research, while the CIA also saw a need for military technology that could collect intelligence on the Soviet Union.

The most important of those new technologies was, as Stalin rightly pointed out, an intercontinental ballistic missile, or ICBM. It would be a categorically different military capability; by the early 1950s, the Soviet Union was building bombers that could carry nuclear weapons to the East Coast of the United States, but they could also be potentially detected and intercepted. In the United States, computer scientists were already hard at work developing computer systems that could link radars together, to allow the military to stop incoming Soviet bombers, but there was in the 1950s no existing technology that could conceivably stop an ICBM attack. Even if a missile were detected by radar, the military would have just seconds to respond, and then there was little to be done to stop it: it would be like trying to shoot a bullet out of the sky.

In the immediate years after World War II, there was initially little enthusiasm in the White House for investing in such long-range missiles. In 1947, President Truman, who had promised to bring federal debt under control, slashed the military's rocket and missile programs. Funding was tight, and it was being fought over. The army, the navy, and the air force all had their own rocket and missile programs, each with justifications, often tenuous, for why that work properly belonged to them. The seeming triumph of American technology was short-lived. The United States had spent millions gathering up German technical talent, but when von Braun proposed research to his Pentagon masters to build more complicated rockets or—his ultimate goal—to design rockets that could travel into space, he was refused. It was a time of "professional gloom" for him and his team.

Yet the Soviets by 1949 had already developed a new ballistic missile, called Pobeda, or "Victory," that could fly higher and carry more than the V-2 rocket. That same year, on August 29, the Soviet Union set off its first atomic bomb on the Kazakh Steppe, ending America's monopoly on nuclear weapons. A little more than a month later, China fell to communism, and in June 1950 North Korea invaded South Korea. Truman, who thought he would demilitarize, was suddenly left dealing with twin threats of a Soviet nuclear and conventional buildup in Europe and a growing communist threat in Asia. The only choice for politicians in Washington seemed to be developing weapons even more powerful than those that had destroyed Hiroshima and Nagasaki.

On November 1, 1952, Herbert York made a call to the nuclear physicist Edward Teller with a brief message. It was "zero hour," York told Teller, who was watching a seismometer at the Radiation Laboratory at Berkeley. Fourteen minutes passed, and then Teller called back with his own coded response: "It's a boy."

That "boy" was Ivy Mike, a 10.4-megaton hydrogen bomb that had just exploded in the clear blue waters of Eniwetok Atoll, vaporizing the island of Elugelab and creating, as Richard Rhodes described it, "a blinding white fireball three miles across." The device, designed by Teller and Stanislaw Ulam, was a thousand times more powerful than the bomb that went off in Hiroshima. York, the young physicist who just seven years earlier had proudly told his father that war was obsolete, was now in charge of recruiting the scientists to design a new class of weapons whose power was so great that at one point it was feared the explosion would ignite the atmosphere and vaporize the oceans. Ivy Mike was a test of the world's first thermonuclear weapon, known as the Super. This new bomb did more than create a new generation of superweapons; it also eliminated one of the last arguments against developing ICBMs. Thermonuclear weapons with yields in the many-megaton range meant that accuracy was no longer critical; with a big enough bang, hitting the target precisely was not as important. And once the thermonuclear weapon could be reduced in size, the military did not need bombers to haul weapons over long distances; it could pack them on an ICBM.

Three days after Ivy Mike exploded, Dwight D. Eisenhower, who had served as the supreme commander of the Allied forces in Europe during World War II, was elected president in a landslide, running on a campaign that focused heavily on battling communism. "World War II should have taught us all one lesson," he declared. "The lesson is this: To vacillate, to hesitate—to appease even by merely betraying unsteady purpose—is to feed a dictator's appetite for conquest and to invite war itself."

By the time Eisenhower took office in January 1953, the Korean War was already drawing to a close, and he was alarmed by the growth in the federal budget. In the past two decades, spending had grown twenty-fold to more than $80 billion, and over half of that was going directly to the Pentagon. To rein in military spending, Eisenhower instituted a policy called New Look, which turned to nuclear weapons as a cost-effective way to offset drawdowns in conventional forces. It was fortuitous timing for rocket enthusiasts. Von Braun and his team had moved in 1950 to Huntsville, Alabama, where they were finally working on a new missile, called the Redstone. In Washington, Eisenhower was met with a flood of reports and panels making the case for rocket technology: both as weapons that could reach the Soviet Union and as a way to carry satellites into space. Rand, a newly established think tank funded by the air force, produced a series of reports proposing an earth-orbiting satellite as a military capability. Because satellites did not yet exist, there was still a question of national sovereignty: Would a satellite that flew over another country, such as the Soviet Union, be regarded as a violation of its airspace?

In 1954, the Technological Capabilities Panel, appointed by Eisenhower to look at the potential of a "surprise attack" by the Soviet Union, offered a solution: the United States would launch a purely scientific satellite as a pretext to establish "freedom of space," which would then pave the way for military satellites. With all three of the military services developing separate technologies, the question was which should get to build the first rocket to space.

As the military services battled over a nascent space program, William Godel in the 1950s was in the midst of a different war in the intelligence world. Back in Washington, D.C., he worked as an assistant to General Graves Erskine, the Pentagon's director of special operations. Godel quickly earned a reputation as the go-to guy for special assignments, particularly those that combined intelligence with science. Whether it was recruiting foreign scientists to work with the Pentagon or formulating plans for Operation Deep Freeze, which established the American presence in Antarctica (and earned him an eponymous plot of frozen water, the Godel Iceport), Godel was known as a man who could get things done.

Godel was also often called in to deal with the turf wars in areas like psychological operations. Frustrated by the lack of coordination for such operations—covert and overt—across government, President Truman in 1951 established the Psychological Strategy Board and appointed Godel as a member. The job brought Godel into periodic battles with the CIA, though many of them were petty. Official correspondence from the time mentions CIA officials clashing with Godel about everything from the CIA director's refusal to attend a Pentagon function for visiting dignitaries to whether the CIA was providing a Hollywood studio with film footage of American prisoners of war held in North Korea. But the infighting was bad enough that Frank Wisner, the head of the CIA's Office of Policy Coordination, banned Godel from his buildings.

It might have been run-ins like those that prompted a security investigation into Godel, something that was not unusual in an era when information dug up from background investigations was used as a blunt weapon to oust political enemies. In 1953, Pentagon security officials interviewed Godel after reports surfaced that his adoptive father had been a Nazi sympathizer. While denying the allegation, Godel also distanced himself from the man who raised him. "I didn't care for him," Godel said. "I had no personal association with him other than as a man who has been very nice to my mother since I left in '38."

The investigation did not stop Godel's upward trajectory in government, however. In 1955, Donald Quarles, then the assistant secretary of defense for research and development, assigned Godel to the National Security Agency, a part of government so highly classified at the time that its existence was not even acknowledged. The NSA had been established in 1952, bringing together the communications intelligence and code-breaking capabilities that had emerged from World Wars I and II. Like the rest of the Defense Department, the NSA was being scrutinized by the Eisenhower administration, which was unhappy with the quality of strategic intelligence. Godel was supposed to help straighten out the NSA's overseas operations and cut back ineffective foreign bases. For Godel, the NSA assignment combined his twin interests in intelligence and technology. In a later unpublished interview, Godel had a simple description of his mission: he was a hatchet man.

In 1955, the year Godel was assigned to scale back the NSA, a copy of his security interview, which included questions raised about his adoptive father's Nazi sympathies, was sent over to the FBI at the personal request of J. Edgar Hoover to review. It is unclear what the FBI chief was looking for, but two years later Secretary of Defense Charles Wilson wrote back to Hoover: "Glad to know you think [Godel's] doing a fine job." Godel's role by then had earned him consideration for a top slot at the NSA.

Godel might have been doing a fine job, but the NSA, like the rest of the defense and intelligence community, was about to become embroiled in yet a new crisis. The same year that Quarles sent Godel to revamp the NSA, he also appointed a panel to decide which rocket proposal would take the United States into space. The problem was that there was no civilian rocket program; only the military services were developing the technology that could launch a satellite into space. The air force's plan was to launch an ICBM into space, and the army proposal would have involved relying on former Nazi scientists working at a military arsenal. The navy's rocket, while the least mature, had the advantage of not being associated with a weapon. In the end, the panel passed over the army's German rocket team and the air force's ICBM, selecting instead the navy proposal, a rocket that was still in development. "This is not a design contest," an outraged von Braun protested. "It is a contest to get a satellite into orbit, and we are way ahead on this."

Von Braun's concerns were ignored, even as over the next two years the navy fell behind schedule. The delays did not spark much concern among America's political leaders, and particularly not for President Eisenhower, who still believed that the United States was ahead of the Soviet Union.

Then, in the fall of 1957, the CIA and the NSA were monitoring Soviet launches of intermediate-range missiles from Kapustin Yar, in western Russia, unaware of a much more important launch that was being prepared in Kazakhstan. Twelve years after winning a scientific gamble on nuclear weapons, Americans were about to face the reality that the horror the six-year-old Michiaki experienced in Nagasaki could soon reach the continental United States. The United States would no longer be invulnerable, and war was anything but obsolete.

CHAPTER 2

Mad Men

On the evening of October 4, 1957, Neil McElroy was enjoying cocktails in Huntsville, Alabama, fresh from a doomsday tour of the United States. McElroy, who was about to become the secretary of defense, was chatting with the army general John Medaris and the German rocket scientist Wernher von Braun during a casual reception held as part of McElroy's tour of the Army Ballistic Missile Agency. It was one of many visits the secretary designate and his entourage were making around the country as he prepared to lead the Pentagon.

Huntsville should have been the least memorable stop for McElroy, who had been traveling the past few weeks in a converted DC-6 transport aircraft reserved primarily for the secretary of defense. Along the way, he was plied with fine liquor and deluxe accommodations, all while getting a crash course in overseeing a military at the dawn of the age of nuclear Armageddon.

The new position was a big change for McElroy. His last job was heading Procter & Gamble, the consumer products company based in Cincinnati, Ohio. McElroy, who had no prior experience in government, was one of the "industrialists" Eisenhower had brought to the capital in the belief that business-style leadership could help straighten out government.

The media had not been kind to McElroy after Eisenhower picked him to head the Pentagon. The native Ohioan had made his name in the nascent field of "brand management," penning a famous letter admonishing Procter & Gamble executives on the importance of promoting the company's soaps to the proper consumer markets so that the products would not compete with each other. "Soap manufacturer Neil McElroy is president's choice to succeed Wilson," *The Milwaukee Journal* declared on August 7. Another report mocked McElroy's experience in advertising, saying that he had been responsible for "vital activities in persuading housewives to buy one bar of soap or another."

Now McElroy and his entourage were being wined and dined across the country by military officials pitching their soon-to-be boss on the importance of their aircraft, missiles, and bases in case of nuclear confrontation with the Soviet Union—all in between plenty of martinis. At Strategic Air Command, near Omaha, Nebraska, they were greeted with a table covered in whiskey, ice, and "fixings," before being shown the control room, where military commanders could launch a nuclear attack. Later, General Curtis LeMay, the head of Strategic Air Command, personally piloted a demonstration of the new KC-135, a refueling aircraft, for McElroy and his staff.

At Edwards Air Force Base in the high desert north of Los Angeles, the group met General Bernard Schriever, the head of the Western Development Division, which was responsible for developing intercontinental ballistic missiles. McElroy and his entourage took an immediate liking to the air force general, who was "extremely able" and could "shoot golf at par."

In Colorado, at North American Aerospace Defense Command, better known by its acronym NORAD, the group was assigned luxury suites at the Broadmoor, whose mountain-view rooms were stocked with bottles of scotch and bourbon. The next day they were briefed on the calculus of a survivable nuclear war, where commanders had to weigh the lives of three million civilians versus protecting a key military site. It was a world, McElroy's aide, Oliver Gale, wrote, "where horror is as much a part of the scene as manufacturing cost is in the soap business."

The final stop on McElroy's itinerary was Redstone Arsenal in Huntsville, a quiet southern town in Alabama whose economy was rapidly shifting from cotton mills to rocket production. General Medaris, commander of the Army Ballistic Missile Agency, was polite but unimpressed by McElroy. The problem with a businessman is that he can "become a sort of czar, surrounded by subordinates who carry out his orders and obey his whims without daring to question his judgment," he wrote in his memoir just a few years following that meeting. "This gives him the illusion that he knows all the answers. He rarely does, outside his own general field."

Neither were McElroy and his staff impressed with the army general, who sported a black mustache and was known for dressing in old-fashioned officer riding breeches. Medaris was a "salesman, promoter, who pushes a bit more than might be considered palatable," wrote Gale, who worked for McElroy at Procter & Gamble and was following him to the Pentagon. Coming from an advertising man, the description was telling. Medaris was trying to sell the services of von Braun and his group of German rocket scientists, who were now based in Huntsville but could not seem to shake their Nazi past. "Von Braun was still wistful about what would have happened if [the V-2s] had all gone off," Gale recorded in his journal, "not because he was sorry that Germany did not win the war (apparently) but because he was sorry his missiles, his achievements, had not been more successful."

Even in Huntsville, the Germans found themselves stymied by the military, starved for funds, and frozen out of the space work they desperately wanted. They were stuck working, yet again, on suborbital missiles. The problem was not scientific know-how but classic bureaucratic rivalry. By the fall of 1957, von Braun's army group had developed the Jupiter-C missile, a four-stage rocket that could have been shot into orbit, if only the army was allowed to launch it. It was not, and so the fourth stage of von Braun's Jupiter-C was filled with sand, rather than propellant, to ensure it would not leave the atmosphere.

Medaris had reason to be skeptical of the incoming defense secretary and his visit. McElroy was replacing Charles "Engine Charlie" Wilson, another captain of industry appointed by Eisenhower. As defense secretary, Wilson threw himself into budget cutting with a passion, carrying out Eisenhower's New Look defense policy, which emphasized advanced technology, such as nuclear weapons and airpower, over conventional forces. Yet satellites, in Wilson's view, were "scientific boondoggles." He did not understand what purpose they would serve for the military. When Wilson had visited Huntsville, army officials tried to impress him with their work, only to have the money-conscious defense secretary interrogate them on the cost of painting wood in his guest quarters.

With McElroy's visit in the fall of 1957, just days away from becoming secretary of defense, it did not seem apparent to Medaris that the new Pentagon chief would chart a different course. As Medaris, McElroy, and von Braun exchanged pleasantries over drinks, an excited public relations officer interrupted the party with news. The Russians had launched a satellite, and *The New York Times* was seeking comment from von Braun. "There was an instant of stunned silence," Medaris recalled.

News of Sputnik was a surprise, but it should not have been. In 1955, the Eisenhower administration announced plans to launch a small scientific satellite as part of the upcoming International Geophysical Year, which would run from July 1957 to December 1958. Not to be outdone, the Soviets countered with their own satellite launch plans. It was always a race, but one in which the United States assumed it had a natural advantage. The Soviet Union could not produce a decent automobile; how could it possibly hope to best the United States in rocket science? In the meantime, American plans for a satellite launch had fallen behind schedule.

However flawed the Soviet Union's consumer goods industry, the regime had an advantage when it came to military and space research. An authoritarian state could focus resources on a specific goal, like a satellite launch, without the bureaucratic wrangling or public pressures that afflicted a democracy like the United States. The Eisenhower administration, prompted by its civilian scientists, wanted to keep its scientific satellite launches separate from its missile programs, even though the underlying technology was nearly identical. That was why the White House opted instead for the navy's Vanguard, much to von Braun's disappointment.

Now, with the soon-to-be defense secretary in front of him, and Sputnik circling overhead, the words began to tumble out of von Braun. "Vanguard will never make it," the German scientist said. "We have the hardware on the shelf. For God's sake turn us loose and let us do something. We can put up a satellite in sixty days, Mr. McElroy! Just give us a green light and sixty days!"

"No, Wernher, ninety days," Medaris interjected.

McElroy had been the guest of honor, but now everyone circled von Braun, peppering the German rocket scientist with questions. Was it really true that the Soviets had launched a satellite? Probably, von Braun replied. Was it a spy satellite? Probably not, though its size and weight, if accurately reported, meant that it could be used for reconnaissance. And what did it all mean? It meant that the Soviets had a rocket with a sizable thrust, von Braun said.

The general and the rocket scientist spent the rest of the evening trying to persuade McElroy to let them launch a satellite. It is likely that the details were well beyond the grasp of McElroy, who had no background in technical issues. The conversation did impart to McElroy at least the importance of the satellite launch, which he might have otherwise missed. At first glance, the satellite did not seem like an immediate threat to the incoming defense secretary. Sputnik weighed 184 pounds and its sole function was to circle the earth, emitting a beep that could be tracked from the ground.

For McElroy, the man most closely tied to the response to Sputnik, the launch was something of a fascinating footnote to a pleasant cocktail party. His aide, Gale, devoted more space to describing a recent evening meal of exotic seafood on the coast of California than he did to the world's first satellite launch. Yet Sputnik was about to trigger a chain reaction that, by the New Year, would engulf all of Washington.

Years later, a myth emerged that the Soviet "artificial moon" immediately prompted people around the country to stare up at the sky in fear and apprehension. "Two generations after the event, words do not easily convey the American reaction to the Soviet satellite," a NASA history covering the time period states. "The only appropriate characterization that begins to capture the mood on 5 October involves the use of the word hysteria."

In fact, there was no collective panic in the first few days following the launch. It was not immediately clear—except to a small group of scientists and policy makers—why the satellite was so important. For those involved in science and satellites, like von Braun and Medaris, the Soviet satellite circling the earth was proof that politics had hampered the American space effort. Yet for most Americans, the beeping beach ball initially produced a collective shrug. That Sputnik failed to shake the heartland to its core was best demonstrated in Milwaukee, where the *Sentinel*'s bold large-type headline on October 5 announced, "Today, We Make History." In fact, the headline had nothing to do with Sputnik but referred to the first World Series game to be played in Milwaukee. News of Sputnik was buried deep in the paper's third section, where the reporter noted merely that news of the unexpected launch had "electrified" an international meeting in Washington to discuss satellites.

In the days following the launch of Sputnik, the Washington bureaucracy moved in slow motion. Eisenhower's attention in the weeks leading up to Sputnik was focused on much more earthbound matters. The standoff over the first attempt to integrate schools in Little Rock, Arkansas, under court order had ended with the president's sending in federal troops. By comparison, the launch of a satellite armed with nothing more than a beacon did not initially seem like something that was going to capture public attention. At a National Security Council meeting held on October 10, Eisenhower listened as his advisers hashed out ideas for responding to Sputnik. Perhaps the administration should emphasize "spectacular achievements" in science, like cancer research? Or the successful launch of a missile that could travel thirty-five hundred miles? Few in the administration seemed to understand what the Soviets had instinctively grasped: the psychological power of a space launch. General Nathan Twining, chairman of the Joint Chiefs of Staff, warned that the United States should not become "hysterical" over Sputnik.

Eisenhower saw Sputnik as a political stunt. He also knew something that the public did not know: in addition to the military's rocket programs, which were public, the United States had been secretly working on the development of spy satellites, which would prove much more important for the strategic balance than a silver ball beeping from the heavens. In the weeks following Sputnik, the administration's policy was simply to downplay Sputnik's importance. General Curtis LeMay called it "just a hunk of iron," and Sherman Adams, Eisenhower's chief of staff, derided concerns over a space race as "a celestial basketball game." The more that the administration tried to dismiss the Soviet accomplishment, the more fodder it gave for political opponents to accuse Eisenhower of allowing the United States to fall behind the Soviet Union. For Lyndon Johnson, the Democratic Senate leader, Sputnik was an opportunity to be fully exploited. In his memoir, Johnson wrote that he got the news of Sputnik while hosting a barbecue at his ranch in Texas. That evening, he walked out with his wife, Lady Bird, to look for the orbiting Soviet satellite. "In the West, you learn to live with the Open Sky," he later wrote. "It is part of your life. But now, somehow, in some new way, the sky seemed alien."

When Johnson looked up in the night sky, what he saw was not Sputnik but a heavenly political gift that would allow him to hammer the Republicans in the months, and possibly years, ahead. "Soon, they will be dropping bombs on us from space like kids dropping rocks onto cars from freeway overpasses," Johnson proclaimed.

Eisenhower, who had so deftly managed his image as a political leader, found himself stumbling. From a technical standpoint, he was more right than wrong. Though the Soviets were somewhat ahead of the United States in booster technology, the United States had a number of strategic advantages that were not known to the public. In addition to the spy satellite technology being developed, the CIA the year before had begun flying a reconnaissance aircraft in the earth's stratosphere. By flying at seventy thousand feet, the Lockheed U-2 spy aircraft was designed to evade detection by ground radar while flying over the Soviet Union and capturing pictures of military bases. The aircraft—and the flights—were top secret. Also secret was that the U-2 flights had already proved that the "bomber gap"—a suspected Soviet advantage in bombers—did not exist. With news of Sputnik, Eisenhower worried about a perceived "missile gap."

Eisenhower refused to be swept up in mass hysteria, however. "Now, so far as the satellite itself is concerned, that does not raise my apprehensions, not one iota," he told a throng of reporters, just days after the Soviet launch. The administration only helped its critics by providing confusing and contradictory statements about the importance of Sputnik. In that initial press conference, Eisenhower claimed that the "Russians captured all of the German scientists in Peenemunde." In truth, the United States through Operation Paperclip had taken the cream of the crop, but the Germans in the United States were stuck filling the fourth stage of their Jupiter-C with sand.

As the weeks passed, the staid articles about Sputnik gave way to sensational coverage. Drew Pearson, the American writer known for his influential Washington Merry-Go-Round column, claimed that "technical intelligence experts" were predicting that the Soviets might try a moon launch on November 7, to commemorate the anniversary of the Bolshevik revolution. "The same missile that launched the 184-pound Sputnik, our experts say, also could shoot a small rocket 239,000 miles to the moon," Pearson wrote. "The Russians might fill the nose cone with red dye and literally splatter a Red Star on the face of the Moon."

Pearson's moon prediction was an outrageous conflation of conjecture and exaggeration, but on November 3, just a month after Sputnik, the Soviets indeed launched a second, larger satellite. Sputnik 2 carried a dog named Laika on a one-way mission to space. It was taken as purported proof that the Soviets would soon be able to launch a man in space (though unlike with Laika the dog, sending a human into space would require the ability to bring the person back safely to earth). The launch sparked panic in the United States and worldwide protests from animal lovers.

Sputnik tapped into a narrative that artfully wove Hollywood, science fiction, and good old-fashioned fearmongering. The public understood that satellites were somehow connected to the ability to launch ICBMs, but the subtleties of terms like "throw weight," or the payload a ballistic missile could carry, were not readily apparent. It took some time, politics, and editorializing, but within a few weeks the American public's initial curiosity and mild apprehension over Sputnik turned to full-blown panic. Eisenhower was right about the science, but he had misjudged the national mood. The administration's response to Sputnik was a mess, but one thing was clear: the solution was going to be formulated by a soap maker from Cincinnati.

McElroy arrived in Washington just in time for peak Sputnik hysteria. The new defense secretary's first few weeks at the Pentagon were marked by an endless parade of military chiefs and presidential advisers, all making suggestions about who should be in charge of space. The air force, not surprisingly, wanted to be in charge of a nascent aerospace force. The navy, which was stumbling with Vanguard, argued incomprehensibly that space was an extension of the oceans. And the army wanted to conquer the moon. Another proposal envisioned creating a tri-service organization. None of the suggestions made a particularly convincing case for ownership or offered a solution to the mismanagement that had led to the current crisis.

One meeting in particular appears to have resonated with McElroy shortly after he arrived at the Pentagon. Ernest Lawrence, the famed nuclear physicist, along with Charles Thomas, a former Manhattan Project scientist and the head of the agribusiness company Monsanto, visited the Pentagon chief and over the course of a meeting that lasted several hours proposed that the secretary establish a central research and development agency with responsibility for all space research. It was a concept that drew on the legacy of the Manhattan Project, the World War II–era government project to build the atomic bomb.

McElroy latched onto the idea, likely because it sounded a lot like the "upstream research" laboratory he had established at Procter & Gamble. Whether the visitors' suggestion sparked the idea—or merely reinforced a thought he already had—is impossible to know. But on November 7, McElroy wrote to his chief counsel to find out if, as defense secretary, he had the authority to set up a research and development agency without seeking new legislative authorities. The answer from counsel was yes, although it was not clear Congress would agree. By the time McElroy showed up on November 20 on Capitol Hill, his idea had a name, and it was called the Defense Special Projects Agency, a space agency that would make sense of the various rocket programs and other space technology ideas. The new agency would consolidate the Pentagon's missile defense technology and space programs while also pursuing, as the defense chief put it, the "vast weapon systems of the future."

Many of the members of the President's Science Advisory Committee were not enthusiastic about this proposal. Fearful of military pressure to hasten an arms race, Eisenhower had purposely selected the panel to represent the interests of the scientific community over military advisers. The scientists on the committee were not necessarily against the Pentagon's consolidating its rocket programs, though they wondered whether it made sense to place ballistic missile defense and space programs all in one agency. As one committee member put it, missile defense was an urgent priority, while there was "no urgency on Mars."

More fundamentally, the science advisers were concerned about placing the space agency under military control. They eventually acquiesced, likely because James Killian, the president's newly appointed science adviser, supported it. The panel did convince the president that a civilian agency, not a Pentagon agency, should ultimately be responsible for nonmilitary space programs. Eisenhower, in his approval of the new organization, made clear that "when and if a civilian space agency is created, these [space] projects will be subject to review to determine which would be under the cognizance of the Department of Defense and which under the cognizance of the new agency."

The reception within the corridors of the Pentagon to the Defense Special Projects Agency was ice cold. The military services viewed it as an attempt to usurp their authority and steal their money. The new agency was a threat to their turf, and their budgets, and they quickly went on a public offensive to undermine support for the proposal. The air force general Schriever told Congress the new agency would be a "very great mistake." If the military wanted to prove that it did not need a centralized agency for rocket programs, its best bet was to prove that it could launch a satellite into space on its own. To that end, in December, all eyes were on Vanguard, the navy satellite that Wernher von Braun had warned McElroy was doomed to failure.

On December 5, 1957, in the midst of Washington battles over the creation of a new research agency, hundreds of reporters and curious onlookers gathered at Cape Canaveral, Florida, to watch the launch of Vanguard. When Sputnik launched in October, John Hagen, the director of the Vanguard program, admitted the navy rocket was five months behind schedule but blamed the Soviet head start on "unethical conduct," as if a surprise satellite launch were the equivalent of cheating at a tennis game. Now, after hurried preparations, Vanguard Test Vehicle No. 3 was ready for launch. Yet the day of the scheduled launch, technical problems kept pushing back the countdown, and America's best hope for catching up with the Soviets became the butt of jokes. The Japanese newsmen called the rocket "Sputternik," the Germans dubbed it "Spaetnik" (a play on the German word for "late"), and the jaded news crews from Washington, D.C., christened it "civil servant," because it "won't work and you can't fire it."

Finally, the next day, December 6, the countdown to launch began. As the count reached zero, Vanguard lifted off. From beaches just two or three miles from the launch site, hundreds of eager people gathered to watch and cheered as shooting flames marked the liftoff, though giant plumes of smoke obscured their view. The few dozen or so official viewers gathered at a hangar not far from the launchpad could see exactly what unfolded: they watched as the navy's rocket lifted a few feet up and then exploded in a massive fireball, toppling over into the sand. In a sad testament to the failed launch, the satellite itself was thrown out of the third stage of the rocket during the explosion and was found not far away, still emitting the beeping signal that was supposed to mark the United States' first foray into space.

The day of the Vanguard disaster, the chairman of the Joint Chiefs of Staff issued a rare note of "non-concurrence" to the establishment of McElroy's proposed research agency—a bureaucratic expression of extreme disagreement. Had Vanguard not just gone up in a literal ball of flames, he might have had a stronger argument. The new defense secretary held firm, and the next month Eisenhower formally approved the creation of the new agency. McElroy agreed to just one small change to his proposal: to avoid confusion with other, similarly named endeavors, like the Office of Special Operations, the new division would be called the Advanced Research Projects Agency, or ARPA.

ARPA was still an idea more than an organization, and not everyone in Washington was optimistic that a new government bureaucracy would be the solution. The frenetic days leading up to the new agency's opening its doors were a mix of highs and lows in the space race. On January 31, 1958, the von Braun team, which had finally been allowed to join the space race, successfully helped launch Explorer 1, based on its Jupiter-C, putting in orbit the first American satellite. That success was quickly overshadowed by the second attempted launch on February 5 of the navy's Vanguard, which broke apart just shy of a minute after launch.

On February 7, ARPA was officially founded with an intentionally vague two-page directive, which established it as an independent agency that reported directly to the secretary of defense. The directive mentioned no projects, or even specific research areas, not even space. "The Agency is authorized to direct such research and development projects being performed within the Department of Defense as the Secretary of Defense may designate," the directive read. The only hint as to the ultimate purpose for this new agency came just weeks earlier during President Eisenhower's State of the Union address: "We must be forward looking in our research and development to anticipate the unimagined weapons of the future."

CHAPTER 3

Mad Scientists

The man who knocked at Oliver Gale's stately home in Georgetown at 7:45 on Saturday evening looked as if he were arriving at an old friend's house to stay the night. He was clutching a rather large suitcase in one hand and a briefcase in the other and smiled warmly when Gale, the defense secretary's special assistant, opened the front door. After confirming Gale's identity, the stranger was ebullient. "Let me dismiss my cab," he announced.

Realizing the man was now stranded at his house, Gale was worried, not for his own safety, but for his time, which was precious. It was January 4, 1958, and the past few months in Washington had been a whirlwind of congressional hearings, many of them focused on Sputnik, and Gale had lugged home a mountain of work that he needed to finish over the weekend.

In the months following the launch of Sputnik, a procession of nuts, opportunists, and salesmen were trying to get to McElroy through Gale to sell their space and missile schemes, which ranged from nuclear-powered rockets to elaborate moon bases. All they needed was a few million (or billion) dollars of taxpayer money. But midwestern politeness is a hard habit to shake, so Gale reluctantly invited the man inside. Once ensconced in Gale's living room, the stranger began to detail his plan for protecting the United States from a barrage of Soviet missiles. Gale listened politely for an hour, decided the man was a complete lunatic, promised finally to put him in touch with someone in the Pentagon, and sent him on his way.

More than three thousand miles away from Washington, D.C., Herbert York, the director of the University of California Radiation Laboratory at Livermore, had a similar encounter, except his visitor was more of a mad genius than a madman. Nicholas Christofilos, a Greek scientist at the lab, burst into York's office yelling, "They're coming!"

Christofilos was "basically frantic," convinced that Sputnik was the harbinger of a Soviet takeover, York recalled. Whether the Russians were coming or not, Sputnik did prove that the Soviet Union was able to launch intercontinental ballistic missiles, leaving the United States helpless to defend itself. Christofilos wanted to do something about it.

The Livermore lab was known as a place that embraced scientists with mad ideas; after all, it was founded by Edward Teller to build the "Super," a thermonuclear weapon. But Christofilos was unique even for Livermore: he had worked for an elevator repair company in Greece before rising to the elite of nuclear scientists in the United States. His path to a nuclear weapons lab began in 1948 when he started writing letters on improving accelerator performance from his home in Greece to the Radiation Laboratory at Berkeley. Scientists at the lab took to calling him "the crazy Greek." Undeterred, Christofilos filed in the United States for patent protection on his ideas and eventually traveled to America. Not only did he succeed in convincing the government scientists that he was sane; he was even given a job at Brookhaven and then later at Livermore.

Christofilos was wild, known for both his drinking bouts and his prodigious ability to work for days without sleep. When he lectured he gesticulated, scribbling out numbers and ideas faster than most scientists could process. The excitement and fear generated by Sputnik and the Soviets possessed Christofilos. The energy he had once put into accelerators and nuclear energy he now channeled to weapons. His ideas were grandiose and bizarre, but usually so genius that they dazzled the physicists around him. What seemed to attract scientists was that the ideas themselves were scientifically sound but required technological miracles to make them work. In late 1957, standing in York's office, he outlined his most fantastical idea yet.

The plan, as York would later describe it, was to create "an Astrodome-like defensive shield made up of high-energy electrons trapped in the earth's magnetic field just above the atmosphere."

This shield would protect the planet against intercontinental ballistic missiles by essentially frying whatever attempted to pass through the band of killer electrons. "His purpose was of epic proportions," York recalled. "He intended nothing less than to place an impenetrable shield of high-energy electrons over our heads, a shield that would destroy any nuclear warhead that might be sent against us."

Christofilos predicted that there were some electrons already trapped in the magnetosphere, a theory that was confirmed weeks later when the first American satellites detected the trapped charged particles (this region was later named the Van Allen radiation belt, after James Van Allen at the University of Iowa, whose instrument confirmed the existence of the electrons). But the practicalities of what Christofilos was proposing were, as even one of his close colleagues called it, "nutty." Christofilos believed that nuclear explosions could inject a much larger number of high-energy electrons in this radiation belt, and those electrons would destroy any missiles passing through the area. In other words, he was proposing an enhanced version of the naturally occurring electron belt. Rather than a Van Allen belt, it would be a death belt.

York loved the idea. The problem was that generating the shield would require exploding nuclear weapons in the earth's magnetosphere. At the time Christofilos first proposed the idea to York, in the late fall of 1957, there was no way to carry out the necessary experiments. Satellites of the type needed to test the idea had not yet been launched, and Livermore, which was part of the Atomic Energy Commission, could not just carry out its own military experiment; that was the job of the Pentagon.

In the early days of 1958, everyone seemed to have ideas for advanced technology, whether lone oddballs, mad scientists, or large defense companies. The pages of trade magazines like *Aviation Week* were filled with advertisements for "rocket stations in the sky" that would "speed man's conquest of space," nuclear-powered aircraft, and missiles striking the moon. It was only three months since Sputnik, and companies were ready to build a space armada. Secretary McElroy and a handful of other senior Pentagon officials were scheduled in January for a dinner set up by the Aircraft Industries Association, a trade group. The meal was to feature four different fine wines and a litany of complaints about how the Pentagon was not sponsoring advanced technologies. McElroy nixed the wine but listened to the concerns, satisfied he had a polite way to fend off the proposals. The lunatics and the opportunists finally had a place to go with their ideas: ARPA.

In January, McElroy shopped around for someone to head the new agency. Earnest Lawrence had suggested his protégé, Herbert York at Livermore, but McElroy wanted someone from the business world with management experience. McElroy that month met with Ralph Cordiner, the president of General Electric, and Sidney Weinberg, the head of Goldman Sachs, to get their suggestions on who should head the new agency. McElroy came back from that meeting with a name: Roy Johnson, a vice president from General Electric—a charismatic businessman with a reputation as a problem solver. After briefly considering the rocket scientist Wernher von Braun for the job of chief scientist, McElroy settled instead on York. ARPA was given an impressive, half-billion-dollar budget for its first year, but the new agency would own no laboratories, would hire no permanent staff, and would not even have its own offices. ARPA, it seemed, was not part of a grand strategy to remake the Pentagon. It was an expedient solution, and a temporary measure, to show that the administration was taking Sputnik seriously.

ARPA's first director exuded the self-confidence of a successful industrialist. At fifty-two, Johnson was described by one newspaper as "urbane and handsome," and according to ARPA's history he "looked every inch like a Fortune cover tycoon." Lee Huff, one of ARPA's earliest employees, recalled the staff being in awe of the urbane business executive. "He'd show up with these gorgeous tans and strut around," Huff said. "He was fun to listen to. He had been in a lot of tough corporate situations solving difficult problems." When Johnson moved into his office in the Pentagon's prestigious E Ring in early February, just a few doors down from the secretary of defense, he thought he was the CEO of the nation's space program. As a business executive who had dealt with appliances and electronics, Johnson knew little about science and less about space. The bigger problem, however, was that he knew nothing about government, much less government bureaucracy.

He got his first taste of Washington on February 13, less than a week after ARPA was established, when he was taken in a Pentagon sedan for a "roast" of Secretary of Defense McElroy held by the local Saints and Sinners Club. Gale, McElroy's assistant, had been nervous about having his boss agree to participate in the fraternal club's event, but the new defense secretary hoped it might ease his way into Washington's social life. Gale's initial fears were confirmed when the lunch started with a cringe-worthy striptease, followed by a skit mocking the Pentagon's leaders. As the defense secretary and the new director of ARPA watched, "McElroy" questioned "General Medaris" on how long it would take for the Pentagon to get a space vehicle to orbit the moon. The "general" replied, "Eight years, one to learn how to do it, and seven to get the decision out of those fatheads in the Pentagon."

ARPA became the nation's first space agency at a time when the United States was several months behind its Soviet rivals. The new agency had inherited the patchwork of overlapping rocket programs that the National Advisory Committee for Aeronautics and the military services had started in the 1950s. The Soviets had already launched two satellites, including a dog, while the United States had only the von Braun team's Explorer 1, which launched just days before ARPA was founded. ARPA was now in charge of all civilian and military space programs, even though its staff on its first day of operations consisted of Johnson.

Johnson was soon joined by Herbert York, the agency's new chief scientist. The appointment of Johnson, a business executive, and York, a physicist, set up what was to be a larger power play over the new space agency. Perennially late for work and meetings, York would show up wearing rumpled suits or even no suit. His corpulence was an affront to Johnson's spit-and-polish business image. For his part, York thought Johnson was an affront to science. "I went over there as chief scientist, [but] I really determined the program," York said later. Johnson, by contrast, publicly described York as his "personal consultant on scientific matters."

The division of work between Johnson and York quickly became clear, however. Johnson was the chief spokesman for ARPA, traveling around the country to proselytize on space to church groups, professional associations, and schools. York was the scientific headhunter, recruiting the technical staff, while also guiding the direction of the various space programs.

In March, Johnson made his official announcement setting up the structure of the new agency. York, along with holding the title of "chief scientist," served as the head of a technical division of about two dozen personnel contracted from the Institute for Defense Analyses, a federally funded nonprofit research center. Those personnel, all essentially on loan to ARPA, were the scientific talent. By contracting them through an outside institution, the agency could afford to pay them more than their normal government salaries while also avoiding the red tape involved in hiring full-time government employees. There were also a few representatives sent over by the military services, such as Robert Truax, a navy captain who worked under official cover to help manage Corona, the CIA's and the air force's top secret satellite program. It, too, had been swept up in ARPA.

Beyond the technical personnel, there was minimal bureaucracy, in large part because ARPA did not even issue its own contracts, instead using the military services to handle the paperwork. Johnson had a deputy, Rear Admiral John Clark, but there were only two other organizational elements to the new agency: Lawrence Gise, a longtime Pentagon bureaucrat, was appointed as head of administration, and William Godel was named head of the Office of Foreign Developments, which would, according to early ARPA documentation, explore promising foreign research. Because it was decided from the outset that ARPA would not have its own contracting staff, paperwork was limited to brief memos, known as ARPA orders. Policies and procedures were ad hoc, largely a result of ARPA having Johnson as its head, according to Donald Hess, one of ARPA's first employees. "Roy Johnson set the pace for the rest of ARPA and we sort of followed the Roy Johnson bible," said Hess.

Indeed, one of the most enduring features of ARPA—and not necessarily something that was intentional in its creation—was its ability to avoid bureaucracy. ARPA could immediately fund projects that the military services might take months, or even years, to start. For example, in early 1958, two scientists at the Johns Hopkins University Applied Physics Laboratory came up with a novel idea for satellite navigation. They had started off tracking the position of Sputnik by measuring the Doppler shifts of its beeps. They quickly hit upon the idea that this same method could work in the inverse: a signal emitted from a satellite could help determine a precise location on earth if they knew the satellite's orbit. Such a satellite system could potentially help submarine-launched missiles determine their exact location. The satellite navigation project, named Transit, was picked up and funded by ARPA, leading decades later to the Global Positioning System.

ARPA's unique position in the late 1950s was the result of happenstance and necessity. The crisis atmosphere post-Sputnik meant that the Pentagon was willing to allow ARPA to make its own rules in the interests of getting things done, and placing a government neophyte like Johnson as the agency's head meant that it was not going to function like a normal bureaucracy. Speaking to a group of scientists from ARPA and the Institute for Defense Analyses, Johnson laid out his view of the agency's mission: "As gun powder succeeded the sword, and as the hydrogen bomb substantially succeeded the rifle, the question confronting us now is what succeeds the hydrogen bomb."

What could succeed the hydrogen bomb, in the view of ARPA's first chief scientist, was a way to defend the United States from a nuclear attack. In April 1958, shortly after arriving at ARPA, York invited Nicholas Christofilos, the "crazy Greek," to Washington to present his idea for a planetwide force field. When Christofilos first proposed his wild missile defense concept the prior year, after the launch of Sputnik, York had not been in a position to do anything with it. Now York was at ARPA, whose purview over military satellites, missiles, and advanced research seemed tailor-made for the force field idea. Thus was born ARPA Order 4, Project Argus, a top secret program to test whether nuclear weapons exploded in the earth's magnetosphere could create a force field that would destroy incoming missiles. It became by far the largest—and most significant—of ARPA's early schemes. "ARPA is the only place that could pick up something like Christofilos's [idea] and support it," York explained later.

Nothing embodied York's vision for ARPA better than the Christofilos missile shield: a highly speculative military scheme based on pure science. ARPA could move quickly, and that was exactly what Pentagon officials wanted because of "the possibility that events in the near future may create conditions unfavorable to the continuation of nuclear tests," Herbert Loper, the assistant to the secretary of defense for atomic energy, wrote to other senior military officials at the Defense Department shortly before ARPA picked up the project. Those "unfavorable" conditions included an expected moratorium on nuclear testing by the United States and the Soviet Union that would go into effect later that year, making any tests of Argus impossible. York made Argus his pet project, even handpicking the launch site, the uninhabited Gough Island in the South Atlantic, after poring over maps. Why York would select a seemingly outlandish idea to promote as one of the agency's first major projects had to do as much with the emerging battle between White House scientists and the Pentagon as it did with ARPA and its nascent plans for space. Johnson had grandiose visions of ARPA as a permanent military space agency, but York saw it simply as a temporary mechanism for supporting scientific research. Argus, he later wrote, "was interesting science," though he acknowledged that the idea was highly unlikely to ever work.

With ARPA in charge of the test program, the Armed Forces Special Weapons Projects prepared to launch three nuclear shots in a ten-day period in August and September 1958. To carry out the tests, the navy created the top secret Task Force 88, which included nine ships and forty-five hundred personnel. The plan was to launch a low-yield nuclear weapon into the upper atmosphere from ships using an X-17A three-stage ballistic missile. From start to finish, the entire operation was unprecedented. The military had only a few months to prepare the tests, instead of the typical time of more than a year. It was also the first launch of a nuclear weapon from a ship at sea, and the only atmospheric test series conducted in secrecy (Trinity, in 1945, was a single test). To keep Task Force 88 secret, the navy was required to come up with elaborate cover stories for the ships involved. The USS *Norton Sound,* selected to launch the first test, was completely split off from the Atlantic Fleet with the pretense that it would be involved in a series of "preliminary tests" for special missile operations in a remote part of the Pacific. In fact, it was on its way to the selected test area in the South Atlantic.

On August 25 and 26, practice rockets were launched from the *Norton Sound*, under the code name Pogo, as a dress rehearsal for the real shot. Finally, at 2:20 a.m. on August 27, amid rough seas, the ship launched the first missile as other ships and circling aircraft stood by to capture the event and record the effects of the blast. Those aboard ship assigned to monitor the missile stared at the night sky wearing high-density goggles meant to protect them in case of early detonation. On the observation aircraft, one pilot was instructed to keep the protective goggles on for a full sixty seconds, to make sure that in the worst-case scenario, an unimpaired pilot could control the aircraft. After the X-17A launched, everyone watched and waited.

At the thirty-six-second mark, the missile reached 100,000 feet and detonated, and the shipboard observers saw a flash light up the clouds. One of the observation pilots reported "a great luminous ball" about forty degrees above the horizon. As expected, the nuclear blast triggered a visual aurora, created by photons of light emitted from the electrically charged particles decaying back to a lower-energy state. For the next half hour, the crew looked on in awe, photographing the brilliant green and blue colors as they swirled into different shapes like a giant kaleidoscope projected into the night sky. The fruits of Christofilos's fertile mind were a spectacular sight, but was it really a force field?

Two more tests were carried out, on August 30 and September 5—also successes, so far as ARPA was concerned. In a memorandum for the president, dated November 3, 1958, and classified top secret, James Killian, the president's science adviser, effusively praised Argus as a "historic experiment, probably the most spectacular ever conducted." Killian's words would soon be echoed in a much more public forum. On March 19, 1959, *The New York Times* revealed the top secret nuclear tests, with the headline that declared it the "greatest experiment." The article brought worldwide attention to Argus and also to its idiosyncratic creator, Christofilos, who became the object of public fascination.

It was never clear who leaked the details on Argus; George Kistiakowsky, who succeeded Killian as Eisenhower's science adviser, suspected it was someone in the Van Allen lab, while York pointed the finger at a navy science official. Whatever the case, it probably did not matter in the end. Christofilos's shield never came to fruition. Despite all the praise heaped on the tests, the experiments ultimately showed that earth's magnetic field was not strong enough to keep the killer electrons in place for long enough to be useful as a giant shield; the "death belt" would decay too quickly. Even when the Argus shield proved impossible, politically and technically, York gave a tongue-in-cheek defense for having pursued the idea. "There could, however, be another earth, another planet with opposing superpowers, where such a shield might actually be possible and make a difference," he wrote.

Soon after he started Argus, York approved another outrageous—or ambitious, depending on one's point of view—project: an interplanetary spaceship. Dreamed up by Theodore Taylor, a former Los Alamos nuclear weapons designer, Project Orion was a grandiose idea for a spaceship powered by nuclear explosions, thousands of them, according to York's description. In the summer of 1958, ARPA agreed to provide about $1 million to General Atomics to fund preliminary design work on Orion. Roy Johnson, for his part, tolerated the project, telling Congress that he thought Orion was a "screwball" idea when it was first proposed just the year prior, "but not quite as screwball today."

Actually, it was a completely screwball idea, even if the science behind it was sound. Just launching Orion required some two hundred nuclear explosions. George Dyson, the son of the physicist and Orion participant Freeman Dyson, described the spaceship as "egg-shaped and the height of a twenty-story building" with "a 1,000-ton pusher plate attached by shock-absorbing legs." Johnson, in his congressional testimony, explained to lawmakers that the shock waves from the nuclear explosions acting against the pusher plate would function like a spring to propel the space plane forward. The tricky part was to do this in a way "so the inhabitants are not killed."

The most obvious problem with Orion was the sheer impracticality of launching a spaceship by setting off hundreds of nuclear explosions. If the ship crashed, it risked massive radioactive contamination. Even if it did not crash, the fallout from the nuclear explosions used to launch and propel the spaceship would leave radioactive fallout in its wake; Freeman Dyson recounted that at one point scientists estimated that a single mission would result in about ten deaths on earth as a result of increased levels of radiation.

After just eighteen months, ARPA ended its support for Orion, though the air force funded it for several years after. In the end, the spaceship never got far beyond a scale model (and President John F. Kennedy was reportedly "appalled" by the mock-up's five hundred nuclear warheads), but it inspired a cult following among those who believed that nuclear fission and fusion were the most feasible method of interplanetary travel. In August 1958, the United States and the Soviet Union agreed to halt atmospheric nuclear explosions, bringing dreams of space shields and atomic spaceships to an end.

ARPA's turbulent first year of existence was caught up in the much larger debate over who would ultimately control space. While the industrialists brought to Washington seemed to hold sway in the administration, Eisenhower was listening to his science advisers, who were pushing for the establishment of a new civilian agency; ARPA's role as the nation's space agency was merely a place holder. The unresolved question then was over whether ARPA would have a longer-term role in military space, and if so, what would military space encompass? Roy Johnson might not have been a scientist, but he had an instinctive understanding of technologies and their applications. The Soviets' focus on engine thrust is what had enabled them to launch Sputnik, and Johnson knew a powerful rocket would be needed to get anything big, like a space vehicle capable of carrying humans, into orbit. ARPA had already proposed a space plane, called the Maneuverable Recoverable Space Vehicle, or MRS-V (ARPA employees pronounced it "Missus Vee"), and Johnson decided the next step was to fund a booster for the space plane.

In ARPA's version of history, two of its technical staff members came up with the idea of clustering between seven and nine engines to produce 1.5 million pounds of thrust, and then Johnson proposed the idea to von Braun's team, offering to fund it. Von Braun was more equivocal about the rocket's paternity. "We were firm believers in the feasibility of clusters, and the question [of] who made the opening statement is a little bit like who started a love affair," von Braun later said. Regardless of who first proposed the idea, Johnson decided, without consulting anyone in the Pentagon or the White House, to fund von Braun's group to work on a super-thrust booster. Such a rocket would, in theory, take MRS-V into space.

On August 15, 1958, ARPA authorized the start of the new von Braun rocket program, called Saturn, with the rather creative explanation that the clustered booster would be used to put up a variety of large military payloads, such as spy satellites. No one really bought the official explanation for Saturn, particularly not Johnson's chief scientist, who knew that the ARPA director's primary goal was to cement the agency's future role in manned space missions. York also knew full well that Johnson was bucking the White House's clear guidance that the mission to put a man in space would belong to a civilian agency, not ARPA or any other defense agency. Saturn became Johnson's obsession, particularly as the agency was about to face its biggest challenge yet.

On October 1, 1958, ARPA was ordered to turn over its scientific satellite programs to the newly formed National Aeronautics and Space Administration, or NASA, the civilian agency that Eisenhower's advisers wanted from the start. Losing the scientific satellites to NASA was no heartbreak for Johnson. Those were nothing but "a sorry string of failures and struggles to throw up relatively dainty" satellites. Johnson fought to preserve the ability of the Pentagon, and particularly ARPA, to pursue military missions in space even as Eisenhower made clear that this would be NASA's mission, not the Pentagon's, and certainly not ARPA's. "If the DOD decides it to be militarily desirable to program for putting man into space, it should not have to justify this activity to this civilian agency," Johnson told lawmakers, in a flat contradiction of the very legislation that Eisenhower with his science advisers had crafted.

Johnson's clashes with the president's scientific advisers, including York, were beginning to divide the agency. Meanwhile, he also faced opposition from the military services, which looked at ARPA as an unwanted competitor to be struck down as soon as possible. "Beset by enemies internally, subjected to critical pressures externally, and starting from scratch in a novel area of endeavor," was how an early

ARPA history described the agency's first few months. ARPA in its early days was at war with other parts of government, and with itself, as York and Johnson competed over power and vision. The agency that emerged bruised and battered from those battles was not a purposeful creation, as was often later claimed, but an accidental by-product of those rivalries.

CHAPTER 4

Society for the Correction of Soviet Excesses

The week after the Soviets launched Sputnik, William Godel was in Oahu, Hawaii, trying to rein in the rapidly expanding National Security Agency, which was in charge of the country's eavesdropping sites. According to heavily redacted top secret documents released under the Freedom of Information Act nearly fifty years later, Godel was visiting the NSA's cryptologic units as a senior member of the Robertson Committee, a top-level panel that had been created at the behest of President Eisenhower in 1957. Godel's trip to Hawaii was part of a broader look at the NSA's presence in the Far East, and the panel was expected to recommend "radical funding reductions" to the NSA's sprawling intercept sites.

As it had the rest of the national security community, Sputnik caught the NSA off guard, underscoring Eisenhower's belief in the need for immediate intelligence reforms. Godel at that point had spent two years at the NSA, when suddenly he was offered a senior position at the newly established ARPA. He was not a scientist, but then again, neither was its newly appointed head, Roy Johnson. In any case, the agency would be organizing, not conducting, scientific research. At ARPA, Godel's appointment and responsibilities were a constant source of mystery. There were rumors that Godel had been passed over for a top intelligence job and was given the ARPA appointment as a consolation prize, but it was never clear what he did and whom he reported to. "I don't know how he got there," said Donald Hess, one of the earliest employees of ARPA.

Even Godel later said he was unsure who proposed his transfer to ARPA. He knew his name had been dropped in the hat for one of the top jobs at the NSA—the director or deputy director—but the ARPA job came up, and he jumped at the chance to be involved with the secret spy satellite work, which was about to be transferred to ARPA. The former marine was a logical choice in many respects; he had experience managing complex science and technology projects from his time detailed to the NSA. He also had his wartime experience rounding up German scientists. "Godel knew quite a bit about recruiting scientists," recalled Lee Huff, a close associate.

Godel said only that he was asked to be a part of the agency to represent the intelligence community, an account confirmed by early employees of ARPA. He was told, he recounted, that he and his assistant should meet with Roy Johnson about joining ARPA. "Since the Soviets are involved, the Intelligence Services must be involved," Godel wrote. He ended up being a fortuitous choice for ARPA. Initially appointed the head of the Office of Foreign Developments and serving as ARPA liaison to the intelligence community, Godel had a profound and wholly unexpected effect on the agency. He was a military strategist who had minimal interest in futuristic technology and even world-class science. Yet had it not been for Godel, ARPA might not have survived beyond 1959.

Herbert York and Roy Johnson did not get along, but for Godel and Johnson "it was love at first sight." The marine turned intelligence bureaucrat had more in common with the business executive than one might expect. Both were intrinsically clever individuals who knew how to manage technical projects. They also both had an aura of mystique as outsiders brought in to resolve a crisis. In Johnson's case, it was at General Electric, where he was often dispatched to solve a specific problem, while Godel was sent to hot spots around the world to negotiate foreign-government cooperation. Johnson admired Godel's ability to push ambitious ideas through red tape. Godel admired Johnson's vision and fight, even if he recognized it was sometimes to ARPA's detriment.

From the start, Godel understood that the United States was involved in a war of public perception as much as it was in a technological competition with the Soviet Union. When ARPA got off the ground in the first few months of 1958, it was managing the same rockets that had been started years earlier by the services, and none of those projects were going to spark the type of excitement that Sputnik had generated. At best, the United States was going to be playing catch-up. The space race was a propaganda war, and it was a war that the United States was losing badly. Godel wanted something that would temporarily steal the headlines away from the Soviets and their orbiting canines. The United States needed to put something really big up in space.

Godel found his response to the Soviet Union's head start in San Diego. Officials at Convair, a division of General Dynamics, pitched ARPA on the idea of launching an entire missile into orbit. Convair was working with the air force on the Atlas, a liquid-fueled intercontinental ballistic missile. One of the keys to the Atlas was a lightweight design that gave it an impressive range and payload. Its "balloon" tank was made of extremely thin stainless steel and was so delicate that it would collapse under its own weight if not fueled; when empty, it had to be pressurized with nitrogen gas. Convair had a stockpile of the missiles, but one Atlas in particular, numbered 10B, was unique, Godel wrote, because it was "one of the early production models in which every critical parameter came together with maximum rather than nominal performance values."

Convair believed that with just a few modifications to the fuel and the nose cone, "the delicate beast," as the engineers called the missile, could actually make it into orbit. At first glance, it did not seem like much of an offer, because the von Braun team's Explorer had already put a satellite into orbit. Convair was proposing to hurl the entire missile into space, which would, if taken at face value, be the world's biggest satellite. Of course, it would only stay up for about two weeks before its orbit would decay, and the missile—together with its modest-size payload—would burn up in the atmosphere. What usually matters in the rocket game is payload, or the cargo the rocket carries into orbit. Typically, the payload once in orbit is separated from the stages of the rocket that carry the engine and the propellant. The payload on the Atlas was very small, but by throwing the entire missile into orbit, even briefly, Godel and Johnson hoped that the public would focus on the overall size. Godel, who sensed the potential for a public relations coup, pitched the idea to Johnson, who liked it so much that on a trip to Convair he chalked his name on the missile.

The project was named SCORE, officially, short for Signal Communications by Orbiting Relay Equipment. For the few select people cleared into the program, the unofficial name was the Society for the Correction of Soviet Excesses, because that was its more important goal. Launching an ICBM into space was an audacious stunt that played on the public's ignorance of satellite technology, beating the Soviets at their own psychological operations game. "The stated objective was to demonstrate the U.S. interest in peaceful uses for outer space," Godel wrote. "In reality, it was a propaganda ploy designed to put a really big, heavy object into space as a means of silencing press, and congressional complaints about small payloads, and rocket failures."

York hated the idea. "They're going to say that it's 'the biggest satellite' and somebody, somewhere is going to say 'nuts, it just isn't,'" York later explained of his objections. "It's a big empty shell with a 100 lb. [payload]." Deputy Secretary of Defense Donald Quarles did not like it either, looking at it as "a publicity stunt rather than good science." York thought it was too transparent a stunt to even work as propaganda. Johnson "felt that this young man from Livermore just doesn't understand about public opinion. I'm not sure he understood it either but I know that he thought I didn't," York recalled. And so, as York put it, Johnson went "merrily on his way with this Score Project."

Johnson and Godel took the idea directly to Eisenhower, who liked it. So, too, did Secretary of State John Foster Dulles. Jerome Wiesner, one of the president's science advisers, who attended the meeting to discuss SCORE, was dead set against it, but Dulles's enthusiasm won over the president. Eisenhower had one caveat: if the United States was going to make a craven shot at publicity, it had better work. "It must remain secret, I want the absolute minimum number of people over there in the Pentagon to know about it, and if it leaks out in any way, the entire project is automatically canceled," Eisenhower warned Godel.

At Eisenhower's insistence, SCORE was shrouded in such extreme secrecy that it required duping not just the public but also hundreds if not thousands of engineers and technicians involved in launching the rocket. To maintain the ruse, Godel recruited Dan Sullivan, a former FBI agent famous for tracking down the bank robber John Dillinger, to serve as ARPA's security manager. The launch was given the cover story of a routine air force Atlas launch so that if it failed, the Pentagon could deny any knowledge of Project SCORE. "Only 88 people, each required to sign an oath of secrecy, even knew of its existence," the ARPA history says, and they were referred to as Club 88.

Godel says it was Eisenhower's idea to put a scientific payload on the Atlas, perhaps a nod to his scientific advisers. And for that, Godel turned to a group of Operation Paperclip scientists at Fort Monmouth, in New Jersey. While the German rocket makers had ended up in Huntsville, Alabama, the communications engineers had been sent to the East Coast. "They had recently proposed that ARPA fund a communications package to be launched on an as yet unselected carrier," Godel wrote. He decided that ARPA could easily fund the Germans' communications equipment without alerting anyone it was going to go on the Atlas.

The army's "communications package," designed by the German scientists but built by RCA, was actually fairly simple, designed to record, receive, and send voice communications. The voice relay would accomplish two goals: first, it would offer a psychological triumph, by allowing the United States to be the first country to broadcast voice messages from outer space, and, second, it would test whether communications from space might be degraded by highly energized particles in the earth's magnetosphere. The next question, however, was what message should the United States relay from space? Godel proposed to Andrew Goodpaster, Eisenhower's close adviser, that the president personally record the message, but Goodpaster disagreed. "You know perfectly well, Bill, that the President doesn't want to have anything to do with this; he simply does not want to be involved, so you put any message in the thing that you want," Goodpaster told Godel. As a result, Wilber Brucker, the secretary of the army, a man nicknamed Ploopie, was selected to be the first voice that humans would hear from space.

SCORE took eight months of preparation, and the more time

passed, the harder it was to keep it a secret. Just forty-eight hours before the launch, members of Club 88 were scheduled to switch out the missile's blunt nose cone with one that was pointier. Now, anyone with even a bit of technical knowledge would know something was up, but the hope was that even if people had their suspicions, there would not be enough time for them to confirm them before launch. The ruse had to be maintained even to the final hours, when a member of the "club" secretly disabled a mechanism designed to cut off fuel to the main engine, which would normally be used if the missile were heading for the ocean.

Then there was one last hitch. When President Eisenhower was briefed on the final preparations for the launch, he decided that it should be his voice on the message from space after all. At that point, however, the army's communications payload was already buttoned up in the nose cone of the Atlas. The launch director, another member of Club 88, gave Godel a choice: take down the whole payload, which would tip off the growing horde of reporters who had shown up to monitor the launch, or rerecord the message remotely. That, too, had a risk. The new message would be sent by radio to the communications payload, and anyone on the right frequency could pick it up. Johnson left it to Godel. "You are the project manager, Bill," he said.

Godel decided that beaming the president's voice—that is, transmitting the new message over an open radio frequency to replace the recorded message inside the capsule—was less risky than physically removing the entire payload. "So, hoping that no radios would be on at two o'clock in the morning, the President's voice went on the air, and erased that of the hapless Secretary Brucker," Godel recalled. Now the only thing left was to launch the recording before anyone could figure out the missile was heading to space.

On January 21, 1959, as the press gathered to watch the launch, a few reporters suspected that something unusual was afoot. Jay Barbree, an NBC television news reporter, already knew exactly what was going to happen, because while hiding in a bathroom stall with his legs drawn up off the floor to avoid detection, he eavesdropped on an air force general and a "spook" talking about the launch and the president's message. Barbree bragged that "good old RCA, NBC's parent company came through" and told him about the classified mission, even playing for him the president's entire message. The reason he did not air the story prior to the launch was not fear of divulging classified information but concern over losing a scoop: if he revealed the plan, the launch would be scrapped and the whole operation denied. If he kept quiet, he would get exclusive coverage at the White House when the broadcast was announced.

The official story was still that the Atlas was about to be launched across the Atlantic Missile Range and land in the ocean, but as the minutes ticked away, the strangeness of the test became more apparent. Those cleared into Club 88 went to ever more elaborate lengths to cover for modifications made to the Atlas, like a missing transponder usually used for range safety. The deception had been so elaborately constructed that even as the final minutes rolled around, the test conductor and the man responsible for pressing the launch button did not know the Atlas was heading into space.

As Roy Johnson sat in the VIP bunker to monitor the launch, Godel stood in front of the launch control monitor. If anything went wrong, it was up to him to give the order to destroy the missile, which would obliterate months of work—not to mention his and Johnson's reputations. At 6:02 p.m., the missile launched, and then everyone waited and watched. Suddenly the missile veered off course. Yet the members of Club 88, keeping up the charade to the end, did nothing. When the range safety officer saw that the missile was not heading into the ocean, he reached over to push the destruct button and had to be restrained. Godel recalled those first 180 seconds, during which the missile hurtled skyward on its path out of earth's atmosphere, as "the longest of his life."

And it worked. No one outside a small circle of people had an idea about the satellite's true purpose. The next day the Cape Canaveral Communications Center picked up the first broadcast of Eisenhower's Christmas message: "This is the President of the United States speaking. Through the marvels of scientific advance, my voice is coming to you via a satellite circling in outer space. My message is a simple one: Through this unique means I convey to you and all mankind, America's wish for peace on Earth and goodwill toward men everywhere."

When word got to the White House that the missile was in orbit and the broadcast had worked, Eisenhower decided to announce Project SCORE's success at a diplomatic dinner that evening. At a hastily arranged press conference following the dinner, the White House played a recorded version of the broadcast for reporters. It did not matter that most people never actually heard the president's voice broadcast from space and instead heard rebroadcasts carried by television and radio news programs. Nor did it matter that the final words were an "indistinct garble" to the reporters gathered for a press conference to hear the actual broadcast from space. In the end, SCORE proved to be the psychological success that Godel had promised.

York, who had so opposed SCORE, had been wrong. So, too, were the White House scientists. The nuances of "payload," versus the weight of the vehicle in orbit, did not appear to matter to the average citizen, who read headlines like "U.S. Orbits Biggest Moon" and "Ours Is Giant Size!" *Life* magazine ran a photo-essay and behind-the-scenes article commemorating the launch and its cloak-and-dagger drama (with no mention of Godel, who remained in the shadows). The day after Eisenhower's message was broadcast to the world from space, Godel, who had spent the last several months working nonstop on SCORE, took an evening off to attend the NSA's holiday reception. When asked where he had been the past few months, Godel joked that he had been "transcribing Christmas messages for the White House."

Project SCORE, though a success, did little to resolve the fundamental tension between the scientists pushing for a civilian space agency and Johnson, who wanted ARPA to remain a military space agency. If anything, it merely antagonized relations further. Johnson had no more interest in York's interplanetary spaceships than York had in holiday messages from space. By the end of 1958, the rift between Johnson and York took a new turn when York was offered the newly created job as the Pentagon's director of defense research and engineering. The new position would outrank the ARPA director. Told of the decision just hours before a press conference announcing York's new position, Johnson was furious.

ARPA still reported directly to the secretary of defense, but York was now in a more senior position. Nor was Johnson's relationship with Secretary of Defense Neil McElroy ever particularly close, and Donald Quarles, the deputy defense secretary, who had been intimately involved with ARPA's establishment, died unexpectedly of a heart attack just two weeks prior to the announcement. "It was awkward," York said, of suddenly leapfrogging over Johnson.

Awkward or not, York went straight to work stripping the young agency of its most important work, the space programs. He now claimed that ARPA had been a stopgap solution to space, while his new office was the long-term solution that would represent the Pentagon on space matters. York's new position as a "space czar" was to move forward with sound projects and to weed out the more insane proposals, such as those from the military proposing moon bases. ("We ought to consider the possibility of moon-based weapons systems eventually to be used against earth and space targets," Major General Dwight E. Beach, the army's director of guided weapons and special projects, told Congress in 1959. "I would hate to think the Russians got to the moon first.") York saw himself as the only way to stop such craziness. "The guy who denied that the moon had to be captured on behalf of the United States had to be in a position above the Service and also had to be believable for intellectual reasons," he said. "It was a mess, it really was."

A May 27, 1959, article in *The New York Times* declared under the headline "Pentagon Lacks Firm Space Plan" that York, now promoted to the Pentagon's head of research and engineering, was the top space official and not Johnson. In a personal note to Gale, McElroy's assistant, Johnson included a copy of the article, on which he wrote that prior to his death Secretary Quarles "had agreed that 'ARPA' would represent DOD in all space matters. This sort of thing 'confuses' even me." That Johnson was essentially calling on a dead man to support ARPA's claim to space was not a good sign for the young agency.

In June 1959, just six months after leaving ARPA, York wrote to Johnson telling him that he was canceling any more funding for the Saturn rocket, Johnson's pet project, because there was no military justification. York might have supported an atomic spaceship, but he was against Johnson's attempt to use Saturn to keep ARPA involved in space. When Johnson protested, York relented on cancellation but insisted that Saturn be moved to the newly formed NASA, which still enraged Johnson. "The date of the transfer of the Saturn program from the Pentagon to NASA may go down in history as the point when the United States firmly committed itself to being a second-class military space power," Johnson told Congress. What Johnson saw as a disaster, York marked as a victory. He later wrote that the transfer of Saturn, along with von Braun's entire team, to the newly created NASA was the "single most important act of my tenure in the Pentagon."

York watched with slight amusement as Johnson's relations with the White House went from bad to worse. Godel was stuck in the middle: he was realistic enough to know that if Johnson was going to fight the White House, it was a losing battle. Saturn lived, but it would belong to NASA, and the president rebuked Johnson. The ARPA director no longer cared, because he was on his way out. "Saturn was the biggest Roy Johnson contribution and he did it over the dead, dying and bleeding bodies of just about everybody," Godel later said.

The immediate question for York, once he was installed as the Pentagon's space czar, was why even keep ARPA? There were already calls for the agency's abolishment, particularly from General Bernard Schriever, who argued the agency's work properly belonged with the military. Other critics called for ARPA to be absorbed by York's new office. By 1959, ARPA's portfolio of space programs had been whittled down to military and intelligence satellites, and it was not clear how much longer it would keep even those. Of those, its most important program was Discoverer, a series of satellite launches that was, among other purported goals, testing life-support systems in space. Godel was one of the few officials at ARPA who knew about Discoverer's true purpose.

On April 13, 1959, William Godel, recently elevated to deputy director of ARPA, stood with a gaggle of journalists on a wooden grandstand ten thousand feet from a launch site at Vandenberg Air Force Base in California. With the public transfixed by the emerging space race between the United States and the Soviet Union, rocket launches of any type were major news events, guaranteed to make headlines. As journalists filled their coffee cups and milled about the rows of typewriters and telephones, Godel briefed them on the impending launch of Discoverer 2, a satellite built by Lockheed Aircraft Corporation. Discoverer had a purely scientific mission, he said, designed to test an "environmental capsule" in space. "You could call the environmental capsule a life support system," Godel told one reporter.

It was all a lie. While Johnson and York were battling it out over space exploration, Godel was spinning stories to cover up secret spy satellites. Discoverer was just a cover story for what would become known later as Corona, a joint CIA–air force satellite program that would take pictures deep inside the Soviet Union. The life-support research was nothing more than an elaborate cover story to deceive reporters, and, more important, the Soviets, and obscure the true purpose of the launches. Discoverer was really the world's first reconnaissance satellite.

The idea of taking pictures from space was, in 1959, still a novel idea. But for five years, the air force had been secretly working on a spy satellite called Weapon System 117L, code-named the Pied Piper, based on a concept proposed by the Rand Corporation. Rand's proposal involved launching a camera on a satellite, which would take pictures of military facilities as it overflew the Soviet Union and then jettison the film, which would be picked up by an aircraft as it floated back to earth. It was technically daunting, but if it worked, it would help resolve questions over the purported missile gap by providing crucial imagery of the Soviet Union. The U-2 spy aircraft could take pictures deep inside the Soviet Union, but it was vulnerable to being shot down, and thus subject to more restrictions. A satellite, on the other hand, could overfly the Soviet Union without being accused of airspace violations associated with an aircraft.

Corona had to be kept secret, however, thus the cover story about the environmental capsule. Even the animals were fake. Inside the capsule of Discoverer 2 were four "mechanical" mice, which were really just small electromechanical devices that mimicked signs of life. The decision to use the mechanical mice was made after two sets of live mice had died, raising the ire of the American Society for the Prevention of Cruelty to Animals. In one launch, the mice died prior to liftoff after they ingested bits of paint sprayed on their cage (believing the mice might be asleep, engineers went up and banged on the capsule, even mimicking a cat's meow). Killing mice in a program that was ostensibly supposed to demonstrate a life-support system was, in general, rather bad publicity, so for the April 13 launch the Pentagon had opted for the mechanical mice. It was to be a good choice, because, as Godel told the reporters, recovering the vehicle was a "remote possibility." In reality, the United States had every intention of recovering the capsule, because it was really designed to contain film, not mice.

At 12:45 local time on April 3, Discoverer 2 launched, and onlookers watched as the missile carrying the mechanical mice passed by the sun, leaving a short-lived trail etched in the sky. The launch was a success, and two hours later ARPA held a celebratory press conference. The next day, however, ARPA announced that plans to recover the capsule had gone awry. The plan was for the capsule to be ejected at a precise time and place so that it would eventually land over Hawaii, where it would be picked up by the air force, but the ground controllers "goofed" and sent the signal at the wrong time. While the worst-case scenario was that the capsule would land in the middle of the Soviet Union, the second worst-case scenario took place. It landed somewhere in the vicinity of Norway's Spitsbergen Islands inside the Arctic Circle, not far from the Soviet Union. Making matters worse, the island was home to Russian mining villages permitted by a treaty signed in 1920.

The mission to find the capsule became the stuff of legend. An air force officer in charge of launch operations claimed that the military got hold of "two guys" in Longyearbyen who saw it come down. Panicked that the Soviets might acquire the capsule, which could reveal critical aspects about the program, the air force launched a recovery mission. Colonel Charles "Moose" Mathison, who, despite not being privy to the full details of Discoverer's true purpose, decided to take recovery matters into his own hands. He hopped on a commercial airliner to Oslo and then persuaded a general in the Norwegian air force to fly to Spitsbergen. Mathison was determined to use Discoverer to garner publicity for the air force and soon reported back that tracks in the snow indicated that the Soviets had indeed nabbed the capsule. An official air force recovery team, headed by Colonel Richard Philbrick, traveled to Spitsbergen and gave local residents colored crayons, asking them to draw what they had seen. The residents drew "a gold bucket and light colored shrouds leading to an international orange and silver parachute—exactly right."

Godel, who was also dispatched to Norway to launch a search from an air base in Bodø, was doubtful of the purported sightings. "The mission was foredoomed to failure," he wrote. No one there had any way to track the incoming capsule, and the chances of someone on a sparsely populated remote island spotting a single capsule were unlikely. In that part of Norway, people barely had electricity, and they were certainly not spending time outside, looking up at the sky. The capsule, Godel concluded, "was buried out there somewhere in a mountain of snow and ice covered with wind-blown coal dust that had not melted since the days of the Mastodon, and no one was going to find it until the end of the next ice age, if then." The capsule was never found.

It took twelve more launches, but on August 18, 1960, film from the launch of Discoverer 14 was recovered by an air force C-119. Within hours, analysts were able to look at the first earth photographs taken from space: grainy images of the Mys Shmidta air base in the Russian Far East. That same week, Gary Powers, the American pilot of a U-2 shot down earlier that year over the Soviet Union while trying to collect imagery, was tried and convicted in Moscow. The era of satellite reconnaissance was launched.

While ARPA, as the Pentagon's space agency, ostensibly had control over the spy satellite program, the management was convoluted. According to an official history of Corona, declassified only in 2012, ARPA was in charge of the funding for the overt elements of Corona, that is, the Discoverer cover story; the CIA was in charge of its real objectives. On paper, at least, Roy Johnson was signing the directives, and there is no indication in official records that the White House regarded ARPA as merely a purse holder.

The battle over control of rockets was an annoying sideshow for the CIA. The air force and the CIA believed that ARPA, which controlled funding for the launches, was attempting to build its own empire in space and taking advantage of the biomedical cover story for its "man in space" work as a "counterweight to the announced NASA program."

In other words, they believed ARPA was using the funds from Corona, which constituted a large portion of its budget, to justify its continuing role in space. The air force was never happy about having another agency manage any of its satellite programs, but now it also had the CIA on its side, accusing ARPA officials of "interference" with Corona.

Those conflicts were soon to end. York was now installed in his new Pentagon position over ARPA, and at his recommendation Secretary of Defense McElroy on September 18, 1959, authorized the formal transfer of ARPA's military space projects back to the services. All of the agency's space programs, including Transit, which would eventually become the basis for the Global Positioning System, were given back to the military services. For ARPA, and Johnson, control of the military space programs was the heart and soul of the agency and, more important, the primary reason for its existence. As quickly as things had gotten started, the agency that was supposed to lead America into space seemed to be falling apart, and Johnson could do nothing to stop it. After a contentious press conference at the Pentagon to announce the new space plans, George Kistiakowsky, the presidential science adviser, wrote that Johnson was "mad as hell and will now resign as director of ARPA."

In November 1959, Johnson announced he was leaving his post to become a "professional artist." He left disillusioned with the agency, with the White House, and with space policy. The military had lost the manned space exploration mission, and ARPA had lost space completely. ARPA's remaining work, ballistic missile defense, was nowhere near as glamorous as space travel. Yet Pentagon officials still insisted they would keep ARPA around to pursue advanced research, and Charles Critchfield, a prominent scientist, was handpicked to be Johnson's successor.

Critchfield was supposed to come to Washington as a salaried employee of Convair, which would continue to pay him his $40,000-a-year salary, and Critchfield would recuse himself of any contracts involving his company. The government would pay him a symbolic $1 per year, a carryover from the "$1 contracts" used during World War II, when companies worked closely with the government. But the war was long over, and lawmakers were growing increasingly skeptical of such arrangements. After Congress raised concerns about the potential for a conflict of interest, Critchfield pulled out of the job, leaving the agency without any director.

ARPA was an expedient solution born in the midst of crisis, and as 1959 came to a close, its brief but chaotic life seemed to be almost over. In less than two years, it had taken the mantle of the country's first space agency and had pushed its agenda for advanced technology aggressively. Yet it had lost more battles than it had won, and it still had neither its own offices nor any permanent employees. That ARPA survived much beyond 1959 could be credited, not completely, but in large part, to Godel, who understood how badly ARPA had bungled relations with the White House and the scientists who held sway there. "We mistreated [the scientists]," Godel admitted in an unpublished interview, "we crushed their nuts."

Just prior to his departure, Roy Johnson wrote a memo to McElroy, looking at options for the agency's future, urging the defense secretary to keep ARPA around as a way to pursue advanced research. The question was what, in the absence of space, should ARPA do? The agency still had work in missile defense, and in a few smaller areas, like propellant chemistry, but nothing that would sustain the interest of top scientists or leaders. Neither space nor science had been of particular interest to Godel. The seasoned intelligence bureaucrat had his sights set not on the moon but on the jungles of Vietnam.

CHAPTER 5

Welcome to the Jungle

In 1950, William Godel stood next to the marine general Graves Erskine looking at a ridgetop in Vietnam, where smoke from an encampment could be seen curling up into the sky. Erskine turned to the French general accompanying them and asked whose forces were on the ridge.

"Le Viet Minh, Monsieur," the commander answered, referring to the communist forces fighting the French.

"Well why don't you send a battalion up there to route them out?" the American general demanded.

"*Mais non, monsieur:* If we sent a battalion up there the Viet Minh would disappear into the jungle," the French general replied. "This way, we know where they are. *N'est-ce pas?*"

General Erskine, who had led the Third Marine Division in its World War II assault on Iwo Jima, was "angered to the point of nausea," Godel recalled. "He got very drunk that night, something he rarely did."

Godel and Erskine were in Vietnam as part of an extended study trip authorized by President Harry Truman, known as the Melby-Erskine mission, that was to look at American economic support for the region. John Melby, the American diplomat who co-headed the mission with Erskine, saw the trip as "a piece of blackmail" the French used to pull America into Vietnam. If the Americans did not assist the French, the communists would win the elections in France and pull out of NATO. The ruse worked well, because Truman later declared that the United States would help the French contain the communists. Four years later, President Dwight D. Eisenhower would expand that logic beyond Vietnam, arguing that losing one country to the communists would be like a "row of dominos," where "you knock over the first one, and what will happen to the last one is the certainty that it will go over very quickly."

Yet the four-month trip in 1950, spent mostly in Vietnam, foreshadowed how perilous American involvement would later become. The situation on the ground shocked officials on the mission, making them realize just how little the United States understood the region. "There was not one single Officer in that Embassy who spoke a word of Vietnamese," reported Melby, who wrote a ten-page single-spaced telegram back to Washington. In Melby's view, the decision to provide economic aid was a "terrible mistake" that would slowly lead to disastrous involvement.

Godel's memory of the trip was equally bleak. The mission convinced him that sending Western ground troops—whether French or American—with their modern technology into Vietnam was a losing proposition. The United States would have to follow a different path if it wanted to avoid getting sucked into the quagmire that had trapped the French in Vietnam. The trip also sparked Godel's passion for Southeast Asia, where he sensed a growing, but ignored, Cold War threat. While top-level planners in the Pentagon were focusing on Soviet nuclear and conventional forces, Godel recognized that the United States was more likely to face small-scale insurgent wars in regions like Southeast Asia and the Middle East. The United States, he concluded, had to learn to win wars that did not involve using nuclear weapons, or battling Soviet conventional forces in Europe. Even more important, it had to find a way to do it without sending its own troops.

In the 1950s, Godel remained heavily involved in Southeast Asia, visiting the region frequently, making friends with high-level Thai, Filipino, and Vietnamese officials. So close were Godel's relations with some of those officials that their children would often spend summers staying at the Godel family home in Virginia, or even longer to attend school in the United States. Godel also worked closely with Edward

Lansdale, one of the nation's most famous counterinsurgency experts, who had extensive experience in Asia.

Lansdale, a former advertising man turned air force officer, had forged his reputation working for the CIA in the Philippines, where he helped President Ramon Magsaysay fight a growing insurgency. Employing a unique mix of American advertising savvy and CIA tricks, the American spy introduced psychological operations techniques that preyed on local superstitions. In one case, he persuaded Magsaysay's government to take advantage of peasants' fears of *asuang*, or vampires, by ambushing a communist rebel, draining his blood, and putting puncture wounds on his neck, leaving the body to be found by his comrades. Whether such maneuvers really helped quash the insurgency, whose support was already waning, is unclear, but it cemented Lansdale's reputation as one of the best-known intelligence operatives in history.

In 1954, Lansdale was assigned to head the CIA's Saigon Military Mission, and he soon insinuated himself into the inner circle of the newly appointed prime minister, Ngo Dinh Diem, repeating the psychological warfare techniques he honed in the Philippines. He ordered the printing of an almanac based on soothsayer predictions that foretold communist defeats. He also helped Diem win a 1955 referendum against Bao Dai, the former emperor, by printing Diem's ballots on red paper, an Asian symbol of joy, while Bao Dai was left with a somber shade of green. Diem took what Lansdale viewed as a subtle but effective tool to sway—or rig—the elections and turned it into a sledgehammer, claiming 98.2 percent of the vote, which reinforced claims of widespread fraud. Lansdale was appalled but did nothing.

Lansdale and Godel shared a similar intellectual disposition toward dealing with the communist guerrillas in Southeast Asia. The goal, they believed, was to focus on defeating insurgencies by building support for the local government, even a flawed one, led by President Diem, whose brutal tactics and authoritarian bent were alienating much of the population. The two men saw in Diem, whatever his faults, a kindred anticommunist spirit and an inspired leader who just needed some American guidance. Despite mounting evidence of brutal repression and rampant corruption, Lansdale and Godel were convinced that Diem was the best hope for Vietnam. Lansdale so frequently reminded Diem that he was the George Washington of Vietnam and the "father of his country" that at one point Diem snapped at Lansdale to stop calling him "*papa*."

Godel was equally taken with South Vietnam's president. After one meeting with Diem, arranged by Lansdale, Godel, then deputy assistant secretary of defense for special operations, pledged his support in a letter to South Vietnam's president. "Please be assured that I shall continue to have a great interest in the affairs of your country and am always ready to be of such assistance as may be within my power," he wrote to Diem in 1956.

Two years after writing that letter, Godel was swept up into ARPA, and it seemed at first that his job at the new agency would be focused on technology and space, not Southeast Asia. But by 1959, the agency was left without a director or a clear mission. As a place holder, Herbert York, who now oversaw ARPA from the Pentagon, installed Austin W. Betts, a hardworking, if scientifically uninspired, army general who had been working in the Office of the Secretary of Defense as the director of guided missiles. York told the general, "You're going to be director of ARPA." Betts saluted and said, "Yes, sir!"

York was very specific about what he wanted Betts to do: almost nothing. York referred to Betts as a "custodian," and the new director got the feeling that the agency was not meant to be around long. "During my one year, I don't believe that we did anything that would have laid big foundations for big future programs," he recalled of his brief time at ARPA. "Herb York made it very clear that he wanted me to play a kind of steady-as-you-go role and not pick up any controversy on any major programs." Godel was an exception for that time period, Betts later recalled, saying his "ideas were a little wild but he was a very bright guy." Godel's wild ideas were about to change the course of ARPA.

In early 1960, as ARPA was drifting into irrelevance, one of William Godel's former colleagues from the Pentagon's Office of Special Operations was brainstorming with fellow military officers about a new era of warfare. In the late 1950s, Samuel Vaughan Wilson, an army officer, had worked together with Godel and regarded him as a man of great drive and energy, if sometimes too ready to put himself, and others, in peril for covert missions. Both men were inspired by Edward Lansdale's vision of counterguerrilla warfare, which required different weapons, techniques, and training. Most important, it required keeping American soldiers out of those conflicts, focusing on making sure local forces could do the job on their own.

The United States was finding itself involved in low-level conflicts around the world—in Vietnam, Cuba, and Lebanon—often advising local governments on how to fight. Wilson, then a lieutenant colonel and director of instruction at the U.S. Army Special Warfare School at Fort Bragg, North Carolina, wanted a name for what military advisers were doing in places like Southeast Asia, where they were working with local government forces battling guerrilla movements. "We were trying to figure out what title to give our work. What would be the brand name, so everyone would know it," recalled Wilson, who later went on to become a lieutenant general.

Wilson and his colleagues at Fort Bragg spent one furtive evening brainstorming, trying to come up with a name. "Counterguerrilla warfare" was the term that had been used up to that point, but Wilson did not feel the term captured what he had in mind. Nor did "counterresistance," because American forces had often worked with "the resistance," particularly during World War II. The communists had already appropriated the term counterrevolutionary. Finally, at 2:00 a.m., Wilson approached the blackboard and scribbled, "Counterinsurgency." One of the men in the room declared, "That's it, let's go home."

That same year, Godel pitched York on a counterinsurgency mission in Southeast Asia. York, whose office now presided over ARPA, agreed to allow Godel to travel across the region to explore cooperative research and development programs for the erstwhile space agency. So with York's blessing, Godel set off to study what he believed would be the future battlefields of the Cold War. Between October and December 1960, he traveled across Asia, looking at everything from military footwear in Thailand to a weather research center in the Philippines. Behind those seemingly modest issues was a larger question: Why were the communists gaining a foothold in the region, and what could the United States do to counter them?

The resulting classified report concluded that the United States was woefully unprepared for coming conflicts in the region. Sophisticated American weaponry given to friendly regimes required American mechanics, which meant in many cases the donated equipment was left unused. The United States, Godel wrote, also had precious little idea how to compete with communist ideological influence in the region and needed to come up with research "that could assist in improving the over-all military capability in the insurrectionary, terrorist and guerilla operations in which we are engaged."

Looking at the recently concluded Korean War, and the brewing conflicts in other parts of Asia, Godel pointed out that the United States was planning military options—conventional and nuclear—for a war in Europe, and yet it was woefully unprepared for the conflict already taking place in Asia. The United States, he concluded, had "devoted inadequate attention to winning the kind of war in which we are engaged. In this war the situation is such that the Free World sustains the very real risk of total defeat and possesses little capability for even token victory."

Godel returned from his trip with a recommendation to establish an ARPA facility in Asia that would be used to experiment with techniques and technologies to fight against guerrillas in the jungles of Thailand, Vietnam, and the Philippines. His plan was much bigger than anything anyone had ever envisioned for ARPA: he wanted to develop the indigenous capabilities of the military forces in Southeast Asia in order to prevent the introduction of American conventional forces. York, who was happy to have ARPA work in an area that did not involve the space race, signed off on it.

When President John F. Kennedy came into office in January 1961, he was faced with a deepening crisis in Southeast Asia. Laos was in the midst of a civil war, the Vietcong insurgency in South Vietnam was expanding, and the Soviet leader, Nikita Khrushchev, had pledged support for "wars of national liberation." A report written by Godel's colleague and mentor, Edward Lansdale, caught Kennedy's eye. Lansdale, who had just returned from a trip to Vietnam, concluded that the Vietcong were determined that year to launch a major offensive and take the south. "This is the worst yet," the president declared after reading how Lansdale described the situation on the ground.

On January 28, 1961, Lansdale, then a brigadier general, was invited to the White House to brief Kennedy and other senior officials on the situation in Vietnam. Lansdale was a hit with the president, who loved the secretive world of spies and derring-do. Just a year earlier, Kennedy had dined with the James Bond author, Ian Fleming, and he even listed Fleming's book *From Russia, with Love* as one of his favorites in a 1961 profile in *Life* magazine. It is no surprise, then, that Lansdale's description of counterguerrilla warfare persuaded the new president, despite considerable skepticism from some of his key military and diplomatic advisers. The president wanted to send Lansdale back to Vietnam as ambassador, but the State Department objected, and Lansdale instead took over as head of the Office of Special Operations in the Pentagon.

From the Pentagon, Lansdale used his new position to continue supporting the Diem regime, working closely again with Godel. On May 11, 1961, the White House approved "A Program of Action to Prevent Communist Domination of South Vietnam." Among other measures, it gave presidential approval for ARPA to set up a counterinsurgency combat center. Lansdale that month wrote to Harold Brown, a physicist who succeeded Herbert York as the director of defense research and engineering, asking him to assemble people "to acquire directly, develop and/or test novel and improved weapons, military hardware, for employment in the Indo-Chinese environment." That team of people, Lansdale wrote, in the highly classified memo, should "immediately initiate planning with the services to dispatch to the field at the earliest possible time a small team capable of rendering initial assistance in this matter to the Viet-Namese armed forces."

The next month, ARPA Order 245 was signed, allotting $500,000 for the Combat Development and Test Center and to fund the start of Project AGILE, an umbrella name for Godel's plan to conduct counterinsurgency research in Southeast Asia. AGILE would soon grow to become the third-largest project in ARPA, in terms of money, but it was even more significant measured by the attention it received from the White House. ARPA, an agency founded to take America to space, was now jumping headlong into the counterinsurgency business.

In an interview years later, Godel said there were two aims to the ARPA program: to help local governments, like that of South Vietnam, learn to fight an insurgency; and to help American personnel fight in very limited engagements, without conventional troops. The idea of

AGILE was to provide "policy options without massive introduction of U.S. troops," he explained. The entire goal was "to let others fight or do it ourselves without massive commitment."

By early 1961, Godel had managed to use the power vacuum at ARPA to carve out a new role for the agency in Vietnam. Reporting directly to Lansdale, he conducted work so secret that even the heads of ARPA, let alone the rank-and-file employees, were unaware of the specifics. A research and development agency might have looked like an odd place from which to launch counterinsurgency warfare, but Godel saw it as an opportunity. It was a young, well-funded agency that operated below the radar screen and had plenty of experience from its space days of operating with a "black budget," or secret money.

Then, on May 25, 1961, President Kennedy set an ambitious goal for the country that brought the space race between the United States and the Soviet Union to an entirely new level. "I believe that this nation should commit itself to achieving the goal, before this decade is out, of landing a man on the moon and returning him safely to the Earth," Kennedy announced. For Godel's colleagues at ARPA, the president's announcement was bittersweet, particularly because the Saturn rocket that would eventually take the first men to the moon was one of ARPA's early projects. The space race was still very much front and center in the nation's Cold War battle with the Soviet Union; it just was not ARPA's battle anymore. Instead, Godel and a suitcase of cash—essentially a down payment for Project AGILE—were on their way to Vietnam to establish a counterinsurgency center. ARPA would stay in that business for the next ten years, eventually expanding it into a worldwide scientific program.

On June 8, 1961, Godel arrived in Saigon with cash and presents. He packed an assortment of small gifts for officials and their families he might meet along the way, including rosewood jewelry cases for women and two dozen Parker Jotter mechanical pencils. For Diem, however, Godel brought something special: a spy camera disguised as an elegant gold-plated lighter. It was an ingenious device, featuring a fast shutter camouflaged by a gold leaf, and a wide-angle lens. The hidden camera could be loaded with 16 mm film and was designed to snap people's pictures surreptitiously while lighting their cigarette. That sort of spy gadget, Godel knew, would delight Diem. He was looking forward to the meeting. He regarded Diem as a good friend with whom he shared many fine evenings of dinner and conversation. Diem, according to an associate, regarded Godel with fear.

When Godel arrived in Saigon with his briefcase of cash, he went immediately to the presidential palace, a grand relic of French colonialism with rooms adorned with porcelain and wood paneling and appointed with overstuffed velvet and silk-embroidered chairs that required constant care to prevent them from sprouting mold in the humidity. As Diem, clad in his usual double-breasted white sharkskin suit, walked into the foreign reception room, he was flanked by two aides: Truong Quang Van, who worked for the intelligence chief, and Bui Quang Trach, a bespectacled military colonel and trusted adviser to Diem.

The gold-plated spy camera Godel brought as a present for Diem was a hit, and everyone admired the elegance of the hidden lens. The gift was just a prelude to the real goal of getting Diem's approval for ARPA's Combat Development and Test Center, which would test, research, and develop tactics and technologies to help the South Vietnamese military defeat the Vietcong. Godel came equipped with an entire laundry list of ideas: "an airborne Volkswagen," a power glider that could fly for hours on a single tank of gas; a steam-engine paddleboat that could carry more than two dozen men, navigating through waters as shallow as a few inches, and run on cane alcohol; chemical defoliants to remove jungle cover; and, more exotically, a "hormone plant killer" that would specifically target cassava, the Vietcong's food source. He also wanted to bring over military dogs that would help hunt Vietcong in the jungles.

President Diem did not necessarily agree with all of the ideas. He laughed at the dog proposal, telling Godel the poor creatures would simply succumb to heat and illness in the jungle and die. But the president agreed to give the dogs a try and claimed to be enthusiastic about Godel's other plans. He approved the creation of Godel's proposed Combat Development and Test Center, and assigned ARPA the old French barracks on Ben Bach Dang Street in Saigon. Diem also tapped his associate and trusted aide Colonel Trach as the Vietnamese head of the new center, which would be run jointly with ARPA.

The president might have liked the combat center, but he greeted Godel's other major proposal at the meeting with more skepticism. Godel wanted to relocate hundreds of thousands of Vietnamese peasants from their villages, which were often and easily infiltrated by the communist guerrillas, and move them into fortified encampments, called strategic hamlets. Those hamlets, in turn, would be fortified with a mix of physical security barriers, such as moats and bamboo-spiked walls, and sensors and alarms, which would be developed by the new combat center. The Vietcong would no longer be able to infiltrate the villages, stealing food, killing government loyalists, and abducting new recruits, and the government, in turn, could attack and destroy Vietcong encampments with impunity, without having to worry about enraging peasants caught in the cross fire.

Godel's proposal for population resettlement was not entirely new. In 1959, Diem had embarked on his own experiment with small-scale population resettlement to separate loyal peasants from Vietcong supporters, called the Agroville program, a plan inspired by the British experience in Malaya. Agroville presumed that if communist insurgents were sneaking into villages, then the best solution was to concentrate the "loyal" villages in a defensible area, allowing the military to root out the insurgents, now deprived of civilian cover.

The Agroville plan collapsed quickly under the weight of corruption and incompetence. Construction of new villages relied on the labor of peasants, who had no motivation to work for free, particularly at the expense of harvesting their own crops. Those who could afford to do so bribed officials to get out of indentured service, and eventually the entire enterprise was abandoned. Diem, fresh from the failure of Agroville, was understandably reluctant to embark on any new resettlement ventures. He did agree that Godel could travel around Vietnam to survey possible sites for the new hamlets and to look at ways to secure villages (Diem accompanied Godel on at least one of the trips, to Pleiku, to study border security). Godel set off with Van, an assistant to Diem's chief of intelligence, giving gifts and cash to village leaders in exchange for information and support.

By the end of Godel's trip in June, the Combat Development and Test Center was up and running. The staff was still small: three American military officers, four enlisted men, and twenty-three Vietnamese personnel. ARPA also planned to send two scientists on a temporary tour of duty. Godel returned to the United States on June 28 and immediately went to work promoting both the combat center and strategic hamlets to policy wonks in Washington. Robert Johnson, a State Department official, reported on a talk Godel gave at the Foreign Service Institute in July 1961 surveying the ways that the United States was helping Diem fight the growing insurgency. "Godel has suggested to Diem that a policy of temporary displacement of populations along the lines of the Malayan operation might serve his purpose well, without encountering the difficulties of the Agrovilles," Johnson noted. "Diem seemed interested."

It took three months between June and August 1961, and around fifty study trips across Vietnam, to eventually convince Diem of the strategic hamlet program. By early 1962, Diem agreed to massive population resettlements along the Mekong delta and the Central Highlands that eventually became the linchpin of the regime's counterinsurgency program, with more than twenty-five hundred hamlets established by August 1962. The breathtaking goal was to concentrate about 90 percent of the rural population into the strategic hamlets by September 1963.

The Defense Department's internal study of the war, known as the *Pentagon Papers,* would later claim Robert Thompson, head of the British Advisory Mission, was the one who proposed and persuaded Diem in December 1961 to pursue strategic hamlets. But by the time Thompson showed up, Godel had already spent months laying the groundwork. Van, the Vietnamese government official who was Godel's traveling companion on his trips in the summer of 1961, made clear who was responsible. "Only one man helped me and my team to instill the idea [of strategic hamlets] to our government," Van told American officials investigating Godel in 1964. "Mr. Godel, and his team." At that point, Van was sitting in a jail cell. Godel soon would be, too.

In Washington, President Kennedy was torn over the Vietnam situation. Though he was heavily influenced by proponents of counterguerrilla warfare, the worsening situation in Vietnam was providing ammunition for those, like Secretary of Defense Robert McNamara, who were pushing to send troops. Unsure of the situation, Kennedy decided to dispatch General Maxwell Taylor and Walt Rostow, an adviser, to Vietnam on a mission that would end up being one of the key turning points of early American involvement in Vietnam. The Taylor Mission, as it was called, arrived in Vietnam in mid-October 1961, just days after a critical National Security Council meeting with the president to discuss the military options. The Joint Chiefs of Staff had suggested that forty thousand troops would be enough to "clean up the Viet Cong threat."

It was against the backdrop of growing calls for conventional troops that Taylor and his group of some half a dozen advisers landed in Saigon. Among those on the trip were Lansdale, the counterinsurgency adviser, and George Rathjens, another ARPA official. Godel, who was already in Vietnam, joined the mission once they arrived. The mood was relaxed—even convivial. Taylor dressed in civilian clothing, and Godel and Lansdale played daily games of doubles tennis against Taylor and Rostow. Over the course of the several-weeks-long mission, Godel tried to impress upon Taylor the importance of developing technologies and tactics specific to guerrilla warfare. Godel showed them the combat center's early work: the military dogs, the sensors, and something called a Q truck, which was an armored truck disguised to look like a commercial vehicle. At one point, Godel accompanied Taylor and Rostow on a flight over Vietnam's waterways, to show them how South Vietnamese river craft were being attacked by the Vietcong while en route to Saigon. Godel explained how ARPA was using Q trucks and Q boats as bait to draw out the Vietcong.

The Q truck was named after Q boats, the decoys that were first used by the British and the Germans in World War I to entrap military vessels. A Q boat was a "wolf in sheep's clothing," disguised to look like a vulnerable merchant vessel but in fact heavily armed, and would attack unsuspecting military ships lured into the trap. Similarly, the armored Q truck would, on the outside, look like South Vietnamese ammunition supply trucks frequently targeted by the Vietcong. On his first trip to set up the ARPA Combat Development and Test Center, Godel had brought cash to build the fake supply truck using armored materials salvaged from old French military vehicles and then welding them onto the inside of a two-and-a-half-ton Japanese commercial truck, the type of supply vehicle commonly used by the South Vietnamese military. The Q truck would not simply entrap a few Vietcong fighters; it would also be used to collect data to learn about the nature and frequency of the attacks. The group liked these novelties, Godel later recounted, but "in all truth General Taylor also thought it was somewhat Mickey Mouse, and that regular forces, regularly equipped, and operating in regular formations with heavy weapons could better do the job."

Taylor liked the technology, but Godel's attempts to sell the general on counterinsurgency fell flat. The trip "was almost superfluous," he later wrote, because those on it already thought they had the answers. Taylor, who made his reputation in World War II, "was at best quizzical about partisan warfare, even when he was reminded that the French and British underground had done good work for his 82nd Airborne Division in France and Germany," Godel wrote. "He was also convinced that the 82nd Airborne could solve the problems of Vietnam with one hand tied behind his back." When Godel tried to impress on Taylor the difficulty of using conventional forces against an army of guerrillas who could launch surprise attacks and then melt back into the jungle, the general was unimpressed. "Nothing that a good double or triple envelopment could not solve, Bill, despite what you say," replied Taylor, who saw conventional troops as the only answer.

Taylor was even less willing to listen to Lansdale, the president's counterinsurgency guru. His only interest, Lansdale recalled, was getting "American genius to work and have an electronic line" that would separate North and South Vietnam, while also cutting off supply routes through Laos and Cambodia. When Lansdale returned from the Vietnam trip, he met with Kennedy to report on his findings, but by that point the president had already decided he wanted something different. "Just stay behind," Kennedy told Lansdale. "I've got something I want you to do." That something was Cuba: the president put Lansdale in charge of Operation Mongoose, a secret project aimed at removing Fidel Castro and the communists from power. That left Godel as the point man for Vietnam counterinsurgency work. Kennedy approved Godel's plan, Project AGILE, going so far as to personally approve some of the equipment, including a new lightweight rifle for Vietnamese troops.

In the summer and fall of 1961, Godel's home near Lake Barcroft, a man-made reservoir in Fairfax County, Virginia, sometimes resembled a real-life incarnation of a James Bond Q lab. Godel brought some of the more benign counterinsurgency gadgets home for a commonsense evaluation. One time, for example, Larry Savadkin, a navy officer assigned to work with Godel, bought sporting pontoons, essentially water shoes, from Abercrombie & Fitch, hoping they might be useful in Vietnam, and Godel let his daughters try them out on the lake. The pontoons would in theory allow a soldier to glide down Vietnam's water canals. Godel's daughters called them "walking on water shoes" or "Jesus shoes." But even in Lake Barcroft, thousands of miles away from Vietnam, it was easy to see they would not work. Every time the girls tried to step forward in the water, they would slip back two steps. "They were not maneuverable," recalled Kathleen Godel-Gengenbach, Godel's oldest daughter.

Many other ARPA novelties could not be brought to Godel's home. ARPA was developing mines disguised as rocks, portable flamethrowers, and thermobaric weapons, which were designed to create intense high-heat explosions to clear jungle foliage. Godel was, quite simply, experimenting with technology and weapons to see what would be effective in the jungle. His most basic premise, which he frequently argued with government officials, was that providing advanced technology, like jet aircraft and helicopters, to developing countries was nearly useless. The jets were difficult for developing countries to service and of limited utility in guerrilla warfare, and helicopters ended up being used to ferry around VIPs, rather than carry troops to combat. What were needed, in most cases, were simple weapons appropriate for jungle warfare.

Godel in the early days was running the AGILE office as his own covert operations shop with a handful of longtime loyalists whom he knew from World War II and had personally recruited into the agency. His team included Savadkin, the navy officer who had bought the "Jesus shoes," and Tom Brundage, a marine colonel, who ran the ARPA program in Saigon. Godel and his team gathered technology from wherever they could find it. For example, he used ARPA funds on a trip to Australia to buy a small, jet-powered drone and gave it to Brundage, in Saigon, making it the first unmanned aircraft deployed in Vietnam.

In July 1961, Godel wrote to Lansdale, updating him on plans for the combat center and providing a laundry list of proposed projects, which included everything from folding bicycles for South Vietnamese soldiers to a "persistent scent identification agent," which would be sprayed by aircraft on Vietcong fighters so that dogs could track them later. Many items Godel listed were for psychological operations, like a combat recorder that would be used for interrogations and a loudspeaker for broadcasting government messages. The list also included study projects such as one for "Means of Using the Montagnards Against the Viet Cong," a reference to the indigenous ethnic group that lived in the Central Highlands and was fast becoming the focus of CIA and military interest.

Over that summer, Godel shuttled back and forth between the United States and Vietnam, often personally delivering weapons to government and pro-government forces. He gave arms to Nguyen Lac Hoa, better known as Father Hoa or the "fighting priest," a Catholic priest and Chinese refugee who had organized a militia to protect his village from communist insurgents. And in October, Godel brought over ten AR-15 rifles to Vietnam to demonstrate the new weapons for Vietnamese soldiers and American military advisers. Godel explained to President Diem that the AR-15 would be a better weapon for jungle warfare, providing Vietnamese soldiers with the confidence to hunt the Vietcong. Diem was impressed. "All I want to know is when will we get them for the Airborne Brigade?" he asked. The new rifle, designed by Eugene Stoner, was lightweight, "something the short, small Vietnamese can fire without bowling themselves over," a summary description of Godel's trip recounted. After an initial positive response from the field, ARPA in December 1961 ordered an additional one thousand AR-15 rifles.

By the fall of 1961, the new Combat Development and Test Center was in full swing, and an assortment of arms and technologies were flowing into the country. Some technologies were so secret that they were funded "off budget," meaning there was no official accounting for them in the ARPA files, Godel later wrote. One, called "Big Ears," was a battery-operated microphone designed to pick up the sound of engines, as an early warning device. Painted to look like jungle vegetation and hung from trees, the sensors proved a failure; the batteries would die after just a week or two. Mixed results aside, Godel used the small-scale projects to convince President Diem of more ambitious technological undertakings. "If the President were convinced that an electronic surveillance system would contribute to the security of even one of his strategic hamlets, and that it might replace a group of tin cans hanging on the barbed wire enclosure, he could also be persuaded to expand the program to other surveillance technologies," Godel wrote.

The fall also marked the start of one of Godel's most ambitious, and controversial, counterinsurgency projects. On the initial list he sent to Lansdale in July, Godel had included a proposal for a chemical plant killer that could specifically target cassava, the Vietcong's subsistence food supply. The "hormone" killer was still theoretical, but Godel also wanted a chemical defoliant for aerial spraying of broad-leaf vegetation. "ARPA advises the chemical could be developed within several months within the U.S.," he wrote.

The idea for chemical defoliation was drawn, like many other aspects of Project AGILE, from the British experience in Malaya, where jungle brush provided cover for guerilla attacks on railways and roads. Similarly, in Vietnam, Godel wanted to use defoliation to clear jungle cover used by the Vietcong. In a September 1961 report that Godel sent describing the newly opened Combat Development and Test Center, he noted that ARPA had already done some early experimentation with defoliants, spreading them by aircraft and vehicle, and was awaiting initial results. If it worked as planned, the idea was to expand its use along the border areas with Laos and Cambodia, where vegetation destruction would be used to increase visibility along key roads and major waterways. But whereas chemical defoliation in Malaya was primarily used for eliminating enemy cover, the plans in Vietnam were much more ambitious. Ambushes were a concern, but another consideration was that subsistence crops were being grown or foraged to support the Vietcong fighters. Defoliation could be used to deprive communist insurgents of valuable food sources—potatoes and cassavas. In other words, the goal was to starve out the insurgents.

President Diem placed a high priority on crop destruction, requesting commercially available products, as well as four helicopters and six fixed-wing aircraft. Godel in his fall trip report noted that ARPA had already funded twenty thousand gallons of defoliant and needed another eighty thousand gallons. Given the urgency of destroying the crops by November, Godel recommended adding napalm bombs to the defoliation campaign, which would hasten crop destruction. In noting the sensitivity of targeting food, Godel said this was being done "at the strong insistence of the Vietnamese Government."

Jack Ruina, who was brought in as ARPA director in early 1961, watched the growing involvement in Vietnam with equal parts horror and confusion. Ruina was, at least in name, Godel's boss, but Godel was answering to the White House and senior Pentagon leaders. Even Harold Brown, the Pentagon official who oversaw DARPA in 1961 and would later become secretary of defense, admitted Godel and his work were a bit of a mystery. "A covert ops guy," was how Brown described him.

Godel and Ruina did not feud so much as simply avoid each other. Ruina, a PhD electrical engineer, wanted ARPA to be a scientific institution and despised everything about Project AGILE and the agency's involvement in Vietnam. Godel even warned members of his team to avoid the ARPA director at all costs. Warren Stark, who worked on AGILE, went so far as to hide behind a door at one point when Ruina walked into the room. Ruina did not care for the growing war in Vietnam. "Harold Brown sort of foisted it on me," Ruina recalled of AGILE. "He said, 'You should do it.' I never liked it, it was full of gimmickry and gadgetry."

Yet Ruina's job as director required him to at least nominally oversee the Vietnam work. He recalled one trip to Vietnam, likely in early to mid-1962, when he had a private meeting with President Diem. At that point, chemical defoliation was well under way, and there was clearly a tension between the goal of eliminating jungle and starving out insurgents. Ruina recalled Diem telling him, "You people did the wrong thing. We asked you to destroy crops."

Ruina, a scientist, had no idea what Diem was talking about and did not much care. "I wasn't involved in operations: what the government was doing in operations, it wasn't my business," Ruina tried to tell the president. But Diem continued to blast the ARPA director. "The ambushing is not important," Diem said. "It's destroying crops." He proceeded to get out a map of Vietnam showing areas of vegetation that were controlled by the Vietcong.

"How do you know which crops are which?" Ruina asked the Vietnamese leader.

"Oh, I know," Diem replied.

The truth was, Diem did not know which crops belonged to the Vietcong and which to villagers, but it probably did not matter to him. Defoliation gave the central government control over the food supply, and control was what Diem wanted. Diem was happy to feign interest in American counterinsurgency strategies if it got the United States to support him, but he was determined to rule the country, and fight the war, his way.

For Ruina, the exchange with Diem merely confirmed his concerns about Project AGILE, which were as much political as technical. He did not see a scientific justification for much of what ARPA did there, and he admitted in an interview more than four decades later that his "political slant" was against involvement in the war. "AGILE made Jack physically ill," Godel later recalled.

Ruina, a disciple of Herbert York's, wanted to create a science agency that served national security, while Godel wanted to build a national security agency served by scientists. The battle over those competing visions would characterize the agency's future.

CHAPTER 6

Ordinary Genius

"He savaged me," a dejected president John F. Kennedy told a *New York Times* reporter in a secret meeting, after a disastrous summit in 1961 with the Soviet leader, Nikita Khrushchev, in Vienna. The new president had hoped the summit would be a chance to demonstrate his ambitious vision for foreign policy and show his strength as a leader. Instead, with the summit's coming on the heels of the disastrous Bay of Pigs invasion of Cuba, the Soviet premier had openly mocked the young president, threatening war over the divided city of Berlin.

During the 1960 election, Kennedy had campaigned against the Republicans by promising a fresh approach to foreign policy. He brandished the alleged "missile gap" with the Soviet Union—the purported lead that the Soviet Union had over the United States in missiles and nuclear firepower—as a blunt political weapon. While the reality and the extent of that gap were hotly debated in Washington, Kennedy had argued, "Whichever figures are accurate, the point is that we are facing a gap on which we are gambling with our survival."

Once in office, however, Kennedy quickly realized the situation was far more complex than he had imagined. Whether Kennedy really believed in the missile gap during the campaign, or simply saw it as a politically expedient bumper sticker, is hard to say, but the political landscape shifted quickly once he was in office. Over the next few months, images from the first successful retrievals of film from the Corona satellite—added to the images coming in from U-2 flights—

would completely dispel any illusion of a missile gap. A little more than two weeks after Kennedy moved into the White House, his defense secretary, Robert McNamara, thinking he was speaking off the record, inadvertently revealed to the nation that there was no missile gap at all.

Now in the White House, Kennedy perhaps had a better idea what President Eisenhower meant when in his farewell address he warned about the influence of the "military-industrial complex" and admonished the country to beware of becoming "captive of a scientific-technological elite." Eisenhower's warning was prescient. The military used tensions with the Soviet Union as the excuse to pressure the new administration to deploy Nike Zeus, the country's first antiballistic missile system. The plan was to launch a nuclear-tipped long-range missile that would explode close enough to an incoming enemy ICBM to destroy it. It was not a particularly good technology, and radar experts knew there really was no way that Nike Zeus could track an incoming missile well enough to intercept it reliably. But that view was not reaching senior political leaders.

Soon after joining ARPA as director, Jack Ruina got word that Kennedy's defense secretary, Robert McNamara, wanted a basic briefing on missile defense. ARPA, after all, had been left in charge of missile defense research, even after the space mission was taken away, and suddenly missile defense was a hot political issue. McNamara apparently wanted to learn the basics of missile defense. Ruina warned that such a briefing could take all day, and McNamara said that the ARPA director would get "all the time we need." Ruina spent the day laying out the technical challenges of a ground-based missile defense system like Nike Zeus. Ruina called the briefing "The Earth Is Round," because that is literally where it started, explaining that the earth was a globe, which meant ground-based radar could only detect a missile within a few thousand miles, providing precious little time to launch another missile to intercept it.

The day before Thanksgiving, Ruina, along with several other government scientists, was invited to the White House to brief President Kennedy on missile defense. The meeting included just Ruina; Harold Brown, the director of defense research and engineering; and Jerry Wiesner, the president's science adviser. All three scientists were critical of Nike Zeus, a doomed system. McNamara was not invited. The meeting went on for several hours, with Kennedy's posing detailed questions to the scientists about the project. Finally, the president's brother Bobby interrupted to say they needed to leave for Hyannis, Massachusetts, the Kennedy family compound. "The helicopter's waiting for you," he told the president. "If you don't get in, we won't get to Hyannis Port in time for Thanksgiving."

The president turned to the scientists and said, "Well, can you guys come up to Hyannis after Thanksgiving so we can continue this discussion?" Who could say no to the president?

The men flew up the day after Thanksgiving to Hyannis, where Kennedy was busy with meetings with various advisers on nuclear security, which included discussion of a nationwide program for fallout shelters. This time McNamara was invited, but Kennedy had already made his decision on Nike Zeus. "I don't think we should go ahead with it, Mac, do you?" Kennedy said to McNamara.

The defense secretary replied, "No, let's not go ahead with it."

And that, Ruina recalled, was the end of Nike Zeus. Now it was up to ARPA to come up with something better. The scientists at ARPA had free rein to apply their intellect to even the most far-fetched sorts of solutions. They wasted no time, delving into antigravity, science-fiction-inspired death beams, and a space net named after a beloved Walt Disney character. ARPA, the agency that almost disappeared two years earlier, was suddenly in the middle of the Kennedy administration's nuclear debates.

Jack Ruina liked to joke that were it not for a chance meeting at a urinal, he might have spent a quiet life in academia. Instead, the Polish-born scientist ended up as the director of an agency engulfed in debates ranging from chemical defoliation to nuclear Armageddon. An engineering professor at the Massachusetts Institute of Technology, he was relieving himself in the men's room, when another professor joined him at the next urinal. The professor told Ruina about a group at the University of Illinois that was doing work on radar tracking, Ruina's specialty, and said that he should talk to them.

Ruina did and was eventually invited to work at Illinois, where he met Chalmers Sherwin, a physicist who would soon become the air force's chief scientist. Sherwin put in Ruina's name for a senior Pentagon job: assistant secretary of the air force. When he arrived in Washington for his new job, Ruina was picked up in a limousine and greeted by his new assistant, a lieutenant colonel. He was taken to his large Pentagon office and introduced to his two secretaries. For Ruina, who was conscripted into the army and never rose beyond corporal, the experience was a shock. The highest-ranking officer he had ever met while serving in the army was a major, and he was a dentist. Now Ruina was suddenly elevated from academic to Pentagon official.

Soon, Ruina moved to a new position, working under Herbert York, who was now the director of defense research and engineering. Ruina was responsible for overseeing ARPA, and he understood what was wrong with the agency. "ARPA is not strong now because it is not supported by Dr. York," Ruina told the agency's senior management. That shocked no one. York did not support his old agency, because it had been a competitor for his space empire. His solution was to make Ruina, who had become a good friend, the head of ARPA.

On January 20, 1961, the same day John F. Kennedy was sworn in as president, Ruina became the third director of ARPA. Suddenly the thirty-seven-year-old electrical engineer found himself the head of a military technology agency with a broad mandate and little oversight. He inherited an agency that was involved in everything from crop defoliation to weather control. It was a strange mix, and nothing was stranger to Ruina than Project AGILE, the Vietnam counterinsurgency work run by William Godel, the deputy director. AGILE was also high profile, having White House support, so Ruina could not cancel it. He chose instead to ignore it.

Despite ARPA's growing work in Southeast Asia, the majority of the agency's budget went to projects left over from its early role as a space agency: the antiballistic missile defense program, called Project Defender, and a broad nuclear test detection research program, which went by the name Vela. A few other minor research projects remained, like propellant chemistry, but none of the others were terribly important from the perspective of scientists or the military. The ARPA of 1961 was barely three years old and had no particular reputation, or even mission: it was simply a collection of science and technology programs no one else coveted. Kennedy changed that, giving the scientists at ARPA what was arguably the world's most important problem: nuclear warfare. Ruina mused that just one more cup of coffee before heading to the men's room and "my whole life would have been different." So, too, might be the world, because the programs ARPA started under his watch would change the course of arms control.

In terms of dollars, the single largest program at ARPA in 1961 was Defender, the research effort aimed at developing technology to protect the United States from intercontinental ballistic missiles. ARPA had been saddled with missile defense research when it was founded, and Roy Johnson had mostly ignored it while he was director. By 1961, with the advent of ICBMs, missile defense was becoming a higher priority for the president and the Pentagon. Ruina disliked the technical inelegance of a human problem like counterinsurgency, but he immediately took an interest in ARPA's work in ballistic missile defense and nuclear test detection. "The only two programs that bordered on major policy issues for the nation, were the [ballistic missile defense] research program and the Vela research program," Ruina said, omitting any mention of ARPA's Vietnam work, which in his view was not real science. "The other research programs were not ones that the secretary of defense or the president were concerned about."

It was not clear, however, that there was any missile defense system better than Nike Zeus, the project the administration had just shelved. ICBMs were still relatively new, and even newer were the theoretical schemes to defend against them. There was no way, using current technology, to track and intercept a missile that took only thirty minutes to circle the earth. ARPA's Defender program faced an even greater challenge: it was very specifically supposed to look at technology that could "protect all the United States," and that lent itself to particularly outrageous and comical concepts involving space-based nets that would trap missiles, and the like. ARPA had to imagine futuristic technology, and pretty soon imaginations ran wild. When it was revealed in 1959 that ARPA had undertaken a broad study project named GLIPAR, short for "Guide Line Identification Program for Antimissile Research," the agency was derided for allegedly considering "anti-gravity, anti-matter and radiation weapons."

By 1961, ARPA was spending about $100 million per year, or half of its entire budget, on missile defense, and yet the program was, according to Ruina, a "god-awful mess." The agency had been inundated with fanciful proposals to shoot down ballistic missiles; one official described 75 percent of the work as outright "crazy." A good amount of this craziness could be attributed to the cutely named ARPA program called BAMBI, short for Ballistic Missile Boost Intercept. The basic idea behind BAMBI was to look at ways to intercept missiles in their initial launch phase. As BAMBI matured, it grew from ambitious to lunatic. One proposal called for orbiting battle stations—large armed satellites—that would shoot out pellets enmeshed in a giant net meant to perforate enemy warheads. No one, however, had figured out how the United States' own satellites or missiles would avoid this net. Herbert York called this satellite swarm a "mad scientist's dream," and not surprisingly he blamed it on Roy Johnson, claiming one of the former ARPA director's final acts was to foist BAMBI on the agency.

Ruina agreed that BAMBI was a "loony idea" that needed to be canceled. "Bambi brought in all the nuts out of the woodwork," Ruina recalled later. Yet when he first arrived at ARPA, he had to at least publicly defend the project to Congress. "Isn't that a bit fantastic?" marveled Representative George Mahon, a Democrat from Texas and frequent critic of ARPA, upon hearing the description of BAMBI. "It is a bit fantastic, true, but so are many things," Ruina replied. "I am sure the Venus probe seemed a bit fantastic 20 years ago. Examining this concept now, the conclusion we come to is that it is not fantastic enough to drop it."

It took two years of fighting off its supporters, particularly from the air force, but Ruina eventually prevailed. In 1963, he told Congress that he had slain BAMBI. Not only was it impractical, he said, but the costs of operating such a system would run on the order of $50 billion a year, about the same amount as the Pentagon's annual budget. Even if BAMBI could be built, it would essentially be at the cost of putting the entire military out of business.

BAMBI illustrated the fine line ARPA had to walk in many areas of technology: If its missile defense solutions were too conventional, then there was no point pursuing them. After all, the idea was to come up with technological solutions that were significantly better than what the military had already done with Nike Zeus. On the other hand, if the solutions were too unrealistic, like BAMBI, then ARPA would be accused of throwing money at science fiction. What it eventually arrived at was something called ARPA Terminal Defense, abbreviated to ARPAT. The scheme involved launching a drone "mother ship" equipped with hypersonic interceptors that would hover some sixty thousand feet above the earth. Ground-based radar would track incoming warheads and try to discriminate warheads from the decoys meant to dupe the system and then pass that information to the mother ship, which would release a hail of dart-shaped interceptors in the vicinity of the incoming warhead. ARPAT was only "kind-of nutty," declared Charles Herzfeld, an Austrian-born physicist recruited to ARPA to run the missile defense programs. Being only "kind-of nutty" apparently gave it an advantage over most schemes, though it still cost about $20 million.

The problem with all the proposals was that, quite simply, none of them worked particularly well. Either they were too expensive, too impractical, or a combination of both. What was needed was a way to differentiate between the scientifically speculative and the technically ludicrous. For that judgment, Ruina turned to a secretive group of elite scientists.

Back in early 1958, the Manhattan Project veterans John Wheeler and Eugene Wigner, along with the economist Oskar Morgenstern, had lobbied to establish a national security science laboratory, a sort of mini-ARPA for scientists. The "Princeton Three," as they were called, had even embarked on an aggressive publicity campaign in the wake of Sputnik to promote the idea, which ended up in *Life* magazine. ARPA was not interested in having a laboratory, but Herbert York agreed to support an exploratory summer study led by Wheeler and gave him a small $50,000 contract for what was code-named Project 137.

Just months after ARPA got off the ground, Project 137 gathered some forty scientists to generate ideas with the modest aim of involving a younger generation in the world of national security. The proposals for that first meeting ranged from looking at chemical sensing based on research into insects' unique ability to find a mate to a method of communicating with nuclear-armed submarines while they prowled the seas using extremely low frequencies. Another idea floated at the first meeting involved using a particle beam to blast ballistic missiles. "If by use of high speed particles the channel can be made straight in the beginning, and the gas within heated so hot that the bulk of the impeding mass is driven aside, then the door is open to sending a quick burst of energy to a great distance," Wheeler wrote.

By the time Wheeler presented the results of Project 137, the idea for a laboratory was vastly scaled down. Instead, he essentially proposed a mini-ARPA for scientists—an organization that would draw on academics for temporary appointments. Even that idea foundered, in large part because Wheeler did not want to leave academia to head the new venture. In 1960, General Betts, as the interim head of ARPA, decided to continue Project 137 as an annual meeting of scientists, called Sunrise. Project Sunrise was not designed to research a specific area of science or technology. Rather, it paid simply to involve young scientists, mostly from academia, in national security issues.

The membership read like a Who's Who of elite physicists. Marvin "Murph" Goldberger, a former student of Enrico Fermi's, headed the group, which included Murray Gell-Mann and Steven Weinberg, both young physicists who would go on to win Nobel Prizes. Members of the group would meet for several weeks over the summer and then report back to ARPA. With help from Charles Townes, who would also later win the Nobel Prize for his work on the maser, the group was established under the auspices of the Institute for Defense Analyses, the research institute that had been supplying much of ARPA's technical talent. Project Sunrise met for the first time in the summer of 1960 and, at the suggestion of Goldberger's wife, promptly changed its name to JASON, after the Greek tale of Jason and the Argonauts.

From the start, the JASONs, as the members were known, were unique among scientific advisers: They were granted top secret clearances, which allowed them access to information that ordinary academics could never have had. Because they were not government scientists, they also had the independence to criticize projects. Though funded by ARPA, the group ran its own affairs and selected its own members. They were brilliant, patriotic, and eager to make some money, a reputation that earned JASON the sarcastic moniker of the "golden fleece." They were also secretive; lists of JASON members were not made public, and the group frowned on scientists advertising their membership in it.

In the early years, the JASONs worked primarily on missile defense—ARPA's Defender project—focusing on problems at the intersection of physics and technology. The first summer, they looked at whether the Soviets might be able to "blind" early warning satellites by setting off preemptive nuclear blasts at high altitude, which could hide the plumes created by ICBMs. (As it turned out, the JASONs found the concern was overblown.) Ruina soon took a liking to the JASONs and their work, describing them as "kind of a truth squad."

The JASONs were not above considering their own off-the-wall ideas, particularly if they came from Nicholas Christofilos, the father of Operation Argus and one of the few national laboratory scientists invited to join the group. His febrile imagination, mixed with genius, entranced the JASONs, who were typically known for their skepticism. Christofilos would dazzle fellow scientists with his frenetic work schedule and ability to generate new ideas and then go on late-night drinking benders in the bars of La Jolla, California, which became the regular meeting place of the JASONs.

After his idea for a planetwide force field fell to the wayside, Christofilos had continued to forward ideas for protecting the nation from Soviet attack, each idea seemingly more outlandish than the last. One proposal involved building an aircraft runway stretching from coast to coast so that American bombers could evade Soviet strikes. It was wryly called a "not good" idea by an early ARPA history. Unperturbed, Christofilos turned his unbridled intellect to a new scheme for a charged particle beam weapon that would obliterate incoming Soviet ICBMs. The concept, part Buck Rogers, part Dr. Strangelove, was classic Christofilos: scientifically brilliant, but requiring a technology almost unimaginably complex.

The particle beam, which emerged from the Project 137 summer study, was picked up by ARPA in its first year of operation. Code-named Seesaw, the particle beam became another entrant into Christofilos's phantasmagoria of megadeath technology. It came to symbolize the type of agency that was being built in the early 1960s under Ruina, and it helped define a unique and enduring relationship between ARPA and the JASONs. At one point, ARPA was planning to spend $300 million on the particle beam weapon and a "death radar sub-system"—an astronomical figure for the time.

Seesaw was not a JASON project, but the group became an integral part of the program over the years, reviewing the results and proposing new avenues of research in a series of secret reports. Ruina found himself, like many other scientists, enthralled with the Greek physicist. He described Christofilos as "fantastic," in part because he lacked any self-awareness that the concepts he proposed were outrageous. "He was not ever frightened of doing an experiment that was beyond what most people could even think of doing," Ruina said. "I mean, he would not hesitate to think: We're going to put up a net to reflect things that's five miles across. Why not put up a big net?" Christofilos's imagination was not tethered to practicality, and that, combined with a sharp intellect, attracted people like Ruina and the JASONs. And that is what allowed Christofilos to think up the missile-killing particle beam.

Particle beams consist of a focused stream of highly charged particles that, when they collide with something, transfer their energy, essentially disintegrating the target. Christofilos was proposing a charged particle beam that would travel through the atmosphere, a feat requiring immense energy and a way to keep the beam focused. It was the type of exotic physics that the JASONs enjoyed doing. ARPA funded Livermore, where Christofilos worked, to study Seesaw using a nuclear fusion reactor called Astron, which was also based on one of his schemes. In turn, the JASONs were heavily involved in reviewing Seesaw; not surprisingly, they always came out in favor of continuing the project, despite the technical hurdles. "Sometimes it was the same scientist each year proving the opposite of what he had proved the previous year," marveled Eberhardt Rechtin, a former ARPA director, referring to the JASON reviews. He joked that Seesaw was "aptly named, because it seesawed each year from 'it's practical' to 'it isn't practical.'" Rechtin said he had doubts about the propriety of the entire Seesaw business, but in the end he supported it, as did other ARPA directors.

Seesaw faced myriad practical barriers: the tunnels needed to generate the beam would have to be hundreds of miles long. The cost of building the tunnels would be exorbitant, and no one knew how to build a power supply that could produce the high power levels needed to produce the beam, which would require a huge amount of electricity, possibly more than existed on the entire grid. "ARPA was generally of a mind that Seesaw was a bad idea," recalled Kent Kresa, a former ARPA official. "It was too expensive and too hard."

Kresa decided one year to sponsor what he hoped would be the last JASON study on Seesaw. Kresa thought that by having the JASONs look at the big picture of what it would take to build a system that could protect a city from an attack—not just the scientific issues of getting a beam through the atmosphere—the scientists would see the folly of the proposal. "If you looked at that, there isn't enough money in the world to be able to protect the United States," he said.

The problem was that Christofilos was on the study, and every time Kresa brought up what he thought was surely a "clear killer" of Seesaw, Christofilos would brainstorm a convoluted solution. "All the other JASONs would say, 'Nick, God that really is neat.'"

"There's a better way to do it," Christofilos announced, when confronted with the costs of drilling the tunnels. He would create the tunnels using nuclear weapons.

"Think of it like a suppository," Christofilos told the JASONs. "We would push it through the rock. As it goes into the rock, it melts the rock, it creates this perfect tube. You just have to keep on pushing it so it's hot enough so it melts the rock. As it goes, you just push it through."

Christofilos did not stop with nuclear suppositories. Another question was how to power a particle beam to shoot down three thousand incoming Soviet missiles. Was there even enough electricity available in the United States to power this weapon? "Nick had the solution. I'll never forget this," Kresa recalled. "He said, 'We're going to do nuclear blasts under the Great Lakes.'"

Kresa was in shock, but Christofilos began to write out calculations, showing the volume of water in the Great Lakes and how much energy could be generated if the lakes were drained in a fifteen-minute period through a set of doors, passing through generators and then into a huge cavity carved out by nuclear explosions. "We're going to put a generator in there, and when the war comes, we're going to drain the Great Lakes, and that will create the energy," Kresa said of the proposal. "Christofilos did a calculation of how much energy it would take, and it was going to work. The JASONs in the room all nodded their head and said, 'My God, Nick, that may work.'"

Seesaw never fired a single shot; the megadeath beam was never built. It was also never extravagantly funded—it certainly never got the $300 million that ARPA once planned—but Seesaw did become the longest-lasting ARPA project and continued at least through the mid-1970s. Everyone seemed to agree that Christofilos was a genius whose ideas had scientific merit, however improbable they seemed. Yet nobody, including Ruina, ever thought the particle beam would work. Seesaw was, by all accounts, a failure, yet early directors still defended it years later as research that embodied the spirit of ARPA in the 1960s: bold and scientifically interesting. "There will not be any payoff; it is not practical," Ruina said, "but there are many good people assigned to it; there is much knowledge being developed from the effort; and it permits freedom of work in a research or laboratory atmosphere."

Projects like Seesaw raised a fundamental question: Was ARPA a science agency with a focus on national security, or was it a national security agency with a focus on science? Just as Jack Ruina hated Godel's Project AGILE, Godel despised Ruina's favorite research projects, which often had only a tenuous link to national security. Godel's main vitriol was directed at the Arecibo Observatory, a radio telescope funded by ARPA under the auspices of the Defender program. Arecibo was ostensibly for use in research related to ballistic missile defense, but everyone in ARPA from Ruina on down to the rank-and-file staff acknowledged it really had nothing to do with national security. It was simply an excellent science facility that would be used by academics to study the ionosphere.

So committed was Charles Herzfeld, the head of Defender, to protecting Arecibo for scientists that when officials from the NSA approached him about using the facility for a secret eavesdropping experiment, he initially refused them. The NSA wanted to use Arecibo to test whether it would be possible to intercept signals bounced off the moon. Herzfeld insisted Arecibo was for unclassified research, though he soon relented, likely at the behest of Godel. And when the

NSA officials decided that Arecibo was not the ideal location for their research, Godel stepped in with a generous, if somewhat shocking, offer: ARPA could arrange to set off a nuclear weapon in the Seychelles for the eavesdropping agency. "A nuclear detonation would be employed, and ARPA guaranteed a minimum residual radioactivity and the proper shape of the crater in which the antenna subsequently would be placed," Nate Gerson, an NSA cryptologist, recounted. "We never pursued this possibility."

One reason the NSA never followed up on Godel's offer to nuke the Seychelles was that such testing was about to become politically impossible to do—thanks to another ARPA project under way.

When President Kennedy took office in 1961, the United States and the Soviet Union were still abiding by the 1958 moratorium on nuclear weapons testing, but there was mounting political pressure on both sides to resume the work. Advocates of nuclear testing, like Edward Teller, argued that no treaty would be able to prevent the Soviets from cheating. In an essay for *Life* magazine, Teller and his colleague Albert Latter laid out their case to the public. Published between a recipe for "tangy tuna-mac" and an advertisement for Air Wick home deodorizer, the essay argued that nuclear testing was necessary to prevent Soviet superiority, even going so far as to suggest that human genetic mutations from radiation, if they occur, may not be such a bad thing. The critical point, the two scientists argued, was that the Soviets would cheat and there would be nothing the United States could do to detect such cheating. "It is almost certain that in the competition between bootlegging and prohibition, the bootlegger will win," they wrote.

Teller could make that argument because there was, in 1961, remarkably little agreement on whether it would be possible to detect a covert Soviet test, particularly one conducted underground. Both sides had seismic sensors that could detect the rumblings caused by such a test, but it was not clear that scientists could reliably distinguish between natural events, like earthquakes, and the explosion of a nuclear device underground. In other words, either side could simply deny there was a nuclear test and claim what had taken place was an earth tremor. This issue would become key to any test ban negotiations between the Soviet Union and the United States. The debate had become such a highly charged issue for treaty negotiations that the common joke was that Kennedy and Khrushchev had both become amateur seismologists.

Not everyone was so convinced of Teller's assertions. ARPA was assigned nuclear test detection under the code name Vela at the end of 1959 as a counterweight to the CIA's and the air force's secret test detection network. ARPA got the work, quite simply, because President Eisenhower did not trust his spooks and wanted an assessment that was independent of the CIA and its assets. The work at ARPA initially foundered, but the election of Kennedy, who was interested in arms control, brought renewed focus and funding to the Vela test detection program. By 1961, Vela had three parts: Vela Uniform, to detect underground nuclear tests; Vela Sierra, to detect nuclear explosions in the atmosphere; and Vela Hotel, which would launch satellites with sensors to detect nuclear tests from space.

In 1961, ARPA began pouring money into academic seismology research in the United States and abroad. The academic discipline of seismology, at the time, was a backwater. Robert Frosch, who was recruited to ARPA to run Vela, recalled going with the director, Robert Sproull, to visit what was supposed to be a state-of-the-art seismic vault, one of the underground bunker-like structures that were used to measure tremors. The two men came out of the vault in shock, feeling as if they had just emerged from a time capsule. The seismologists there were using pen recorders and primitive galvanometers, an analog instrument used to measure electrical current. Vela began to change that with an influx of funding for seismology that was almost unimaginable in scale for most areas of science. The military's need to distinguish earthquakes from nuclear tests brought seismology "kicking and screaming" into the twentieth century, according to Frosch. At one point, he said, he funded almost "every seismologist in the world, except for two Jesuits at Fordham University" who refused to take money from the Pentagon.

Frosch's ambitious idea for advancing both seismology and nuclear test detection was to build a novel system that would identify the vast majority of Soviet earthquakes, resolving once and for all the debate over distinguishing earth tremors from nuclear tests. Frosch's project, called the Large Aperture Seismic Array, or LASA, was a massive nuclear detection system that comprised two hundred "seismic vaults" buried across a two-hundred-kilometer-diameter area in the eastern half of Montana. For it to work, more than a dozen of these enormous sites would have to be constructed around the world to monitor the Soviet Union. There had been smaller arrays, including one in the United Kingdom, but no one had ever built an array of LASA's size or scope or knew if it would really improve detection. The air force hated the idea, and seismologists, Frosch said, considered his idea "mildly crazy." Frosch saw it as a way of putting ARPA's flexibility to the test. "If you had an idea, you didn't have to go through two years of getting permission and three years of getting the contract people to make a mess of it," he said.

ARPA's leadership gave it their blessing. In the end, it required negotiating with some fourteen different utility cooperatives and dozens of Montana landowners, who were less than thrilled to have the federal government installing nuclear detection devices on their private property. Frosch recalled ARPA getting a complaint from a landowner, who spotted personnel one morning working on a seismic vault. "When I was sitting there eating my breakfast yesterday, I could see someone digging on my land," the owner reportedly said. "I don't like to see someone on my land when I'm eating my breakfast."

What was amazing about LASA, according to Frosch, was the scale of the work, which was completed in just eighteen months, a schedule unimaginable for government projects that typically take years, if not decades. When ARPA needed to have a center where all the seismic data could be collected and analyzed, the agency ended up renting space in downtown Billings, where data from the array was routed to an IBM computer. "It was literally opened up in a storefront in Billings, and in the back room we had computers and we were the center of the array," Frosch said.

ARPA also began funding the placement of seismograph stations around the world that were operated by scientists. The ARPA-funded seismograph network was designed not to replace the military's classified system but merely to expand the science of test detection. It was, however, a powerful counterweight to the CIA and the air force, who up to that point had a monopoly on advice to political leaders about what was theoretically possible to monitor in a test ban. Naturally, the CIA and the air force regarded ARPA as "a bunch of incompetents," and worse, incompetents who were "insisting that their work be placed in the public domain."

The Worldwide Standardized Seismograph Network, as the ARPA project was called, was massive in scale and scope. Instead of recording results on paper, these new stations would convert results onto 70 mm film, which could be more easily transported and shared among researchers. That was one revelation for seismologists; the second was the extraordinary reach of the seismographic stations, which stretched from Alaska to Tasmania, often in remote or exotic locations. In Trieste, Italy, mountain climbers helped place the seismometers; other locales involved using dogsleds and rickshaws. ARPA also encountered setbacks, like in the South Pole, where instruments ended up frozen inside a block of ice.

According to Lee Huff, who co-authored the early ARPA institutional history, it fell to Godel and his team to secure the agreements to install some of the initial seismic arrays. They began to fan out across the globe to negotiate ARPA's new system, from Thailand to Iran. "Godel had already a lot of experience putting in networks of listening devices of one kind or another around the Soviet Union. He did some of that for the NSA, literally, going himself out into the countries and getting it done," Huff said. "He was an old hand at that."

For the most part, getting a country, be it India or Iran, to agree to the seismic stations was not difficult at the time. ARPA was essentially offering to build—for free—seismological stations that would be operated by the host country, and local scientists only needed to agree to operate them and share the data. The network would eventually involve some 125 stations in more than sixty countries. "You could do things a lot easier than you can today, and they were simply letter agreements," said Jon Peterson, a scientist for the U.S. Coast and Geodetic Survey, a federal agency that worked with ARPA on the stations.

As ARPA built up a worldwide detection network, the program highlighted a growing tension between secret and open research that was about to become critical for arms control and national policy. The air force and the CIA refused to release data from their network of sensors. "Everybody thought that they were too secretive, they always gave you data that you wanted, but they wouldn't open up their books," Ruina said. The bête noire of the nuclear detection world was Carl Romney, a scientist who worked for the Air Force Technical Applications Center, or AFTAC, the agency responsible for nuclear test detection. Romney was widely regarded as the nation's leading expert in test detection. Many in the field called him brilliant, but he was also widely vilified by critics for his role in blocking every attempt to move forward with a test ban. "Romney never tried to mess up the data, he would only deliberately misinterpret the data," claimed Jack Evernden, a seismologist who worked for Romney, before later moving to ARPA.

Whether deliberate or not, the problem with secret data, as Ruina pointed out, was that "nobody could argue with it; they could just question it." The secret data problem came to a head in 1962, when the United States carried out a test called Aardvark, a part of the first series of tests conducted completely underground. Aardvark, a forty-kiloton nuclear device intended for nuclear artillery, produced reliable seismographic data on a nuclear underground explosion, and Romney suddenly realized he had been wrong about a critical national security issue. He had been arguing that it would be difficult to distinguish small underground nuclear tests from earthquakes, which would make verifying a nuclear test ban treaty difficult, if not impossible. Now, with the Aardvark data, he knew he had been wrong on a key point. During a July 3, 1962, meeting, Romney announced that the new seismic data led him to conclude that distinguishing between tremors and small nuclear tests might not be as difficult as he had previously thought. McNamara was upset and decided the Pentagon had to issue a press release, because if the new data leaked—which often happened—it would look as if the government were "withholding information that would tend to ease the inspection problem in a nuclear test ban."

Ruina called it an "honest mistake," but one that would have been avoided if other scientists had been given access to the classified data that Romney jealously guarded. "This is what can happen when you have one person interpreting data, there's no peer group reviewing it, and there's nobody duplicating the experiment," the ARPA director wrote in a three-page letter, blaming the mistake on secrecy. ARPA's work finally began to sway President Kennedy, according to Glenn Seaborg, who as chairman of the Atomic Energy Commission played a key role in test ban negotiations. Kennedy paid close attention to ARPA's test detection research, relying on its results in deciding how to go forward with the treaty. "VELA seemed to indicate that the detection capability was better than had been thought by American experts in the period from 1959 to 1961," Seaborg wrote in his memoir detailing the negotiations.

On October 7, 1963, a month after Ruina left ARPA, President Kennedy signed the Limited Test Ban Treaty, which halted nuclear testing in the atmosphere, in outer space, and underwater. The first launch of ARPA's Vela Hotel satellite, to detect space-based nuclear tests, took place just days later. It was, according to all accounts, a spectacular success, proving the earlier naysayers wrong. Though the treaty did not prohibit underground testing, the ARPA work was credited with helping reach any agreement with the Soviet Union. "There are three reasons for the partial test ban," said Robert Sproull, who took over as director after Ruina left. "One that Mr. Kennedy wanted it, one that the Soviet Union wanted it. And one that ARPA made it possible for the Senate to ratify it. And all three of those were required."

By 1963, ARPA's Vela program had established a counterweight to the intelligence community and, in the process, helped achieve a limited test ban by proving that detection of atmospheric and underground nuclear tests was indeed possible. Vela would also over the next few years have an equally significant impact on the science of seismology. As measurements from the Worldwide Standardized Seismograph Network started coming in and being shared among academics, the seismologist Lynn Sykes was able to use the data from ARPA-funded stations to more accurately track the location of oceanic earthquakes. Whereas seismologists once placed the earthquakes all over the ocean floors, Sykes was now able to prove that in fact the tremors were occurring along mid-oceanic ridges. Plate tectonics, which had previously been a highly controversial theory, could now be substantiated with data from ARPA's network.

In 1968, Sykes, along with his fellow seismologists Bryan Isacks and Jack Oliver, published a seminal paper that finally paved the way for the acceptance of plate tectonics. "Seismology and the New Global Tectonics" relied heavily on years of results drawn from the ARPA network, including data that showed how waves crossed within and between continental plates. As Sykes put it, Vela "almost instantaneously transformed seismology from a sleepy, poorly supported scientific backwater to a field flooded with new funds, professionals, students and excitement."

Sometimes science and policy meet up, as they did with arms control in the early 1960s; other times they do not, demonstrating that technology alone is not enough to solve problems. After the Limited Test Ban Treaty of 1963, enthusiasm for further negotiations waned. Following Kennedy's death in 1963, Lyndon Johnson expressed little interest in pushing forward with a comprehensive test ban. It would take almost three decades for the United States to declare a moratorium on underground nuclear testing, in 1992, which followed a unilateral Soviet moratorium the year prior. The Senate never ratified the Comprehensive Nuclear Test Ban Treaty, even after the agreement was adopted in 1996. Nonetheless, Vela is widely—and properly—credited as a technical and political success for its early contributions to a limited test ban and then for helping move along negotiations for a comprehensive ban, even if belatedly. In the process, Vela also "revolutionized seismology," concluded Lee Huff.

ARPA's legacy in missile defense, on the other hand, was far more tenuous. The Arecibo Observatory that briefly interested the NSA proved to be an excellent science facility but contributed almost nothing to the military. The megadeath particle beam known as Seesaw did not produce anything of particular value for the Pentagon. And ARPA's advocacy for exotic directed-energy schemes for missile defense inadvertently planted the seed that would blossom decades later under President Ronald Reagan as a fantastical global shield. ARPA could, however, take credit for killing off some "loony" ideas like BAMBI's killer spiderweb and dissuading senior leaders from pursuing doomed efforts like Nike Zeus.

Yet ARPA's work in the nuclear world produced new science and advanced technology, enabling political change. It represented a "triumph" in technology and national security, according to Stephen Lukasik, who was recruited to ARPA in 1966 to work on nuclear test detection and later became the agency's director. It was a triumph because the science matched the politics, he argued. Ballistic missile defense and test ban treaties were at the top of the White House's agenda, and ARPA's scientific progress in these fields marched in lockstep. It proved that verification was scientifically feasible and that missile defense was technologically infeasible. ARPA was successful, according to Lukasik, "not because we were geniuses—we were just ordinary geniuses—but because the country was ready for test bans, the country was ready for nonproliferation treaties, the country was ready for ballistic missile defense treaties, and the country was ready for limitations on theorized nuclear weapons and cruise missiles."

In other words, ordinary geniuses transformed an entire scientific field and helped open the door to arms control, not just because ARPA scientists had the freedom to do what they wanted, but because these ordinary geniuses were working on problems of national importance. The question was what ARPA could accomplish if it employed an extraordinary genius.

CHAPTER 7

Extraordinary Genius

"We have some big trouble," President John F. Kennedy told his brother Attorney General Bobby Kennedy early in the morning of October 16, 1962.

A few hours later, the younger Kennedy was staring at pictures of Cuba that had been taken by the U-2 surveillance aircraft. "Those sons of bitches, Russians," he said, sitting in the White House with a group of officials dedicated to overthrowing Fidel Castro.

The pictures showed the telltale signs of Soviet missile launchers. The CIA had used a massive computer—it took the better part of a room to house it—to calculate the precise measurements and capabilities of the missiles installed. Their dismal conclusion was that the missiles had a range of more than a thousand miles, making them capable of reaching Washington in just thirteen minutes. This revelation touched off a crisis that lasted almost two weeks. As the standoff over Cuba intensified, American military forces reached DEFCON 2, just one alert level before the start of a nuclear war.

As military and civilian commanders clamored for information on a minute-by-minute basis, computers like the air force's IBM 473L were being used for the first time in the midst of a conflict to process real-time information on how, for example, to allocate military forces. According to a top secret Pentagon report on military command and control, the Cuban missile crisis demonstrated how operational data, now available on computers, "were recognized by more and more

Joint Staff officers," and "informal requests for outputs increased." Yet even with the growing availability of computers, sharing that information among military commanders involved a time lag. The idea of having the information travel between connected computers did not yet exist.

After thirteen days of scrambling forces to carry out a potential attack, the Soviet Union agreed to remove its missiles from Cuba. Nuclear war was averted, but the standoff also demonstrated the limits of command and control. With the complexities of modern warfare, how can you effectively control your nuclear forces if you cannot share information in real time? Unbeknownst to most of the military's senior leadership, a relatively low-level scientist had just arrived at the Pentagon to address that problem. The solution he would come up with became the agency's most famous project, revolutionizing not just military command and control but modern computing.

Joseph Carl Robnett Licklider, who went by the initials J.C.R., or simply Lick to his friends, spent much of his time at the Pentagon hiding. In a building where most bureaucrats measured their importance by proximity to the secretary of defense, Licklider was relieved when ARPA assigned him an office in the D Ring, one of the Pentagon's windowless inner rings. There, he could work in peace and hopefully avoid William Godel, who always seemed to be sticking his nose in everyone's business or, worse, would try to get Licklider involved in what was inevitably a cockamamie Vietnam project.

One time, for example, Godel wanted Licklider to evaluate a proposal for using mass hypnosis in Vietnamese villages to increase support for the South Vietnamese government. Licklider managed to duck out of an initial meeting with the company promoting the idea, but finally, when pressed to review the proposal, he penned a memo to the head of Project AGILE providing a diplomatic, but direct, response. "I do not want to be negative merely because it is controversial," Licklider wrote. "I do want to argue, however, that you should investigate very thoroughly the accreditation of any people or organization you employ in the field of hypnosis and that you have the work monitored—if not actually done—by recognized national authorities in medicine and psychology." ARPA, it appears, took Licklider's concerns to heart and stayed out of the hypnosis business.

There were other disturbances to his work, like the time a military contractor dropped by to show off a gun that shot micro-rockets—another Vietnam-inspired novelty. The weapon was a semiautomatic pistol that fired .49-caliber micro-rocket projectiles; it was being tested by ARPA in the United States for possible use in Vietnam. For reasons unknown, the visitor chose to demonstrate it in Licklider's office. "These things got going and the place was left a shambles," he recalled. Then there was ARPA's role as a cover for the black budget; Licklider was forced to fund one project so secret that he was never told its real purpose. Years later all he could say was that his office paid "for digging a hole in Lafayette Square," presumably a reference to a secret project conducted near the White House grounds.

Those annoyances aside, Licklider was mostly left to do his own thing. It helped that most people at ARPA in those years did not know, or understand, what exactly Licklider was doing. In theory, Licklider was in charge of two research efforts at ARPA, one in the behavioral sciences, which apparently made him the resident expert on hypnosis, and another in the obliquely named command and control, which initially consisted of taking ownership of an expensive, but now unneeded, air defense computer. Those responsibilities belied his more ambitious reason for being at ARPA, which was to transform how people interacted with computers.

Among his colleagues in the computer world, Licklider was known as a fervent believer in big ideas, with an easy sense of humor and undying love of corny puns. At ARPA, which was new to the computer world, his colleagues described him as nice but quiet. "I knew that he didn't get into anyone's business, except his own," recalled Donald Hess, who worked at the time as a senior financial administrator for the agency. "It was as though he wanted you to stay away from him."

That was only half-true. Licklider did not engage in small talk, but it was usually because he wanted to avoid getting drawn into discussions of topics like hypnosis, not because he did not want anyone to know what he was doing. In fact, Hess recalled Licklider inviting ARPA employees to a meeting at the Marriott hotel near the Fourteenth Street Bridge, between the Pentagon and the Potomac River.

Licklider had set up equipment to show how someone in the future would use a computer to access information. As Hess recalled, there was a demonstration of how people would have a computer console in their kitchens and use it to access a recipe from a network of connected computers. Most of the technology needed to create what the ARPA history called Licklider's "messianic" vision was years away. Licklider, as the chief proselytizer for interactive computing, wanted people first to understand the concept. Licklider was trying to demonstrate how, in the future, everyone would have a computer, people would interact directly with those computers, and the computers would all be connected together. He was demonstrating personal computing and the modern Internet, years before they existed.

There is little debate that Licklider's reserved but forceful presence in ARPA laid the foundations for computer networking—work that would eventually lead to the modern Internet. The real question is why? ARPA was a military agency, so surely the network was not intended just to exchange casserole recipes. By the 1990s, however, popular news accounts were frequently repeating as accepted truth that the Internet's origins could be traced back to ARPA's work on creating a military communications system that could survive nuclear war. This account sparked a counter-narrative by ARPA-funded scientists who insisted that computer networking was pursued primarily for its civilian applications. The truth is more complicated, and it is impossible to divorce the Internet's origins from the Pentagon's interest in the early 1960s in the problems of war, both limited and nuclear. The Internet would likely not have been born without the military's need to wage war, or at least it would not have been born at ARPA. Tracking the origins of computer networking at ARPA requires understanding what motivated the Pentagon to hire someone like Licklider in the first place. It started with brainwashing.

"My son! My son! Bless the lord," shouted Bessie Dickenson, standing on the ramp of Andrews Air Force Base, in Maryland, as the child she had not seen in more than three years got off the plane. It was 1953, and their reunion was short-lived; twenty-three-year-old Edward Dickenson, sometimes described as a "mountain boy" by the press, would soon be court-martialed for cooperating with the enemy. Dickenson was one of almost two dozen POWs from the Korean War who initially opted to stay in North Korea, throwing in his lot with the communists. Dickenson soon changed his mind and returned to the United States, where he was initially welcomed, then called a traitor. At his court-martial, defense lawyers argued that the young man, who hailed from the almost fictional-sounding Cracker's Neck, Virginia, was a simple country boy who had been "brainwashed" by the communists during his years in captivity. Unmoved, a panel of eight officers convicted him, and he was sentenced to ten years in prison.

"Brainwashing" was a new term in the early 1950s, first introduced and popularized by Edward Hunter, a spy turned journalist who wrote about this dangerous new weapon that could sway men's minds. The communists had been at work on brainwashing for years, but the Korean War was a turning point, Hunter argued. In 1958, he told the House Committee on Un-American Activities that as a result of brainwashing tactics, "one in three American prisoners collaborated with the Communists in some way, either as informers or propagandists." The communists were racing ahead of the United States in mental warfare, Hunter claimed, and their "new weapons are for conquest intact, of people and cities."

Brainwashing soon made its way into the popular imagination with the publication in 1959 of Richard Condon's best-selling novel, *The Manchurian Candidate*, in which a POW, the son of a prominent family, returns to the United States as a sleeper agent, trained for assassination (the lasting popularity of the idea is seen in the modern series *Homeland*, which depicts an American POW "turned" by al-Qaeda).

Whatever the truth of actual brainwashing incidents, the battle for people's minds loomed large in the late 1950s and was the subject of serious Pentagon discussions. The United States and the Soviet Union were engaged in an ideological—and psychological—battle. Eager to exploit the science of human behavior, as it had physics and chemistry, the Pentagon commissioned a high-level panel at the Smithsonian Institution to recommend the best course of action. The Smithsonian's highly influential Research Group in Psychology and the Social Sciences was established in 1959 and tasked to advise the Pentagon on long-term research plans. While the panel's full report was classified,

Charles Bray, who headed the group, published some of its unclassified findings in a paper titled "Toward a Technology of Human Behavior for Defense Use," which outlined a broad role for the Pentagon in psychology. "In any future war of significant length, there will be 'special warfare,' guerrilla operations, and infiltration," Bray wrote. "Subversion of our troops and populations will be attempted and prisoners of war will be subjected to 'brainwashing.' The military establishment must be prepared to assist in promoting recuperation and cohesiveness within possibly disorganized civilian populations, while attempting to shift loyalties within enemy populations."

Psychology during the Cold War had fast become a darling of the military. "By the early 1960s the DOD was spending almost all of its social science research budget on psychology, around $15 million annually, more than the entire budget for military research and development before World War II," wrote Ellen Herman in her survey of the field. Of course, the Pentagon's interests, and the Smithsonian panel's recommendations, were about more than just brainwashing. Bray wrote of applications ranging from "persuasion and motivation" to the role of computers "as a man-machine, scientist-computer, system."

The Smithsonian Institution panel eventually recommended to the Pentagon's director of defense research and engineering that ARPA conduct a comprehensive program that would include both the behavioral and the computer sciences. That recommendation was translated by Pentagon officials into two separate assignments handed down to ARPA: one in the behavioral sciences, which would include everything from the psychology of brainwashing to quantitative modeling of society, and the second, in command and control, which would focus on computers. Though the Pentagon treated ARPA's command-and-control and behavioral sciences assignments as distinct, the archives of the Smithsonian panel make clear that its members viewed the areas as deeply related: both were about creating a science out of human behavior, whether it was humans interacting with machines or with other people. Who better to lead those twin efforts than a psychologist interested in computers? On May 24, 1961, Bray wrote to Licklider, a PhD psychologist who was working for Bolt, Beranek and Newman Inc., in Massachusetts, to ask him if he might be interested in a job at ARPA. The position would be to head up the "Behavioral Sciences

Council," Bray explained. "It will be obvious to you, I believe, that the position has immense potential for good and bad," Bray wrote. The work would be "onerous and exhausting" and like most government positions of the time would not pay well either, between $14,000 and $17,000.

Licklider's initial field of specialty was psychoacoustics, the perception of sound, but he had become interested in computers while working at MIT's Lincoln Laboratory on ways to protect the United States from a Soviet bomber attack. There, Licklider had been involved in the Semi-Automatic Ground Environment, or SAGE, the Cold War computer system that was designed to link twenty-three air defense sites to coordinate tracking of Soviet bombers in case of an attack on the United States. The SAGE computer would work with the human operators to help them calculate the best way to respond to an incoming Soviet bomber attack. It was, in essence, a decision-making tool for nuclear Armageddon, and it spawned decades of popular culture notions of doomsday computers, from the movies *War Games* to *Terminator*.

The reality was that by the time SAGE was actually deployed, it was rendered almost obsolete by the advent of intercontinental ballistic missiles. Still, for scientists like Licklider, who worked on SAGE, the experience transformed how they looked at computers. Prior to SAGE, computers were big mainframes that used batch processing, meaning programs were entered one at a time, often by punch card, and then the computer did the calculations and spit out answers. The idea that someone might sit daily in front of a computer—and interact with it—was unfathomable to most people. But with SAGE, operators for the first time had individual consoles that displayed information visually, and even more important, they worked directly with those consoles using buttons and light pens. In other words, SAGE was the first demonstration of interactive computing, where users could give commands directly, and time-sharing, where multiple users could work with a single computer.

Based on his experience from SAGE, Licklider saw a future where people would interact with computers through personal consoles at their desks, rather than having to walk into a large room and feed punch cards into machines that would crunch numbers. In other words, Licklider envisioned the modern conception of interactive computing. What seems so obvious today was a revolutionary concept in the early 1960s, when computers were still large, exotic creatures housed in university laboratories or in government facilities and used for specialized military purposes. That vision meant doing away with batch processing, where a single user worked on a computer for a single purpose. Instead, users at remote consoles would be able to tap into the resources of a single computer, performing different functions almost simultaneously. Licklider's 1957 article, "The Truly SAGE System; or, Toward a Man-Machine System for Thinking," was one of the first manifestos outlining this new approach, and marked him as a leader of a group of scientists who wanted to transform computing.

In 1960, Licklider took this thinking a step further with the publication of what would become a seminal paper on the path toward the Internet. The article, titled simply "Man-Computer Symbiosis," was not the work of an ordinary computer scientist, as demonstrated by his opening lines. "The fig tree is pollinated only by the insect *Blastophaga grossorun,*" he wrote. "The larva of the insect lives in the ovary of the fig tree, and there it gets its food. The tree and the insect are thus heavily interdependent: the tree cannot reproduce without the insect; the insect cannot eat without the tree; together, they constitute not only a viable but a productive and thriving partnership. This cooperative 'living together in intimate association, or even close union, of two dissimilar organisms' is called symbiosis."

This symbiosis between man and machine was fundamentally different from a present dominated by batch-processing computers; it also differed from the hard-core artificial intelligence enthusiasts, who were pinning their hopes on thinking computers. Licklider suspected that true artificial intelligence was much further away than some people thought and that there would be an interim period dominated by this symbiosis of man and machine. The picture he painted was of people using a network of computers, "connected to one another by wide-band communication lines and to individual users by leased-wire services."

Military applications were certainly high on Licklider's list: after all, his ideas were prompted by SAGE, and his essay addressed the needs of military commanders. Yet his vision was also much broader, and in his paper he included corporate leaders in need of quick decisions and libraries whose collections would be linked together. Licklider wanted people to understand that more than any specific application, what he was describing was an entire metamorphosis of man and machine interaction. Personal consoles, time-sharing, and networking—the article essentially spelled out all the underpinnings of the modern Internet. But all it was at that point was a vision; someone had to develop the underlying technologies to make it happen. When Licklider was offered the ARPA job in 1962, the position was low pay, high stress, and at an obscure agency that was barely four years old. Employees of the agency were all temporary hires, with the expectation that they would leave after just a few years. He agreed to take the position for one year, because it offered him the opportunity to make his vision of computer networking a reality.

In 1960, the same year that Licklider's manifesto on computer networking was released, Paul Baran, an analyst at the Rand Corporation in California, published "Reliable Digital Communications Systems Using Unreliable Network Repeater Nodes." The paper was Baran's proposal for using a redundant communications network to ensure that the United States could still launch its nuclear weapons after an initial attack. His description, like Licklider's, bore a lot of similarities to the structure of the modern Internet.

Years later, as people began to explore the origins of the Internet, a debate emerged over who could rightfully be considered the originator of the idea. The problem with pinning the Internet to any one person, or idea, is that a number of people were thinking about networking computers in the 1960s. The real question is who was in a position to actually translate this vision into a nuts-and-bolts reality. Rand was a possibility: Although it was more of a think tank than a research agency, it shared ARPA's flexibility. The air force might assign Rand broad national security problems to address, and then its analysts had relative freedom to propose solutions that might be picked up by the military for funding. Rand analysts, for example, had theorized the first spy satellite, which led the air force to pursue Corona. Rand also gave great intellectual freedom to its workers, boasting some of the top nuclear theorists of the twentieth century, like Herman Kahn, whose ruminations on "wargasm," his term for all-out nuclear war, made him great fodder for caricature.

Baran, however, was thinking about practical solutions to nuclear war. And in 1960, the same year that Licklider published his landmark paper "Man-Computer Symbiosis," Baran was working with colleagues at Rand on simulations to test the resiliency of the communications system in case of nuclear attack. "We built a network like a fishnet, with different degrees of redundancy," he recalled in an interview with *Wired* magazine. "A net with the minimum number of wires to connect all the nodes together, we called 1. If it was crisscrossed with twice as many wires, that was redundancy level 2. Then 3 and 4. Then we threw an attack against it, a random attack."

Picture the communications network as a series of nodes: If there is just one connection between two nodes and it is destroyed in a nuclear attack, it is no longer possible to communicate. Now imagine nodes with multiple connections to other nodes, providing an alternate path of communication if some nodes are taken out. The question for Baran was, how much redundancy is enough? Through simulations of an attack, Baran and his colleagues found that if you have three levels of redundancy, the probability that two nodes in the network could survive a nuclear attack was extremely high. "The enemy could destroy 50, 60, 70 percent of the targets or more and it would still work," he said. "It's very robust."

Baran later explained his thinking as being aimed squarely at concerns over the hair-trigger alert that both the United States and the Soviet Union maintained with nuclear weapons. Having the ability to survive a nuclear attack would, in theory, make deterrence more stable by taking away the temptation for the other side to launch a first strike. "The early missile control systems were not physically robust," he said. "Thus, there was a dangerous temptation for either party to misunderstand the actions of the other and fire first. If the strategic weapons command and control systems could be more survivable, then the country's retaliatory capability could better allow it to withstand an attack and still function; a more stable position."

For the idea to work, the network would have to be digital, rather than analog, which would degrade the signal as it traveled. It was an ambitious, new idea, and the problem was that Rand, which Baran joked stood for "research and *no* development," could not create such a system on its own.

Rand could not build the network, but the air force could, and its leaders were interested in Baran's idea. Before work started on it, however, a bureaucratic reorganization pushed the project over to the Defense Communications Agency—a stodgy Pentagon bureaucracy that Baran suspected was stuck in the analog world. Better to end the project, he figured, than to see it botched. "I pulled the plug on the whole baby. There was no point. I said, 'Just wait until some competent agency comes around.'" That competent agency would end up being ARPA.

Licklider arrived at ARPA the same month that the superpowers almost went to war over the missiles in Cuba. For senior Pentagon officials, it seemed obvious that ARPA's work on command and control was about nuclear weapons. William Godel, then the deputy director of the agency, recalled that ARPA's new assignment was supposed to look at alternatives to Looking Glass, the code name for the military's "nuclear Armageddon" aircraft that flew around-the-clock on alert. At the Pentagon, Harold Brown, the director of defense research and engineering, thought that he had assigned ARPA to work on problems dealing with the command and control of nuclear weapons. Brown, who wrote the assignment, recalled being influenced by one of his deputies, Robert Prim, a mathematician from Bell Labs. Prim was heavily focused on technologies for the command and control of nuclear weapons, including research that eventually led to Permissive Action Links, security devices for nuclear weapons. Brown was unhappy with the pace of developments in the military services, so he assigned command-and-control research to ARPA in the hopes it might come up with something better.

The need to come up with something better to control nuclear weapons loomed large in the fall of 1962. Just a few weeks after he started work, Licklider attended an air-force-sponsored conference on command-and-control systems held in Hot Springs, Virginia, where the Cuban missile crisis had been front and center on people's agenda.

Yet the meeting had been lackluster, with no one really coming up with any creative ideas. On the train returning to Washington, D.C., Licklider and a Massachusetts Institute of Technology professor, Robert Fano, struck up a conversation, and soon a host of other computer science colleagues on the train got involved. Licklider used it as another opportunity to proselytize for his vision: creating a better command-and-control system required coming up with an entirely new framework for man-machine interaction.

Licklider was well aware of the Pentagon's interest in command and control of nuclear weapons. One of his early program descriptions for work on computer networking referenced the need to link computers that would be part of the nascent National Military Command System used to control nuclear weapons. His vision was for something much broader, however. When he met with Jack Ruina, the head of ARPA, and Eugene Fubini, one of Harold Brown's deputies, Licklider pitched them on interactive computing. Rather than focusing strictly on technologies to improve command and control, he wanted to transform the way people worked with computers, which meant moving away from batch processing toward time-sharing and eventually networking. "Who can direct a battle when he's got to write a program in the middle of that battle?" Licklider asked rhetorically.

The new ARPA research manager was determined to show that command and control could be something more important than just building a computer to control nuclear weapons. When he would meet with Pentagon officials and they would start to talk about command and control, Licklider shifted the conversation to interactive computing. "I did realize that the guys in the secretary's office started off thinking that I was running the Command and Control Office, but every time I possibly could I got them to say interactive computing," Licklider said. "I think eventually that was what they thought I was doing."

Pentagon officials did not quite understand what Licklider was talking about, but it sounded interesting, and Ruina agreed, or at least he agreed that Licklider was smart, and the specifics were not important. When the secretary of defense "asked to see me about something, he never asked me to see about computer science," Ruina said, "he asked to see me about ballistics defense or nuclear testing detection. So those were the big issues." Licklider's work "was a small but interesting program on the side."

Ruina, an engineer, was even less interested in the behavioral sciences, Licklider's other assignment, which was allocated just $2 million a year. Ruina dismissed the entire field as Freudian ruminations. "Tell me what has happened in the last twenty years in behavioral sciences that you would think of as a breakthrough in the sense of giving us new concepts, and thinking, and important contributions, and . . . did it come from any government contract—cut-pipe work, or was it a guy who is more of a novelist like Tolstoy who was able to do great human insights without having to get a government contract to do it?" Ruina asked Licklider. "He said he would think about it, and I remember he came back and could not produce anything that was very interesting, and I said, 'Yes, that was my concern about that program.'" Licklider ended up spending most of the behavioral science money on human-computer interaction, rather than anything related to social science, which suited Ruina just fine.

When Licklider arrived at ARPA, he found an organization that contained a mix of geniuses and mediocre bureaucrats. The agency had recently started its involvement in Southeast Asia, which was still being carried out mostly in secrecy, something that concerned Licklider. "There was a kind of a cloak and dagger part of it," he recalled in a later interview. And, more troubling, Godel, the head of all that work, "was always trying to get control over what I was doing," Licklider said. "I could never tell what he was doing, so that part made me nervous." Mostly Licklider was left alone, however. ARPA was still a young agency with few established precedents to fall back on. Ruina had eliminated the ARPA Program Council, one of the few bureaucratic formalities. Newcomers like Licklider were essentially making up the rules as they went along, creating what would later be regarded as the hallmark of ARPA: freewheeling program managers given broad berth to establish research programs that might be tied only tangentially to a larger Pentagon goal.

Licklider's immediate problem was dealing with the "white elephant," a prototype for a new computer that was built for the SAGE air defense system. The expensive behemoth, designated the AN/FSQ-32, was a prototype for an upgraded version of SAGE, which the Pentagon had recently canceled. By 1960, the Pentagon's primary concern was the threat of ICBMs, not manned Soviet bombers. No longer needed, the computer was essentially dumped on ARPA's doorstep, at least administratively, along with the costs associated with the contractor responsible for it, System Development Corporation, an offshoot of Rand. The computer was a "great asset," Licklider recalled, but it was being used for batch processing. For Licklider, an oracle of time-sharing, that was a waste, and with a cost of nearly $6 million the white elephant took up the largest portion of Licklider's new $8 million budget for command and control. He could not kill the project, so he used the SAGE computer as an opportunity to solicit ideas from computer scientists who shared his vision. He slowly shifted the money to "centers of excellence" in computing.

The most ambitious of those contracts took the name Project MAC, short for Machine-Aided Cognition or Multiple-Access Computer, a wide-ranging $2 million grant to MIT. Project MAC covered the span of interactive computing, from artificial intelligence and graphics to time-sharing and networking. ARPA provided MIT with a great deal of autonomy, so long as it used the money for the goals prescribed by the agency. Licklider, who focused on vision over reputation, also took a risk on more unknown scientists, like Doug Engelbart, at the Stanford Research Institute. By the time Licklider was done handing out contracts, his centers of excellence stretched from the East Coast to the West Coast and included MIT, Berkeley, Stanford, the Stanford Research Institute, Carnegie Tech, Rand, and System Development Corporation.

In April 1963, just six months after joining ARPA, Licklider dashed off a memo to the people he was funding, in what would become one of the more famous missives of his time at the agency. He addressed it to the "Members and Affiliates of the Intergalactic Computer Network," a tongue-in-cheek way of telling the ARPA-funded researchers they were part of a broader community working toward a common goal. "At this extreme, the problem is essentially the one discussed by science fiction writers: 'How do you get communications started among totally uncorrelated 'sapient' beings?" he wrote. The six-page memo went on to state explicitly what he had in mind. "It seems to me interesting and important nevertheless to develop a capability for integrated network operation," he continued. "If such a network as I envisage nebulously could be brought into operations, we would have at least four large computers, perhaps six or eight small computers, and a great assortment of disk files and magnetic tapes units—not to mention the remote consoles and teletype stations—all churning away."

It was the clearest articulation yet of his vision for interactive computing, and vision is what mattered in 1963, because much of what Licklider was building was a foundation of research, not an actual computer network. The lack of anything concrete to show from the initial research was also potentially a liability, because few in the Pentagon at the time really understood the full potential of computers. When Ruina left in 1963, his replacement, Robert Sproull, a scientist from Cornell University in Ithaca, New York, almost canceled Licklider's entire program. After the heyday of ARPA's first year, when it managed space programs and had half-a-billion-dollar budget, funding for the agency had been cut almost in half by the mid-1960s, to $274 million. Sproull was under orders to trim $15 million from ARPA's budget, and he immediately looked for programs that had not appeared to produce much in the past two years. Licklider's computer work ended up at the top of his list, and the new ARPA director was on the verge of shutting it down.

Licklider dealt with the threat of cancellation with his typical calm. "Okay, look, before you cancel this program, why don't you come around and look at some of the labs that are doing my work," Licklider suggested. Sproull went with Licklider to three or four of the major computer centers around the country and was duly impressed. Licklider kept his funding. Asked decades later about whether he was the man who "almost killed the Internet," Sproull laughed and said, "Yes."

By the time Licklider left ARPA in 1964, the term "command and control" for its computer science work was abandoned in favor of a new name, the Information Processing Techniques Office, cementing its focus on computing and shedding its old nuclear identity. His investments were already bearing fruit, small and large. At MIT, the ARPA-sponsored time-sharing system spawned the first e-mail program, called MAIL, written by a student named Tom Van Vleck. At Stanford Research Institute, the previously unknown Engelbart had experimented with different tools that would allow users to interact directly with computers; after trying out devices like light pens, he eventually settled on a small wooden block, which he called a "mouse."

Ivan Sutherland, a brilliant young computer scientist who had already forged an impressive reputation for his work in computer graphics, replaced Licklider. But at the time he was brought into ARPA, Sutherland was just a twenty-six-year-old army lieutenant, recruited because no one else qualified wanted the job. ARPA's unusual system of having only temporary employees made the position difficult to fill. Government salaries were low, and there were no provisions at that point for temporary academic appointments. Sutherland, however, had no choice. "I was in the army, and I got some orders which said, 'You are hereby ordered to proceed to the Pentagon and take this job,'" he recalled.

Sutherland wanted to follow in Licklider's intellectual footsteps but found himself facing resistance from computer scientists. He tried to get the University of California, Los Angeles, to create a network using three of its computers, but the researchers involved did not see how it would benefit them. University professors were scared that networking computers would allow others to tap into their coveted computer resources. Steve Crocker, then a graduate student at UCLA, recalled the battles for computer time. "There came a moment when the tension was so high that the police had to be called to tear the people apart who were about to fight with each other," he said. When ARPA tried to carry out that first computer-networking project at UCLA, it ran into similar resistance. The head of the computer center "decided that being under the gun from ARPA to produce on some timely basis was not consistent with the way a university ought to operate" and "pulled the plug" on the ARPA contract, Crocker recalled.

Sutherland called the thwarted computer-networking project "my major failure." It was not actually a failure, it was just too soon. Shortly after Sutherland left, his deputy, Robert Taylor, took over. Taylor did not have Sutherland's or Licklider's reputation, but he did have vision and determination. In 1965, Taylor approached Charles Herzfeld, the new ARPA director, in his E Ring Pentagon office, and laid out his idea for a computer network that would link geographically dispersed sites. Herzfeld had long had a keen interest in computers. As a graduate student at the University of Chicago, he attended what he described as a life-changing lecture by John von Neumann, the famed mathematician and physicist, who talked about the Electronic Numerical Integrator and Computer, or ENIAC, the World War II computer that was built to speed up the calculation of artillery firing tables. Later, at ARPA, Herzfeld befriended Licklider, whose proselytizing at ARPA about brain-computer symbiosis had an equally profound effect on Herzfeld. "I became a disciple of Licklider early on," he recalled.

Taylor was not proposing Licklider's small-scale laboratory experiments of a couple years earlier. Taylor wanted to create an actual cross-country computer network—something that had never been done before and would require significant new technology, investment, and arm-twisting of researchers.

"How much money do you need to get off the ground?" Herzfeld asked.

"A million dollars or so, just to get it organized," Taylor replied.

"You've got it," Herzfeld replied.

And that was it. The conversation to approve the money for the ARPANET, the computer network that would eventually become the Internet, took just fifteen minutes.

The Internet might have no single progenitor, but Licklider's ability to carry out his vision from Washington, while Baran's perfectly good idea died on the vine at Rand, underscored ARPA's unique position in the 1960s. The ARPANET was a product of that extraordinary confluence of factors at the agency in the early 1960s: the focus on important but loosely defined military problems, freedom to address those problems from the broadest possible perspective, and, crucially, an extraordinary research manager whose solution, while relevant to the military problem, extended beyond the narrow interests of the Defense Department. An assignment grounded in Cold War paranoia about men's minds had morphed into concerns about the security of nuclear weapons and had now been reimagined as interactive computing, which would bring forth the age of personal computing. It was a strange journey.

As for the ARPANET's link to nuclear Armageddon, the truth is convoluted. The ARPANET was not built as a nuclear command-and-control system, but it was inspired by Cold War fears of nuclear annihilation and Soviet domination. Harold Brown's interest in nuclear command and control, the Smithsonian Institution panel's concerns about guerrilla warfare and propaganda, and Licklider's interest in man and machine were all factors that contributed to computer networking. In an interview years later, Paul Baran aptly compared the creation of the Internet to building a cathedral. "Over the course of several hundred years: new people come along and each lays down a block on top of the old foundations, each saying, 'I built a cathedral.' Next month another block is placed atop the previous one. Then comes along an historian who asks, 'Well, who built the cathedral?' Peter added some stones here, and Paul added a few more. If you are not careful you can con yourself into believing that you did the most important part. But the reality is that each contribution has to follow onto previous work. Everything is tied to everything else."

As the director of the agency that started computer networking, Ruina understood the difficulties of crediting any one person, or influence, for what would eventually be ARPA's most famous project. In later years, Ruina was interviewed frequently about his role in the development of the Internet. He often repeated some variation of the same theme: he really had little idea what Licklider was doing, other than he thought it was something good. "I did nothing for the Internet except hire the guy who did it," Ruina said.

Ruina's statement reflected a larger truth about the research that would pave the way for the Internet: computer networking did not occupy much of the agency's time or budget in the 1960s. Licklider was able to do what he did precisely because it was below the radar. The irony of what would later be touted as the agency's greatest accomplishment was that it grew up in the shadow of what was about to become a much bigger focus of the agency's activities: Vietnam. In 1962, as Licklider quietly began to build up the nation's computer research—laying the groundwork for the Internet and personal computing—another part of ARPA was secretly helping to lay the groundwork for a disastrous war in Southeast Asia.

The extraordinary freedom given to ARPA in the early 1960s to tackle military problems was supporting fruitful research into computer networking and nuclear test detection, but in Southeast Asia the agency was pursuing something equally ambitious, but with far darker consequences. For the next decade, ARPA's involvement in the Vietnam War would have profound consequences for the agency. Its ambitious vision of solving insurgency was creating a body of research that would be resurrected decades later in Afghanistan and Iraq, leading to technology that would eventually reshape the very way the United States fights its wars.

CHAPTER 8

Up in Flames

By late 1961, the United States was on a path of escalating involvement in Vietnam, taking ARPA along with it. The Taylor Mission to Southeast Asia in October 1961 had laid the groundwork by calling for more troops and support for South Vietnam's government. The Kennedy administration was now committed to defeating communism in Southeast Asia, and so the debates in Washington after that point were largely over how best to achieve that goal. It was decided that much of that support would be done in secret.

"The growing U.S. military involvement in Vietnam was kept secret, partly because it violated the Geneva agreement," which placed limits on outside involvement in the country, like the number of foreign military advisers, "and partly to deceive the American public," wrote Stanley Karnow in his history of the Vietnam War. Karnow, then a journalist in Vietnam, describes in late 1961 seeing from his hotel in Saigon an American aircraft carrier in the harbor loaded with several dozen helicopters strapped to the deck. Yet when he pointed the sight out to the American military spokesman he was having drinks with, the officer told him, "I don't see nothing."

ARPA's presence in Vietnam, under the rubric of Project AGILE, was a vital part of this secret escalation, and its field office was an unprecedented merging of science and technology with counterinsurgency operations. Whether it was manipulating food supplies, experimenting with forest fires, or moving whole villages, ARPA's Project

AGILE was conducting the world's first large-scale scientific experiments in counterinsurgency. One of the first and most significant of those experiments was chemical defoliation, which was under the personal control of William Godel. ARPA started in the fall of 1961 with experimental spray tests using an adapted UH-1 Huey, called a HIDAL, short for Helicopter, Insecticide Dispersal Apparatus, Liquid, as well as fixed-wing aircraft. At that stage, the focus was still on experimenting with different chemicals on test plots. "This program has been undertaken with due regard to possible adverse psychological effects and at the strong insistence of the Vietnamese Government. Local district and village heads have coordinated the program," Godel wrote in a secret September 1961 report sent to senior leaders at the Pentagon, State Department, and White House.

That experiment appeared, at least in Godel's mind, to have been successful, and President Diem himself began requesting chemicals to destroy manioc and rice plants. The initial campaign would require four H-34 helicopters and six C-19 aircraft to spray the defoliant. Time was of the essence; it was late fall and some rice would already be ready for harvest, so the report also recommended using air-delivered napalm bombs on the grain fields. The plan was to start with a thousand-mile section of "zone D," not to destroy food, but to eliminate ground cover for the Vietcong. The extreme sensitivity of chemical spraying required a policy decision from the highest levels. That came on November 30, 1961, when President Kennedy secretly authorized the use of chemical defoliation in Vietnam.

Five days later, on December 4, 1961, Godel summoned James W. Brown, a scientist from the U.S. Army Chemical Corps Biological Laboratories at Fort Detrick, Maryland, to discuss the start of ARPA's secret defoliation experiment. The location of the experiment was described, in a later memo, merely as a "friendly country." Godel explained that a crop destruction operation had been approved, and a preselected area would be sprayed with a defoliant, followed by napalm to burn down foliage. Brown would be in charge of initial spray operations and ensuring government authorities in South Vietnam would be ready to move in once the spraying was complete. The sensitivity of the operation was clear: Godel instructed Brown that he was "to be ignorant" if local government authorities asked him for information about the chemical agents or about protective measures. Brown should tell them nothing, a classified account of the meeting noted.

On January 7, 1962, three C-123 aircraft touched down at the Tan Son Nhut Air Base just as the sun was setting over Saigon. The C-123, a rugged, twin-engine short-haul transport aircraft, was popular for covert missions because of its ability to take off and land from primitive airstrips. The White House had initially weighed stripping the planes of military markings and flying them under the guise of civilian aircraft. That idea was rejected, but the mission was nonetheless classified. To avoid detection, the aircraft were parked under guard on a ramp normally used by President Diem's counterinsurgency squadron, an elite unit led by Nguyen Cao Ky, a flamboyant military commander known for his penchant for purple scarves and close links with American covert operations.

The crews had been briefed on the secret mission and had signed statements promising not to reveal where they were going or what they would be doing. Designated simply Tactical Air Force Transport Squadron, Provisional 1, the aircraft would later be known by their project code name, Operation Ranch Hand. The security measures were not unwarranted: one morning the American crews found that someone had sabotaged the planes and slashed the throat of a Vietnamese guard. Within a week of arriving in Vietnam, the aircraft began flying missions over Bien Hoa–Vung Tau highway, also known as Route 15. Flying just over the treetops, the aircraft, outfitted with spraying rigs, released their chemicals in a steady stream over the forest below. Inside the aircraft, the barrels were marked with colored bands to denote the type of chemical compound they contained. The bands on the barrels of those first aircraft were purple, denoting "Agent Purple," a fifty-fifty mixture of two herbicides: dichlorophenoxyacetic acid and trichlorophenoxyacetic acid. Other agents included Pink, Green, Blue, White, and what would eventually be the most widely used defoliant in Vietnam, Agent Orange.

In many respects, Godel was running AGILE as a one-man show, bypassing even ARPA's director, Jack Ruina, who was happy to ignore the whole mess. Godel reported to senior Pentagon and White House officials through the Special Group for Counterinsurgency and treated AGILE as his personal domain. Godel had, in fact, a very clear idea of what he was trying to achieve. He explained his rationale for the various projects as a "weapons system" approach, meaning each weapon or technology was selected not as an end unto itself but for a specific purpose: to enable the South Vietnamese to fight on their own. "The development and use of patrol dogs and defoliants, for example, is designed to bring Vietnamese troops out of their Beau Geste forts into active pursuit of the enemy," Godel wrote.

All of Godel's AGILE projects were tied together. Chemical defoliation would cut off food sources and eliminate ground cover for the Vietcong in the jungle. The South Vietnamese military would be provided, courtesy of ARPA, a host of new technologies—such as new lightweight guns—to fight the Vietcong in the jungles. Finally, the linchpin of the strategy was the strategic hamlet program, which would resettle farmers and their families to areas that could be secured, cutting off opportunities for the Vietcong to terrorize, recruit, and resupply. Or so went the plan.

The idea of population resettlement in Southeast Asia traces its tragic roots to the British-ruled Malaya, which faced a communist insurgency in the 1940s. Robert Thompson, permanent secretary of defense of Malaya, drew up a wide-ranging counterinsurgency plan, which involved a complex set of psychological operations, such as flying "loud mouth" aircraft that ordered insurgents to give themselves up, chemical defoliation to eliminate cover for rebels, and fortified villages, a small-scale version of what Godel and other officials ended up proposing in Vietnam.

Just as Edward Lansdale brought his lessons from the Philippines to Vietnam, counterinsurgency experts like Thompson offered lessons from Malaya. And like the Philippines, Malaya offered a seductive case study, because it was regarded as a successful defeat of insurgency. There were key differences between the two regions, however. In Malaya, the communists were ethnic Chinese, a distinct group from the Malay, so when the population was resettled, it was relatively easy to keep the Chinese out of the new villages. In Vietnam, however, there was no easy way of distinguishing the Vietcong from peasants loyal to the government. Another important distinction was size: the communist insurgents in Malaya probably consisted of fewer than ten thousand fighters, while the Vietcong by 1962 would number at least eighty thousand.

The strategic hamlet program was a massive undertaking by any measure; it involved transplanting South Vietnam's rural population to newly constructed villages. It required a government capable of explaining the policy and then providing the infrastructure and security needed to build and secure the hamlets. The presumption on the American side was that the villagers wanted to be protected from the Vietcong, so they would willingly, and even happily, move to areas that were protected, but the reality on the ground was far more complex. The creation of the strategic hamlets began in Binh Duong province, where the South Vietnamese government launched the program in March 1962 under the name Operation Sunrise. Few peasants were willing to move voluntarily to fortified encampments far from their farmlands. The South Vietnamese government pushed forward anyway, with Diem's brother Nhu, a man far removed from the reality of peasant life, heading the process. By the fall, the government in Saigon claimed that 3,225 strategic hamlets had been built "and that over 33 percent of the nation's total population was already living in completed hamlets."

Yet ARPA from the start understood that those numbers belied a more complex reality on the ground. The "strategic hamlet," according to a 1962 ARPA report on rural security, "is a condition of the mind and heart, not an impenetrable fortress." It was another way of saying that there was no clear blueprint for what a hamlet would look like, because little in the way of resources was provided by South Vietnam's government for these new encampments. The specifics were unclear, but the goal was not. The same ARPA report described the hamlets bluntly as "machinery for formal government control" over the population. The "effective" strategic hamlet "is one in which every citizen and his activities are generally known and strangers and unusual activities are easily observed," the report stated.

While the South Vietnamese government was claiming great success, those working with Project AGILE were dubious. Reports of forced labor, disgruntled farmers, and empty villages were filtering back to Washington through the Combat Development and Test Center in Saigon. In 1962, Gerald Hickey, a Vietnamese-speaking anthropologist newly hired by the Rand Corporation to work on ARPA's Project AGILE, began visiting some of the strategic hamlets around South Vietnam, including in Cu Chi. What he saw was not encouraging. Farmers complained of forced labor to help fortify the hamlets, which took them away from their fields. Agriculture output was dropping precipitously in some areas, and for those already struggling to eke out a living, the uncompensated labor was devastating. As for the elaborate fortifications, which included bamboo spikes and warning devices, the Vietcong, at least around Cu Chi, simply dug tunnels to infiltrate the hamlets.

When Hickey attended a "theatrical" ceremony for the official completion of the Cu Chi hamlet, the scene was markedly different. The hamlet had been hastily cleaned up, and the villagers had been told to stay inside. Diem's brother Nhu arrived at what seemed in almost every respect to be a Potemkin village. Gripping a cigarette holder, Nhu presided over a ceremony bereft of the residents who were not trusted enough to attend. Hickey's report to ARPA on the strategic hamlet program, which was resoundingly negative, was not greeted with enthusiasm in Washington, either from civilian or from military leaders. Harold Brown, one of many Pentagon officials Hickey briefed, literally turned his back on Hickey and a colleague during their briefing, swiveling his chair away from them. A marine general slammed his fist on the desk and told the Rand consultants that the peasants would be forced to help the program.

In April 1962, as the strategic hamlet program was collapsing in Vietnam, ARPA hired the Rand Corporation to hold a seminar in Washington, D.C., gathering the world's top experts in counterinsurgency, drawing on military officers with experience in Malaya, the Philippines, Kenya, and Algeria. Those experts included names that would become synonymous with counterinsurgency, like Edward Lansdale and David Galula, a French officer. They spent four days discussing their experiences quashing insurgencies using a variety of techniques, including forced resettlement. They were upbeat. In Algeria, women liked the "varied social life" in the new settlements, Galula declared. Other attendees praised what they described as successful resettlements in Malaya and Kenya. The counterinsurgents in Washington, D.C., seemed oblivious to what was going on in Vietnam.

Officials back in Washington might not have wanted to hear it, but the strategic hamlets in Vietnam, the very linchpin of counterinsurgency, started coming apart at the seams almost as soon as they were created. "That it was the major facet of the Regime's counterinsurgency efforts from 1961 on is undeniable," Lawrence Grinter wrote in an unpublished analysis of strategic hamlets buried in the archives of an ARPA contractor. "That it failed miserably is beyond question."

By the fall of 1963, Project AGILE was collapsing on all fronts as the insurgency in South Vietnam grew in size and scope amid mounting general opposition to Diem's regime. Counterinsurgency experts like Edward Lansdale, who supported Diem, were oblivious to the key role of Ngo Dinh Nhu, Diem's brother. If the kindly Diem was the public face of the South Vietnamese government, Nhu was the éminence grise, working behind the scenes as the president's closest confidant and strongman. Nhu, a French-educated self-styled intellectual, was an enigma for Western advisers mostly because he avoided meeting them. An opium addict with pretensions of being a philosopher, Nhu wielded de facto power over the regime's paramilitary forces.

The strategic hamlets, run by Nhu, were in shambles, and the Mekong delta was now dotted with empty villages filled with roofless huts and broken-down barbed wire, like abandoned prisons whose inmates had long ago escaped. Many of Godel's other ideas, like his plans to introduce military dogs to South Vietnamese forces, were also failing. The patrol dogs had been one of the first projects Godel introduced under Project AGILE, in the hopes they would help South Vietnamese soldiers fight more effectively in Vietnam's dense tropical foliage. Unable to adapt to the jungle, Godel's dogs had largely grown sick and died or were eaten. Diem had warned him that would happen. Then there was chemical defoliation. The spraying was supposed to be done in secret, but it was quickly exposed and proved a boon for North Vietnamese and Soviet propaganda that accused Americans of poisoning crops, which was true.

Even some American officials were wary of the defoliation. Seymour Deitchman, a Pentagon engineer, noticed the dead trees and brown plants in the ARPA compound in Saigon. Empty barrels containing Agent Orange had been stored amid the low-slung colonial buildings on the compound, and the vapors had managed to kill off vegetation in and around the ARPA offices. At one point, Colonel Trach, the Vietnamese head of the Combat Development and Test Center, started selling the barrels, only to get in trouble, not because the barrels were dangerous, but because he failed to give a cut of the profits to his higher-ups. Deitchman was concerned; no one appeared to have given much thought to the health consequences of using the chemicals. "Well, do you know what effect that is going to have on the people caught up by it?" Deitchman asked Godel in one meeting at the Pentagon. The response, Deitchman recalled, was a four-letter curse and an angry dismissal. "ARPA was going to win the war," Godel told Deitchman. "That was its role."

Some of Godel's projects also ran into unexpected political problems. The AR-15s that Godel brought to Vietnam were supposed to demonstrate their utility for South Vietnamese troops as a lightweight, low-recoil weapon that would be better suited for the jungle than either the World War II–era M1 Garand or the M14 rifle, the standard-issue U.S. Army rifle at the time. A limited number of AR-15s were fielded to American Special Forces and Vietnamese soldiers in 1962. Godel admitted later that he also "wanted to stick a finger in the Army's eye," because its leaders had resisted any use of the new rifle.

The experiment looked like a grand success, based on the initial ARPA field report. "The suitability of the AR-15 as the basic shoulder weapon for the Vietnamese has been established," according to a field test report from August 1962. "For the type of conflict now occurring in Vietnam," American military advisers found the AR-15 to be "superior in virtually all respects" compared with other weapons tested. In response, the U.S. Military Assistance Advisory Group in Vietnam then requested twenty thousand AR-15s. Similarly, the Pentagon's Systems Analysis Group, using the ARPA results, published its own study, which concluded "that the AR-15 is decidedly superior in many of the factors considered. In none of them is the M14 superior. The report, therefore, concludes that in combat the AR-15 is the superior weapon." Yet when advocates of the AR-15 back in Washington used the positive reports to try to persuade the Pentagon to buy the new rifle for American soldiers, it ignited a political firestorm among the

White House, Congress, and the Pentagon. The army did not want to be told what weapon to buy, and its leadership believed that the AR-15 was not as lethal as the M14.

The army conducted its own report in 1963, concluding that the M14 was a better weapon. For the next three years, the army, which was accused by critics of bias, would battle advocates of the AR-15 in the Pentagon and Congress, until it was finally forced to purchase the gun. Even then, the army ordered modifications to the original design, including ball propellant ammunition, which was later blamed for causing the rifle to jam during firefights. The debates went back and forth: soldiers voiced frustration over the jamming and the need to constantly clean the weapon, and AR-15 advocates blamed the army for modifying the design. In the meantime, the original purpose of the ARPA tests—to quickly field a rifle better suited for the Army of the Republic of Vietnam—was delayed as a result of the bureaucratic infighting the weapons had sparked in the Pentagon. As a result, it took six years for the AR-15 to be fielded in large quantities to South Vietnamese soldiers, who would not get the new rifles until 1968, after the Tet Offensive. Godel had wanted the AR-15 to help the Vietnamese soldiers battle an insurgency and instead saw his idea get bogged down over an entirely different question: the fielding of a new weapon for American troops. Eventually, all three U.S. military services adopted the AR-15, designated the M16, and ARPA officials frequently touted it as a success in later years. Godel's assessment was quite different. "It was a blatant, blithering failure," he said.

Even if some of AGILE's technology—the "gimmickry and gadgetry" as the ARPA director Jack Ruina had derided it—had worked, it is not clear they would have been effective in their overall goal of waging counterinsurgency, which was posited on having an effective South Vietnamese government. Diem's autocratic regime was plagued by nepotism and weighed down by corruption and incompetence, and its legitimacy was marred by brutal crackdowns, particularly against South Vietnam's Buddhists.

Patience with Diem, and with the counterinsurgency proponents, waned back in Washington. Secretary of Defense Robert McNamara, who had made his reputation in operations research, never embraced counterinsurgency, at least not the type forwarded by men like Lansdale, who by 1963 had largely been sidelined, at least from matters relating to Vietnam. And in Saigon, American military leaders were also getting frustrated with ARPA, which they felt was stepping on their turf. ARPA reported not to the commanders in charge of operations in Vietnam but instead to Harold Brown back in the Pentagon. According to an official army history, the military services looked at ARPA as an unwanted competitor, and they "regarded the ARPA field unit with distrust."

ARPA was now in a position similar to its nascent days in 1958 and 1959. Thanks to William Godel, the agency had successfully carved out a role for itself in an area that had become front and center in American policy. And for the second time in its history, ARPA found itself surrounded by enemies who wanted to dismantle its growing empire, only instead of space it was in Vietnam.

At 4:30 p.m. local time on November 1, 1963, a panicked President Diem called up the American ambassador in Saigon, Henry Cabot Lodge Jr. There was shooting at the presidential palace, and a coup was under way: What was the position of the United States?

"I do not feel well enough informed to be able to tell you," Lodge replied. "I have heard the shooting, but am not acquainted with all the facts. Also it is 4:30 a.m. in Washington and the U.S. Government cannot possibly have a view."

"But you must have some general ideas," Diem, now frantic, replied.

South Vietnam's president had reason to question the American ambassador; he had never trusted the Americans, and though he had no way of knowing for sure, he correctly suspected that the United States supported the coup now under way. Rebels in the South Vietnamese military took control of Saigon and were quickly closing in on the palace. ARPA personnel in the city, including Charles Herzfeld, who was visiting from Washington on his first trip to Vietnam, watched the fighting until late in the night from the rooftop bar of downtown Saigon's Caravelle Hotel, occasionally ducking at the sight of machine gun tracers. That evening, Diem and his brother escaped from the surrounded palace through a secret underground tunnel carrying a briefcase stuffed with American dollars and sought refuge in Cholon, a Chinese section of Saigon.

Back in Washington, Diem's supporters had other problems. Edward Lansdale, the patron saint of Project AGILE, was forced to retire from the air force, having already been pushed out of his role in Southeast Asia counterinsurgency over the past two years (following the CIA's Bay of Pigs fiasco, Lansdale helped lead Operation Mongoose, the series of failed attempts to assassinate the Cuban leader, Fidel Castro). News of the coup reached Godel shortly before a meeting that would have a profound effect on his career, and on ARPA.

John Wylie, the Pentagon bureaucrat who oversaw the funds used to finance the agency's classified Vietnam operations, called William Corson, a marine major assigned to ARPA in charge of distributing the cash. Wylie wanted to meet with Corson and Godel that morning. Strangely, Wylie did not want to talk at the Pentagon, instead insisting on the nearby Twin Bridges Marriott Motor Hotel, a popular meeting point for military officials, thanks to its location just five minutes away from the Pentagon and its nice views of the Potomac River and the Washington Monument. It was not the place, however, where one would normally hold a serious financial meeting about funds used for a classified Vietnam program. Wylie's request was odd, but Corson chose not to argue with a senior Pentagon official.

Corson was a mysterious figure. Assigned to ARPA as Godel's special assistant in June 1962, he was also working as a secretary on the Joint Defense Department–CIA Committee on Counterinsurgency. The young marine officer seemed to despise ARPA and all the work it was doing. He believed that the defense scientists he was working with were "vultures," whose novelties, like crop defoliants and man-hunting devices, were doing little to address the underlying causes of insurgency. In a meeting to discuss proposals for counterinsurgency work, Corson derided all of the ARPA studies on offer, saying they were "doomed to failure." Regarding a proposal to look at CIA–Special Forces activities, Corson quipped, "This has about as much value to ARPA as engaging a marriage counselor to provide guidance to Elizabeth Taylor and Richard Burton about marriage."

At ARPA, Corson was designated a "Class A" agent. In the spy business, Class A agents were used to finance operations that required cash for special purposes, such as paying informants. In this case, Corson was responsible for disbursing cash related to ARPA's classified activities in Southeast Asia. The marine major later claimed he had little knowledge of what he was supposed to do, other than function as a "paymaster."

Over the next year, Corson would issue a series of cash advances to Godel, as well as Wylie, receiving receipts that might contain descriptions as vague as for "ARPA directed activities." Later, when returning from Vietnam or other countries, Godel would file a full expense report, and the cash receipts would be destroyed, a way of acknowledging that the advance had been canceled. The system seemed to work well enough until early November 1963, when McNamara suddenly demanded a full accounting of the cash accounts. As auditors started to comb through the Class A records, they found a shortfall in the funds. Wylie, as the man ultimately responsible for ARPA's cash, knew there was going to be trouble. It was the sort of thing that could send people to jail.

It was raining that Friday morning when Wylie picked up Corson in his Pontiac from the Pentagon's river entrance and drove him over to the Marriott. The two men sat in awkward silence in the hotel's breakfast area until Godel showed up a half hour later in the official ARPA sedan. Godel's arrival did not improve the mood; he was preoccupied with the coup in Vietnam, and Wylie was in a panic over the cash accounts. As Godel and Wylie began to discuss the state of the accounts, Corson claimed ignorance. "You can be around people that are in the intelligence business," Corson recalled, "you know they are speaking English, you know the meaning of every word, but the conversation was extremely elliptical."

There were problems reconciling the cash accounts, and Godel was getting increasingly frustrated with Wylie, who was in charge of the money. "If you are in trouble, you should go to your boss right now and tell him," Godel finally snapped at Wylie, "and if you are in deep trouble, serious trouble, then you ought to get a lawyer, a good one right away."

With little resolved, Wylie departed alone, and Corson and Godel drove back together to the Pentagon. On the ride back, Godel asked Corson how much he owed to the Class A account, meaning how much he was indebted based on the cash advance he had received for his most recent Southeast Asia travels. Corson told him he had an outstanding balance of about $3,000. After lunch, Godel went with Corson to David Mann Jewelers, a private store in the lobby of the Pentagon that sold everything from watchbands to engagement rings. Godel, who knew the owner, cashed a personal check for $2,000. The remaining $1,000 he took from the safe in his office and settled his account with Corson. Thinking the matter was closed, Godel returned to following the grim situation in Vietnam.

Back in Saigon, Diem and his brother had sought refuge in a French church in Cholon. Believing a deal had been worked out to allow them passage out of the country, the two brothers got into an armored personnel carrier sent by the coup's leaders. The two brothers were shot and stabbed repeatedly as the vehicle headed back to military headquarters; pictures of their mutilated corpses eventually made their way into the news. When word of their gruesome deaths got back to the White House on November 2, President Kennedy "rushed from the room with a look of shock and dismay on his face." The White House might have green-lighted the coup, but by all accounts Kennedy's shock over the brothers' deaths was genuine.

For Godel, who believed deeply in Diem, the death was a devastating personal blow. Then, within days, on November 5, 1963, an order came down from Secretary of Defense Robert McNamara: Class A agents were to turn in all of their cash and liquidate the accounts. A surprise audit followed. When auditors showed up at Wylie's office, he first asked them to come back later, and when they refused, he offered up a $4,000 personal check and withdrew a $100 bill from his wallet to make up for missing cash. What started as an audit ballooned into a criminal investigation. Agents turned their attention to Godel, who was used to spending cash on covert operations with few or no questions asked. In the Pentagon's Office of Special Operations, Godel's penchant for spending money on covert operations without proper approvals had gotten him reprimanded, but nothing serious. Now he was at the center of an FBI investigation.

Godel gave a series of contradictory answers to investigators from the FBI and the inspector general, some of which he later attributed to confusion over his multiple trips to Vietnam, and some to their confusion over how the accounts were handled. He was also hesitant, he recalled at trial, to discuss with auditors what was really going on with Project AGILE. He was trying, he later explained, to dissuade the auditors from thinking the work was "in some way covert intelligence operations, highly technically defined, and which I did not want associated with this program."

On November 22, 1963, another event changed the course of Project AGILE. While traveling in a motorcade in Dallas, Texas, President Kennedy, whose enthusiasm for covert warfare had enabled AGILE's broad mandate in Vietnam, was assassinated. The Friday night Kennedy was killed, three of ARPA's senior officials—Robert Sproull, the director, and his two deputies, Robert Frosch and Charles Herzfeld—had a somber dinner together. The nation was mourning its president, and Frosch recalled that the evening consisted of the three normally loquacious men eating in silence, with each occasionally interjecting the repeated statement "The president's been shot." Not present at that dinner was William Godel, who within months would be fired from his position at ARPA and indicted on criminal charges of fraud.

It was late on a Friday evening at the federal courthouse in Alexandria, Virginia, when the jury finished deliberating. The case against William Godel, the former deputy head of ARPA, went to trial in May 1965, as American troops began pouring into South Vietnam, the very thing that ARPA's counterinsurgency work was supposed to forestall. Godel's original justification for AGILE, which was to equip indigenous forces to keep American troops out of the conflict, was gone. That goal had been overtaken by events much larger than ARPA or Godel. The counterinsurgency campaign was now a conventional war.

The criminal trial was supposed to be a straightforward case of government embezzlement, but the deliberations had dragged on for almost an entire week, and the jurors were getting anxious. Initially deadlocked, they were admonished by the judge. "This is an important case," he told them earlier that afternoon, urging them to try to reach a verdict. "The trial has been expensive in time and effort and money to both the defense and the prosecution."

The jurors had already been at the courthouse for three weeks, listening to testimony. Earlier in the day, a power main had broken, and the lights started to flicker. If they could not reach a decision that evening, they would have to spend the night in a hotel near the courthouse. None of them had a change of clothes or even a toothbrush. Everyone wanted to go home.

As the jury deliberated in Virginia, tens of thousands of students gathered on the other side of the country, in Berkeley, California, for a Vietnam Day "teach-in" featuring activists like Norman Mailer, whose speech, "Hot Damn, Vietnam," blasted the war and the Johnson administration. ("Things were getting too quiet in Vietnam. If there was one thing hotter than Harlem in the summer, it was air raids on rice paddies and napalm on red gooks.") The conflict in Southeast Asia was only just starting to attract national attention at the time of the trial. The week before the jury deliberations, the Vietcong had ended what had been an over two-month pause in the ground war, attacking, as *Time* magazine wrote on May 21, like "distant thunder that precedes a monsoonal line squall." The aerial bombardment of North Vietnam, in the meantime, was also ramping up with Operation Rolling Thunder, "the hard, hot application of U.S. air power."

The judge repeatedly tried to keep the testimony away from Vietnam. This was a fraud case, he insisted, not a trial of American involvement in Southeast Asia. But because Godel was accused of embezzling money intended for secret work in Vietnam, it was hard to separate the two issues. The government alleged that Godel, and his co-defendant John Wylie, had taken a series of large cash advances and then conspired to file fraudulent expense reports against those advances, pocketing the money. It sounded like a simple allegation, but as the jurors listened to the testimony, it became both more complex and more confusing. The jury was told about a series of financial transactions—almost all cash—that had taken place over two years, and trying to untangle the accounting was almost impossible. ARPA had essentially used a system designed for paying covert operatives or, as Godel described it at trial, "to support what we called special operations, that is to say intelligence or otherwise special activities for which a Congressional record of expenditures was not for various reasons desired." The trial dragged out over several weeks, as jurors listened to stories about spy cameras disguised as cigarette lighters, meetings with the Vietnamese president, Ngo Dinh Diem, and the creation of a mass resettlement policy.

Godel's friends found the fraud charges preposterous. He had five daughters, a wife, and a successful career; the idea that Godel would sacrifice everything for a few thousand dollars was not believable to them. "I never met a finer man," said Roy Johnson, the first director of ARPA, who came to Godel's defense at the trial. Even associates who were not quite as gracious in their assessment found the charges unbelievable. "Bill Godel wouldn't bother with anything so small as that!" Kenneth Landon, a former colleague, declared. "Now if you were telling me that he had absconded with half a million dollars, I would be impressed that that might be possible because Bill Godel always thought big."

Those who knew John Wylie, the fifty-eight-year-old Pentagon financial bureaucrat on trial with Godel, offered a less charitable view. Wylie had a fondness for luxury boats and expensive cars. In the months leading up to the trial, he suddenly claimed mental illness, entering a hospital for a month of treatment before being declared fit to stand trial. During the trial, Wylie appeared nearly comatose, staring down at his feet, though observers noticed that once outside the courtroom, he appeared just fine. Wylie never testified.

The case against Wylie was clear. When the Pentagon auditors investigated in 1963, they found he had taken cash from the ARPA accounts to pay for a yacht. In Godel's case, however, there was little evidence to suggest he took the money for personal expenses. Rather, the question was whether Godel spent the money the way he claimed he did. Meaning it came down to whether the jury believed that the types of cash transactions he was describing, such as $2,000 for an armored "Q truck," were credible. The prosecutor, Plato Cacheris, who would later rise to fame as a defense lawyer for spies like the former FBI agent Robert Hanssen and the CIA officer Aldrich Ames, insisted the trial was never about Vietnam, just money. "It was a couple guys trying to make a few bucks," Cacheris later recalled.

Shortly before 10:00 p.m., the foreman sent a message to the judge. The twelve members of the jury, three women, all housewives, and nine men, managers and clerks of local small businesses, had finally reached a decision. Wylie was found guilty of embezzlement and fraud. Godel was acquitted of embezzlement but found guilty of making false statements and conspiring with Wylie. For Godel's family and friends, it made no sense that the jury would clear him of stealing money and yet find him guilty of working with Wylie, a man Godel could barely tolerate. Both men were sentenced to five years in prison.

In the next year, conventional military involvement in Vietnam, which Godel had fought so hard to prevent, continued to escalate rapidly. From his federal prison cell in Pennsylvania, Godel watched in horror at the mounting death toll. His tools for counterinsurgency, like defoliation, were warped into weapons of a conventional war, something that was never supposed to happen. Over the course of the Vietnam War, more than ten million gallons of Agent Orange, the most widely used defoliation compound, would be sprayed on Vietnam, exposing tens of thousands of American military personnel and countless more Vietnamese to cancer-causing chemicals. Agent Orange became synonymous with the entire chemical defoliation program and, for many people, symbolic of the flawed war effort. Of all the things Godel brought to Vietnam, chemical defoliation left the deepest and most tragic legacy.

Godel left ARPA believing all his efforts had ended in failure: he had not succeeded in stopping the insurgency from escalating into war, and he had not stopped the introduction of conventional troops to Vietnam. "The goal of AGILE," according to Godel, was "winning that war, at that time, in those circumstances, and have a legacy at the end—a model of workable answers for the 'next time.'" That did not happen, and when Kennedy died, the idea of counterinsurgency as a way of preventing foreign entanglements died with him. AGILE was created by Godel, enabled by Kennedy, and supported by Diem. Now Kennedy and Diem were dead, and Godel was in prison. Yet rather than ending, AGILE, like the Vietnam War, just got bigger. "We ended up trying to win the war with technology," Godel said. Asked ten years later in an interview commissioned by ARPA to list any true successes from AGILE, Godel gave a one-word answer: "None."

Despite the failure of AGILE, Godel had instilled in many of his disciples, including Charles Herzfeld, the desire to apply the tools of science to warfare. Even with Godel gone, Herzfeld, who was about to become director of the agency, embraced the former intelligence operator's legacy and continued to expand AGILE. As for the conviction of his onetime mentor, the only thing Herzfeld would say on the subject was "stuff happens." Herzfeld's vision for ARPA, and AGILE, was more expansive than any previous director's. For Herzfeld, the world was a giant laboratory.

CHAPTER 9

A Worldwide Laboratory

At a press conference on July 28, 1965, President Lyndon Johnson announced that American military forces in Vietnam would rise to a staggering 125,000 men. Monthly draft calls were at 35,000 that year. "This is a different kind of war," Johnson said. "There are no marching armies or solemn declarations. Some citizens of South Vietnam, at times with understandable grievances, have joined in the attack on their own government. But we must not let this mask the central fact that this is really war. It is guided by North Vietnam, and it is spurred by Communist China. Its goal is to conquer the South, to defeat American power, and to extend the Asiatic dominion of communism. There are great stakes in the balance."

Those stakes, in the view of the administration, were also increasingly global. The Cuban Revolution that in 1959 overthrew Fulgencio Batista and brought Fidel Castro to power was soon followed by a decade of third world insurgencies. From Southeast Asia and Latin America to the Middle East and Africa, a motley collection of insurgents fought powerful central governments, often quite successfully. Whether it was under the banner of social justice, Marxism, communism, or simply national liberation, insurgencies battling American-backed regimes were regarded by officials in Washington as a disease to be rooted out.

For ARPA, this also meant expanding its mandate beyond Vietnam. Project AGILE was no longer limited to guerrilla warfare in Southeast

Asia. At hearings ARPA officials began to describe the agency's mission in terms of creating a global laboratory for counterinsurgent warfare. Reading from an ARPA program description during a hearing in 1965, Texas Representative George Mahon noted that ARPA was working on "simulating the behavior of nations and individuals in the laboratory conditions and in comparing the two."

"Is there any practical way to simulate the behavior of a nation in a laboratory?" Mahon, in apparent disbelief, asked ARPA's director, Robert Sproull.

"I believe so," Sproull answered confidently.

Sproull privately expressed doubts, but that year Charles Herzfeld was elevated to director, and he embraced this global vision. War and science were critical parts of Herzfeld's upbringing, and he was a steadfast believer in the latter's ability to influence the former.

Herzfeld was born to a prominent Austrian family of Jewish heritage that had converted earlier that century to Catholicism amid rising anti-Semitism in Vienna. In 1938, when Herzfeld was a teenager, Austria was annexed by Germany, and the family fled to Budapest, taking a circuitous route common to escaping refugees, before finally arriving in the United States, where Herzfeld's uncle, a prominent physicist, was already settled. Once there, Herzfeld followed in his uncle's footsteps, pursuing a career in science. Although accepted to Harvard for graduate school, in 1945, he chose the University of Chicago, where Manhattan Project veterans like Edward Teller and Enrico Fermi were heading. After graduate school, however, Herzfeld discovered that he was more interested in managing scientific research than conducting it. "I was no Mozart, but perhaps I was a fair Toscanini," he wrote.

In May 1961, Herzfeld was working in the government's scientific bureaucracy when he got a call from Jack Ruina inviting him to visit ARPA to discuss taking over the agency's missile defense work. Over the course of two days, Herzfeld was briefed on the breadth of the agency's research, ranging from ballistic missile defense to counterinsurgency. Herzfeld liked the idea of applying technical solutions to complex, big problems, and he was charmed by William Godel's "towering intellect." Yet he was still undecided about taking the job until, in the middle of a business trip to Europe, he got word that Soviet-controlled East Germany was putting up a wall to divide Berlin. "I thought it was a declaration of war," he said.

When he returned to the United States, Herzfeld called Ruina and told him he would take the job at ARPA. Ambitious and brilliant but prone to arrogance, Herzfeld was put in charge of Defender, ARPA's missile defense program, which was half of the agency's budget. He quickly rose to assistant director, overseeing the agency's work in counterinsurgency, missile defense, and computer science. Though he would not formally become director until 1965, he was effectively already running large parts of the agency a year earlier. With his deep Austrian accent and elite scientific pedigree, Herzfeld carried himself with the confidence of a senior statesman. He embodied ARPA at its height: when it succeeded, as it did with the ARPANET, it succeeded big and helped change the world. But when it failed, it failed big, and that changed the world, too, and not always for the better. "The only thing worth doing in an ARPA way is big problems," he said. "Precisely because little problems you can give to the bureaucracy."

Whatever area he worked in, Herzfeld thought big. In missile defense, he argued for deploying a system, optimistically arguing the technology was ready. "I think one could do reasonably well with $10 billion, maybe $12 or $14 billion, if you stretch it over a period of five years, it isn't all that much money really," he said. But interest in missile defense was already waning by the time Herzfeld took over ARPA, and with no new test ban treaty on the horizon so, too, was ARPA's nuclear test detection work.

Almost from the start, however, Herzfeld gravitated to AGILE, explaining that it appealed to his personal history. "I've been thinking about various forms of warfare for most of my life. I don't enjoy it, but it's what I wound up doing," he recalled in an interview years later. "I found it was a really important problem. It's easy to say, 'Oh I don't want to get my hands dirty.' The next question is, who would you rather do it, if you don't? Who does it?" Some ARPA directors, like Jack Ruina, did not like the messiness of AGILE. "I learned to love it," Herzfeld said.

In Vietnam, Herzfeld saw a place where ARPA could apply its scientific expertise to the problems of counterinsurgency. He threw himself into Project AGILE, sometimes dreaming up ideas of his own and then traveling to Vietnam to test them. "I enjoyed that sort of fieldwork," he recalled of his trips visiting the agency's counterinsurgency work abroad. "I'm more of a doer than a thinker, and the bottom line is I think a lot about things I do. I got a charge out of doing things that changed the environment. It was exciting, it was fun."

For Herzfeld, Vietnam was a place where scientific ideas could be tested. At one point, he teamed up with Robert Frosch, the head of Project Vela, to put the agency's experience in nuclear test detection to work in guerrilla warfare. ARPA had experience detecting things underground, like nuclear explosions. Vietnam did not have nuclear weapons, but the insurgents had their own form of underground warfare: an extensive network of tunnels that the Vietcong used for resupply, communications, and even living. Detecting those underground tunnels had proved almost impossible. Herzfeld "cooked up the idea" of detecting the tunnels by looking for sick trees.

"The question was, if you dig a tunnel near a tree, does it get sick, and if it does, does that show?" Herzfeld said of his idea. "We did experiments in the United States, mostly in Virginia. We checked the trees by taking infrared pictures of them. Sure enough, if you dig a tunnel through the tree's root system, the tree gets unhappy and it shows up on infrared; it looks different than healthy trees. Great triumph we thought." Then ARPA took the infrared tunnel detector system to Vietnam to test, as Herzfeld put it, "the ground truth" of the idea. "We took pictures of trees in South Vietnam with infrared," he said. "It turns out that one-third of all the trees in Vietnam are sick. Okay, great idea, but not feasible."

Herzfeld was interested in history and politics, but most of all he was interested in using science to change the world. In the process, he expanded AGILE from a counterinsurgency effort designed to help indigenous forces into a global program to support American conventional forces and even to protect the president of the United States. "We were expected to solve the problems we were assigned, not just work on them," Herzfeld later said of his philosophy. Under Herzfeld, AGILE would grow to be the most ambitious global counterinsurgency research program ever conducted, and at times the strangest, turning entire countries into test beds. "I was involved in this really going big scale," he recounted. "Boy, it was different, and not all the results were happy."

Warren Stark's road to administering a global counterinsurgency laboratory began at a Washington cocktail party. A Harvard Business School graduate, Stark was one of the legions of young men and women inspired by the 1960 election of the youthful John F. Kennedy, whose inaugural address—"ask not what your country can do for you; ask what you can do for your country"—was a literal call to action, or at least to Washington. Stark was not exactly sure what he could do for his country, but he had an Ivy League pedigree and some experience in business and had served as an enlisted soldier in World War II, so he decided to move to the capital and find out.

His Harvard connections would serve him well. Stark worked the Washington cocktail scene and soon landed what sounded like a plum job: the American ambassador to Costa Rica offered him a position at the embassy with the U.S. Agency for International Development. Excited, Stark phoned his wife in Florida and told her to sell the house, sell the furniture, sell everything; they were about to move to Costa Rica, a country he knew almost nothing about. There was just one hitch. He first had to get a top secret security clearance, a process that could take many months.

In early 1963, still waiting for his clearance and working Washington's social scene, Stark got a party invitation from Larry Savadkin, an old family friend. Savadkin's record was impressive: In World War II, he had suffered head injuries during the German bombing of the destroyer USS *Mayrant*. In search of something safer, Savadkin transferred to the submarine USS *Tang*, which was sunk by one of its own torpedoes on a secret mission in the Formosa Strait, hunting Japanese convoys. One of a handful of survivors, he spent a brutal year in Japan as a prisoner of war. Savadkin in the early 1960s was working on counterinsurgency with William Godel at ARPA. He introduced Stark to Godel, and the former marine took an immediate liking to the bright young Harvard MBA. When Stark told Godel about his plans to go to Costa Rica, Godel countered with a different proposal. "I'm thinking of opening an office in Latin America," Godel told Stark. "Would you do some consulting for me?"

"To be honest with you, I don't know anything about Latin America," Stark told Godel. "I've never been there, other than the Caribbean." Stark also did not know anything about ARPA, but with his security clearance still in limbo he agreed to consult for Godel. His first assignment was to write a political report on Latin America. He rushed out and bought a book on Latin America by Lincoln Gordon, the American ambassador to Brazil. "I plagiarized a lot of stuff," he admitted later. "I ended up doing some research. I wrote him a report, and Godel liked it very much. He asked me if I would like to join his group in ARPA, which I did not know very much about. I told him I really wasn't all that interested. He kept asking me and asking me. I was attending meetings, and he was sort of embarrassing me."

Stark finally made Godel his own offer: "Look, whoever gets me my clearance first, that's where I'll go." For Godel, that was apparently easy. In a week, Stark received his top secret clearance. Stark had no idea how Godel did it, but he finally had a government job.

At ARPA's headquarters in the Pentagon, Stark found himself in the middle of Project AGILE's rapid expansion. After the Cuban missile crisis of 1962, the Pentagon decided that insurgency was a global problem and required a global program. The Pentagon's spending on counterinsurgency operations had grown from $10 million to $160 million between 1960 and 1966, and much of that funding went to ARPA. Even though some of the work that ARPA was sponsoring was unclassified, the very existence of the ARPA field offices was so sensitive that in public records of congressional hearings, the names of the countries that ARPA was involved in, other than Vietnam and Thailand, were blacked out.

The agency had opened up a small field office in Panama, whose jungles could be used to test equipment heading to Vietnam, but the best place to test counterinsurgency technology was Thailand, an American ally that was also facing a rising communist insurgency and that had environmental conditions roughly similar to Vietnam's. The insurgency in Thailand was focused in the northeast, where communists were able to exploit ethnic and economic divisions. For the United States, the insurgency there offered an ideal proving ground for tactics and technology: Thailand shared features of Vietnam's geography, including dense jungle and a network of waterways that proved effective for both commerce and warfare. The ARPA office in Thailand would eventually grow to hundreds of employees, making it a central hub of the agency's global counterinsurgency network. "Thailand was basically a laboratory to conduct research on projects that would end up being used in Vietnam," explained Stark, who was put in charge of much of the AGILE research in the region.

Stark arrived at ARPA eager and enthusiastic but with, by his own admission, no knowledge about Southeast Asia, which was clearly going to be the focus of his work. He would travel frequently around Thailand, sometimes in the ARPA-owned Caribou aircraft, which could take off from little more than a small clearing, and sometimes by helicopter, catching glimpses of Thai life in the provinces. The conditions in remote Thai villages, such as the lack of electricity and plumbing, shocked him. During one of his first visits to Thailand, Stark went with another ARPA official to a local hostess bar, where a Thai woman's company could be bought by the hour. He paid for an hour, not out of romantic interest, but so that he could have someone who would tell him something about Thai culture, about which he knew nothing.

Despite the admitted lack of knowledge that senior officials, like Stark, had about the region, AGILE had an underlying philosophy that was breathtaking in scope: the agency ultimately envisioned creating a global database of people, politics, and places that could then be applied as a model anywhere in the world where the United States needed to battle insurgency. The "people and politics" branch of this model was embodied in a project called the Remote Area Conflict Information Center, or RACIC, whose objective was "to establish an information system encompassing a broad area of military and sociological information from which state-of-the-art surveys, interdisciplinary analyses and studies, and specific technical information requirements can be derived."

The "places," or physical environments, or "bio-ecological classification of military environment," would be studied under a program called "Duty," which would create a database of geographic and environmental information of regions where the United States might have to conduct counterinsurgency campaigns. "With a proper approach to the collection of environmental data from a limited number of selected areas in the world, it may be possible to arrive at generalizations which will have predictive value," Robert Sproull, the director of ARPA, explained to Congress in 1965. "We have underway a contract, which suggests that the conditions of jungle, meteorological conditions, hydrology, and so on, can be classified into approximately 12 general categories. If on further investigation this proves to be the case, we should be able to extract from our experience in Vietnam, as an illustration, a judgment as to the applicability of certain types of equipment and material to similar jungle areas elsewhere."

Duty, of course, was heavily influenced by ARPA's involvement in Agent Orange and chemical defoliation. "The need for ecological and physiological studies of vegetation is clear if we are to determine the 'why' with respect to the effects of herbicides and then convert this information into predictive procedures," Stark wrote in one memo, explaining the need to collect and classify environmental data. In other words, ARPA was coming up with models of places and people. Need to defoliate jungles in Latin America? ARPA could determine what sort of herbicide would work best for the local vegetation. Fighting guerrillas in the mountains? ARPA would know what equipment would be the most effective at high altitudes. Need to create a "hearts and minds" campaign for a region infiltrated by pro-Castro communists? There again, ARPA could draw on its database of social movements.

The start of that database was Thailand, where ARPA could with relative freedom collect data and test equipment. One of Thailand's most attractive features as a war laboratory was its jungle, and so it is not surprising that its earliest use was as a proving ground for jungle communication. The project originated with the president's brother Bobby Kennedy. According to Stark, Bobby Kennedy was chairing a meeting of the Special Group for Counterinsurgency, when he asked in frustration, "We can communicate by radio between the Earth and the Moon. How come we can't communicate a couple of hundred yards in the jungle?"

The answer was attenuation. As radio waves propagate over wet, dense vegetation, they lose their strength. For soldiers operating in Vietnam's jungles—whether American advisers or South Vietnamese soldiers—this presented a major problem. The proposed solution was SEACORE, short for Southeast Asia Communications Research, which was designed to address this problem through a series of data collection experiments in Thailand.

At first glance SEACORE seemed to be a project tailor made for ARPA. Here was a fundamental military problem—radio communications in the jungle—with a specific scientific challenge: attenuation caused by tropical growth. But SEACORE was troubled from the start. ARPA essentially adopted a U.S. Army program started in Panama and transported it to Southeast Asia. What started in 1962 as an eighteen-month study with a $2.5 million budget ballooned in cost and dragged out over years. An ARPA consultant hired to review SEACORE concluded in 1964 that the work "had little potential."

Even with that negative assessment, the work stretched out over seven years and involved moving an entire laboratory to Thailand. ARPA even paid to build a faux village in the Thai jungle to carry out the doomed communications experiments. The best proposal SEACORE could come up with was a half-wave dipole placed on three elephants, which would "maintain a much more efficient antenna capability along the trail." The idea was novel but not terribly practical for American or Vietnamese forces. The easy fix was simply to put antennas on top of tall trees.

Though the program cited some incremental improvements, like the development of a high-frequency radio that was used in Vietnam and around the world, its largest legacy was a raft of publications and annual reports. "While SEACORE may have had some marginal achievements, e.g., leaving some enhanced Thai military communications research capability, it appears to have neither substantially affected Thai military communications capabilities nor to have made a major contribution to the state of the art of communications in a jungle environment," the ARPA history concluded.

Smaller projects fared little better. One example was a "sunlight projector" developed by ARPA for psychological operations teams. As the name implied, it was a movie projector that used a mirror—a shaving mirror in this case—to focus sunlight on movie film to project it on a screen. The idea was that this sort of ruggedized equipment could be easily carried into small villages to play government propaganda films. An American psychological operations team actually tested the device in 1965 in a small village. The results were not good: the projector ended up focusing too much sunlight and burned a hole through the film.

Not all of the Thai programs were a failure. The Mekong River surveillance system was a rare bright spot. The project was designed to come up with a comprehensive way for Thai authorities to control the flow of insurgents and smugglers along the Mekong. ARPA provided boats, radar, and training. Another successful project, called the *Junk Blue Book,* achieved something of a legendary status among boat enthusiasts and military personnel alike. Published in 1962, the original *Junk Blue Book* cataloged every type of civil watercraft (of which there were many) that traveled along South Vietnam's waterways. ARPA then followed up with another version, for Thailand. According to Charles Herzfeld, the goal was to see if one could link the construction of junks to the region they were built and in that way help the military determine whether a particular boat was from a hostile area. Herzfeld called it "All the South China Sea Junks," a play on the compendium *All the World's Fighting Ships,* which he described as one of his favorite books from childhood. The *Junk Blue Book,* which was to be used by the military to help identify potential smugglers or insurgents, also captured a piece of history that is used by boat enthusiasts even today.

That rare success aside, Stark came to realize that much of the military's—and ARPA's—problem was that it simply did not understand the societies or culture of Southeast Asia. The agency's staff were scientists or policy wonks, like himself, with little prior experience in the region or cultural knowledge. As Stark walked down the street in Thailand, Thai children would often call him *farang kee nok,* or "bird shit foreigner." Stark would answer with *farang kee Tai,* or "bird shit Thai."

In northeast Thailand, Stark was overseeing ARPA's Rural Security Systems Program, which was applying systems analysis—a method that looks at each part of a large organization or endeavor—to every aspect of counterinsurgency, from policing to economic development. Pentagon-style systems analysis was even taught to the Thais, sometimes with humorous results. American military officers, known for elaborate slide briefings, would explain the Rural Security Systems Program with an infographic comparing different elements of the project to the pillars of a Greek temple, a favorite icon among defense technocrats (it was used, for example, in the Institute for Defense

Analyses logo). The Thais attempted to mimic the Americans, but because a Greek temple was an unfamiliar symbol in Thailand, they used a diagram of a chicken to depict the various elements of systems analysis.

The contractors running the Rural Security Systems Program had little experience in Thailand, and the idea of reducing society down to a set of variables that could be manipulated proved quixotic. One former senior ARPA official called it "laughable" and "really bad." ARPA's employees also tried their best to parrot the language of systems analysis, though they often did not do much better than the Thai chicken. Stark remembers giving one briefing on Thailand, trying to show how the entire program worked. "I remember taking this big sheet of paper, and in the middle I had village security and I had little circles; it must have been twenty different programs and how they all intermixed and really fit into this whole thing," he recalled. "[The official] liked it. He said, 'That all makes sense.'"

No, Stark thought to himself, it really did not make any sense. Stark was beginning to think the entire program was garbage, yet ARPA management disagreed: they promoted Stark to a senior management position and sent him to the White House to have his picture taken with President Lyndon Johnson.

ARPA's program needed professional help, so Stark in 1967 turned to scientific experts, convening a high-level advisory board that included social scientists. He took the experts to Thailand for a weeklong visit to review ARPA's programs and make suggestions on improving them. Stark recalled one eminent Yale economist expounding for ten minutes on variations within Thai villages in the size of houses and quality of clothing people wore. Asked what could account for such discrepancies, Bob Kickert, ARPA's resident anthropologist who specialized in Thailand, rolled his eyes and replied, "Wealth." (Kickert, one of the few agency employees with actual expertise in the culture of the region, would later quit after drinking himself silly on gin during a year spent in a remote village near the border with Burma.)

It got worse. Stark asked the JASONs, the genius-packed group that advised ARPA, to look at counterinsurgency strategies. The group had been involved in Vietnam, but mostly informally, by suggesting possible technological novelties for the battlefield, many impractical. In 1966, for example, Nicholas Christofilos, the JASON who once suggested a planetary force field, forwarded a proposal to his friend John S. Foster Jr., the director of defense research and engineering, for detecting hidden weapons caches and munitions by suspending a two-hundred- to three-hundred-foot wire between two helicopters flying parallel to each other. The helicopters would fly over an area where suspected Vietcong might be lurking, and weapons caches could be detected by measuring the frequency change of inductance in the wire caused by the presence of ferromagnetic materials. Herzfeld greeted the idea with enthusiasm. "The approach appears feasible and is worth pursuing," he wrote in a brief note to Foster, who agreed. Foster called it "a unique approach that merits further consideration."

Christofilos's idea was forwarded to a program manager at ARPA, who pointed out that, although a novel idea, "the suspension of a wire between helicopters will present an intolerable safety hazard to pilots, besides being most difficult to achieve." The Christofilos proposal appears to have been tested but, not surprisingly, never used in combat.

The JASONs were mostly physicists, but Stark thought their reputation for brilliance meant they might have some ideas on running a counterinsurgency program in Southeast Asia, so he gathered them for a series of conferences in Cape Cod in the summer of 1967 to discuss the war. They did have ideas, though not particularly good ones. Murray Gell-Mann, a physicist, whimsically suggested ARPA look at the effects of various security tactics, like cutting off the ears of insurgents. The minutes of that meeting, with the "ear cutting" comment, were later leaked to the press, leading to campus demonstrations against the JASONs. Just a couple years later, Gell-Mann won the Nobel Prize for his theory of quarks, confirming that brilliance in physics does not necessarily translate into other fields.

Stark knew the situation on the ground in Vietnam was getting worse by the day, and it was not clear that science, or scientists, were going to help. In fact, it would soon become clear they were making it worse. Counterinsurgency was premised on helping the local government fight communist rebels so that the United States would not have to commit its own troops, but in Vietnam, foreign assistance to an inept and corrupt government was instead fueling the insurgency. As counterinsurgency work grew, the ranks of the Vietcong swelled with new recruits. Yet ARPA's attention had already moved beyond Southeast Asia.

On the afternoon of June 3, 1963, Ruhollah Khomeini was driven through the unlit streets to the Fayziya madrassa in the holy city of Qom, Iran. Followed by a procession of students along the way, the popular imam was greeted by an overflowing crowd, but no electricity. In recent months, the government had been ramping up the White Revolution, a secularization campaign that was aimed at undermining the clergy's power. The authorities had cut off power to the entire city of Qom in anticipation of the imam's speech; his lectures had become so popular they were traded on cassettes. Khomeini still spoke; his microphone was hooked up to a generator.

The timing was auspicious; it was Ashura, the day when Shiite Muslims mourn Imam Hussein, the grandson of the Prophet. Imam Khomeini had been escalating his rhetoric in recent months against the shah of Iran, whose regime he regarded as a puppet for the West. This speech was a watershed, however. Khomeini had decided to take aim directly at the shah, the monarch who had ruled Iran since a CIA-engineered coup in 1953 overthrew the popular and democratically elected prime minister, Mohammad Mosaddegh. "You helpless creature," Khomeini said, addressing the shah directly, "you don't realize that on the day when a true outburst occurs, not one of these so-called friends of yours will want to know you."

Khomeini was arrested shortly after his speech, sparking a round of protests and arrests. The shah's rule, and Khomeini's popularity, were exposing cracks in Iranian society along religious and class lines, exacerbated by nationalist concerns about foreign meddling. An insurgency in Iran was growing, just as it had in Vietnam, and in other parts of the world, and ARPA saw an opportunity.

In December 1963, Rand issued a secret report commissioned by ARPA titled "Support Capabilities for Limited War in Iran," which looked at the potential of either a direct Soviet invasion of Iran or

Soviet support for a local insurgency. It concluded that the United States should focus on improving Iran's own defense capabilities. The next year, ARPA formally opened its first Middle East office, called the ARPA Research and Development Field Office, Middle East. Based in Beirut, Lebanon, it was, in theory, supposed to work with local governments to help strengthen their research and development capabilities. The key focus for the new office was not actually Beirut—that was just a convenient place to run the program—but Iran.

"Iran has been the primary focus of project operations," wrote a Pentagon official in a classified memo explaining ARPA's Middle East office. This focus, the letter goes on to state, was due to several factors, including the Iranian leadership's "receptivity" to the types of projects that ARPA was proposing. Just as Thailand was to be the "soft side" to test counterinsurgency techniques for Vietnam, Iran was the new counterinsurgency proving ground for the Middle East. It had all the right conditions: a friendly government, a military in the midst of modernization, and a low-level insurgency. The Pentagon official cited "Iran's attractiveness as a laboratory to study aspects of remote area conflicts."

Iran was the next frontier: an American ally facing challenges of legitimacy, internal unrest, and looming concerns about Soviet interference. In fact, expanding to the Middle East, according to secret memos, had been a goal of AGILE almost from the start of the program.

One of ARPA's largest projects in Iran was working with the Imperial Iranian Gendarmerie, which was heavily involved in counterinsurgency. In 1942, H. Norman Schwarzkopf Sr., chief of the New Jersey State Police, was sent to Iran to help to remake the gendarmerie "in the mold of the New Jersey State Police." Schwarzkopf, whose son would lead American forces in 1991's Operation Desert Storm, was credited with remaking the gendarmerie into an effective counterinsurgency force that put down tribal rebellions in Kurdistan and Iranian Azerbaijan.

Just as ARPA had tried to help the South Vietnamese military use sensors to detect the movements of the Vietcong, it attempted to apply some of this same methodology and technology to Iran. In Iran, though, the infiltration of the country's borders had little to do with arms smuggling and more to do with the market for illegal drugs. "The biggest challenge was opium smuggling," Herzfeld recalled. "I spent a fair amount of time on that."

For several years, ARPA paid to teach the Iranian forces to use counter-infiltration technology, like seismic intrusion detectors, break wires, and heat sensors. While focused primarily on the gendarmerie, ARPA also held demonstrations for other Iranian security agencies, including the notorious SAVAK, the shah's secret police. Herzfeld recounted that one popular method for drug smuggling at the time was to use water trucks: the front of the truck would be carrying water, and a back partition would contain opium. ARPA demonstrated how, by using infrared sensors, it was possible to spot temperature differences in the trucks that indicated if they were carrying opium. Eventually, however, ARPA was told by the Iranian government to stop its work on detecting heroin smuggling because it had "gotten too close to the top level traffickers," Herzfeld recounted in his memoir. "That ended that."

The problem with the ARPA counter-smuggling program, like almost all of the agency's work in Iran, was that it presumed the Iranian government wanted to solve the same problems the United States wanted to solve. But the Iranian government was rife with corruption and dysfunction; members of the shah's family were believed to be involved in the opium trade, and the shah himself lifted the ban on opium cultivation in 1969. Declassified reports from the AGILE files note that as with other elements of ARPA's work in Iran the gendarmerie rarely if ever followed through with what was suggested. Nonetheless, ARPA persevered for several years, trying to teach the gendarmerie how to use technology for aerial and ground surveillance, as well as along coastal areas. One ARPA report described the surveillance work as "trivial and floundering" and claimed the Iranians were not actually providing any support.

In Lebanon, ARPA funded the American University of Beirut under a project called Factors in Regional Change. Supported by ARPA, the university conducted studies on everything from a confessional census of Lebanon, an extremely sensitive topic in a country that apportioned power based on religion, to students' political attitudes. The university looked at ARPA's money as supporting basic research, not tied to any specific military goal.

Herzfeld, however, believed the Middle East office provided valuable intelligence for the military. "They gave us lots of insight about what was going on in the various then small insurgencies in the Arab countries," he said of the Lebanon office. In Beirut, for example, ARPA at one point planned to fund the university to conduct what Herzfeld described as "an enormous project" to analyze the written records of Muslim families in the region. "Absolutely everything is written down. It goes back at least a thousand years," he said. "We thought we'd mine that." The goal was to see if the family records might provide hints to potential conflict and insurgencies. "Can you detect, or get a guess early on, who is likely to go fight against whom," Herzfeld said. "I guess one way you look at it: What might have been the sources of future insurgencies. Which families?"

ARPA's achievements in the Middle East never matched its ambitions. In theory, the Beirut office covered all of the countries in the Middle East and North Africa, but ARPA only managed to actually run programs in Iran, Lebanon, and Ethiopia. Official reports back to Washington would often cite successes that sounded odd for an "advanced projects" agency. "We have recently completed development of a field ration so that Ethiopian military forces now no longer must live off the land," one report enthused.

Whether there was a rhyme or reason to ARPA's work was unclear, but even those within AGILE seemed to recognize that the Middle East office lacked a larger purpose. And yet it was expanding to new regions. Herzfeld spent several years—and many trips—traveling to India to forge a research and development partnership with the country's military establishment. In 1962, communist China and India went to war over a disputed border region, with much of the fighting taking place in mountainous areas. India had learned a great deal about fighting in high-altitude conditions, and Herzfeld thought the United States could learn from India's experience. In the end, the White House quashed the nascent partnership over rising tensions in the region, a rare defeat for the rapidly expanding AGILE program.

ARPA's staff often knew even less about the Middle East than they did about Southeast Asia. Donald Hess, the senior financial administrator for ARPA, recalled being sent to Capitol Hill to discuss the agency's budget request for funding its Middle East work. "We wanted to open another AGILE base somewhere in the Middle East," he said, though he could not recall where. "It looked to be small enough that nobody cared." The lawmakers, however, began to pepper him with questions. Why was ARPA opening the office? Who told ARPA to do this? What was the office going to do? Hess admitted he had no idea, other than some vague notion that the White House wanted it. No one, at least among ARPA's rank and file, really seemed to understand the overall goal.

Stark, the AGILE program manager, traveled multiple times to Lebanon, but he admitted recalling more details about his food and the hotel than he did about any of the work. At one point, he said, the agency was advising both Israel and Lebanon on how to defend one from the other. "We did a study for the Lebanese on planning to evacuate villages in southern Lebanon in the event of an Israeli invasion. What would be a good way to protect the people?" Stark said. "At the same time, we were working with the Israelis to protect their borders."

It was not that the two projects were inherently contradictory but simply that ARPA did not seem to have any larger goal other than planting its flag in the region, and that, according to Stark, came to characterize much of what he saw throughout AGILE, from Thailand to Lebanon. It was expansion for the sake of expansion.

When he first started at ARPA, Stark, like many, had been enamored of William Godel and his intelligence exploits; the two men grew to be friends, and his family sometimes spent Sundays at the Godel house in Virginia. Stark had been enthusiastic about the idea of creating a worldwide counterinsurgency laboratory. At one point he even forwarded a proposal to Godel for expanding AGILE domestically, warning of "an impending insurgency situation in the United States resulting from racial unrest." This conflict, he said, was similar to the types of insurgencies that AGILE was helping combat around the world. "It is my opinion that the military should start thinking about this situation in its most drastic form," Stark wrote, citing leaders like Malcolm X. "A contingency planning group in DOD could begin to think about the various forms such an insurgency could take, the requirements for preventative military action, and the abundant problems associated with this potentially large scale conflict." There is no evidence that ARPA actually followed up on Stark's suggestion, however.

The closest AGILE ever got to applying its counterinsurgency work at home in the United States was a secret assignment that followed the assassination of President Kennedy. On the Monday after Kennedy was shot, ARPA, as part of Project AGILE, launched a research program focused on presidential protection. The work was given the formal name Star, short for Strategic Threat Analysis and Research, a secret project to help protect the president of the United States. Privately, however, Harold Brown, ARPA's boss, began calling the work "Operation Barn Door," because, as Frosch said, "that's what you [close] after the horses have already escaped."

With its move into presidential protection, ARPA had now carved out a mission in almost every critical area of national security. Its new work was problematic from the start, however. The newly sworn-in president, Lyndon Johnson, was going to be running for reelection soon, and he was not interested in anything that would make it look as if he were hiding behind bulletproof glass. "We couldn't possibly let the project be known," recalled Robert Sproull in an interview years later. "As soon as it became known, it would be killed, and that was complicated because ordinary security isn't enough. You know the Pentagon leaks like a sieve. So, how are we going to keep it quiet?"

It did not take long, as predicted, for word of the project to leak out. In response to a congressional inquiry about Project Star, Herzfeld, the deputy director, wrote to lawmakers to say that ARPA only provided $15,000 to Aberdeen Proving Ground to hasten development of some armor materials for the presidential limousine. "The Defense Department has, of course, no responsibility with respect to the use of the automobile by the President," Herzfeld wrote. The letter was at best misleading: it failed to mention that ARPA was helping to design a new armored limousine for the president, not to mention half a dozen other related research projects that would likely have raised eyebrows in Congress.

Putting Star in the AGILE office was a way of conducting the work in secrecy; information about the project was shared on a strict "need to know" basis. Like other parts of AGILE, Project Star consisted of a string of failed or rejected ideas. One proposal, for example, suggested using chemical weapons to protect the president. "There also exists a need for a system which would make an unfriendly crowd become friendly almost instantaneously," one note in the files stated. "This goes beyond the desire to divert a crowd, as could be done by the prompt and generous use of cash money. The possible use of gasses, sound, lights and other chemical biological or psychological agents to achieve such a change as well as other attributes they might possess for crowd control will require further study." Though there are a number of references to psychological agents in the Star records, it does not appear that any ideas were actually tested in the lab, let alone on people, at least in the United States.

Another proposal examined under Star was to have a continuous stream of air flowing in front of the president's speaker stand. It was thought this airflow might slightly deflect bullets or other projectiles, at least enough to protect the president from a direct hit. Simple calculations showed that the airstream would have a minimal effect on anything except tomatoes, and even then the ARPA-funded analysts predicted the thrower would be able to correct his or her aim by the second or third throw.

Other ideas of more technical promise—such as creating a metal screen of rotating rods that would deflect bullets while still allowing visibility of the president—were never pursued, probably for practical reasons. Star also included brainstorming sessions that produced ideas that ranged from the mind-numbingly obvious, like having the president wear body armor, to the exotically far-fetched, like a "mirage producing system" that involved heating the air or gas around the president to change the index of refraction (in other words distorting light, which would make it harder for an assassin to take aim). Some suggested tactics surely failed the proverbial snicker test, such as having the president continuously "move about when in the car" or use armor disguised as a sunshade, which might require "spreading false weather reports" to justify its use.

Despite high ambitions, the only new technology that appeared to result from Star was a comical high-powered squirt gun, which was designed to disable a single person in a crowd and could be carried and used much like a regular gun. The nonlethal weapon would fire a high-power stream of liquid containing capsicum, the active ingredient in tear gas, disabling a potential assassin. The liquid squirt gun, though built, suffered from the practical limitations of such weapons. Getting the water to go out in a relatively concentrated stream beyond twenty feet was difficult (and if someone was indeed an imminent threat to the president, the more likely gun of choice for any Secret Service agent would be a regular service weapon). Though the squirt guns were delivered in 1965, it appears that they were never used and eventually misplaced.

ARPA's final contributions included its work on the upgraded presidential vehicle, modest improvements to the presidential transport helicopter, some two dozen studies covering various aspects of presidential security, and a never-used squirt gun. Beyond armoring the presidential vehicle, the only substantive Project Star innovation that Harold Brown could recall being implemented was having an airstream directed at flags positioned behind the president. "The rationale was that a waving flag background could confuse a shooter's aim," Brown recalled. "How that has worked out in theory or practice I have no idea."

Project AGILE's approach of treating the entire world as a living laboratory had a fundamental problem: real people were living in those laboratories, and some of the agency's ideas looked at best callous or at worst sinister. ARPA pursued a "people sniffer" that would, as the name suggested, hunt Vietcong fighters by sensing the presence of ammonium from urine. The "Airborne Concealed Personnel Detector" was placed on Huey helicopters and flown over the jungles of Vietnam. ARPA even funded a study to see whether the common green bottle fly, known for its ability to smell from hundreds of yards away, could be used to detect humans. To be fair, ARPA often turned down some of the more gruesome ideas, like a proposal from Hughes Aircraft Company for "nonlethal decay mechanisms," which involved spreading tainted grain and human parasites that might lower the morale of the Vietcong.

Yet of all ARPA's experiments, defoliation proved the most troubling. While the air force's operational responsibility for chemical defoliation overshadowed ARPA's role (and ARPA happily let its paternity be forgotten), the agency's involvement with Agent Orange and the other "rainbow herbicides" extended well into the war, with the agency's studying its effects extensively in Vietnam and Thailand, at times with Herzfeld's encouragement. An ARPA document that includes a "Follow-Up to Dr. Herzfeld's Trip Report" from November 1965 notes a request for an update to a study on "Human effects of crop destruction and the potential use of crop destruction," which is described "as a large scale countermeasure" to the Vietcong. A handwritten report labeled "Dr. Herzfeld's Trip Actions" assigned the herbicide study to Stark as a "high priority."

Stark, however, was beginning to have serious doubts about the entire approach to counterinsurgency "laboratories," whether in the Middle East, Asia, or the United States. After his friend and mentor Godel was fired and imprisoned, he began to feel ARPA's programs were becoming unmoored. Stark's disillusionment with the science of counterinsurgency had been deepening for some time. His trip reports over the years grew longer, and more dismal, noting the constant renaming of failed projects, like strategic hamlets, which had become the "new rural life hamlets." In 1967, he finally quit in disgust. "During my five years at ARPA, I spent $100 million of the taxpayers' hard earned money," Stark later wrote. "I'd like to have it all back."

Whatever the concerns of personnel like Stark, ARPA's experiments were about to have a fundamental impact on the way the American military was conducting the war in Vietnam. Back in 1964, Godel had laid out a proposal to stop the flow of Vietcong fighters and arms into South Vietnam by "sealing" the border using "novel" technologies, a nearly monumental task given the country's combination of mountainous and jungle terrain. Part of the goal was to cut off the Vietcong's resupply route, known as the Ho Chi Minh Trail, which wound its way from North to South Vietnam, snaking at times into Laos and Cambodia.

The idea of experimenting with some sort of virtual barrier using modern technology—because a physical barrier would be impossible across that geographic expanse—had been around since the early

1960s. Maxwell Taylor had proposed the idea to Edward Lansdale and William Godel during the 1962 study trip. Lansdale had no interest in the project, so ARPA under Godel pursued it. The resulting proposal called for deforesting 80 to 90 percent of a crucial 180-mile portion of South Vietnam's border with Laos that formed a part of the Ho Chi Minh Trail. A handwritten list of proposed technologies was indeed novel, if in some cases horrifying. The barrier would require 100,000 "throw-away" shotguns, 250,000 rocket pistols, one million tetrahedrons (ground-based spikes, also known as caltrops), two million mines disguised as rocks, twenty thousand bomblets loaded with chemical defoliants, and an unidentified amount of "insect attractants" (as opposed to insect repellant). Perhaps most disturbingly, the proposal suggested twenty-five thousand "biological weapons systems," without specifying what those weapons might be or how they would be used.

The border-sealing proposal was initially rejected at the time as too expensive. Seymour Deitchman, a Pentagon engineer and counterinsurgency adviser to Harold Brown, argued that pursuing the proposal "in South Vietnam requires, I believe, a major strategic decision to pursue the war in this manner, because extension of border security would have to be accompanied by a significantly larger combat operation in the border area than is now contemplated."

Elements of the proposal did move forward, nevertheless. In March 1965, for example, ARPA conducted what could be considered the agency's most ambitious experiment in its Southeast Asian laboratory. The experiment was carried out as a secret mission conducted by the 315th Air Commando Group, targeting part of the Boi Loi Forest, which the Vietcong had been using as cover. A classified history of Strategic Air Command called the raid "one of the most unusual uses made of B-52s in South Vietnam." That was because the purpose of the raid was not simply to bomb the Vietcong but rather to spark an out-of-control forest fire using a combination of defoliants and incendiary bombs that would eliminate ground cover for insurgents.

The bombing raid was an experiment in using forest fire as a weapon, an extension of ARPA's defoliation work. It also demonstrated how a small-scale effort to clear vegetation and prevent ambushes had ballooned into a wide-scale program with a life of its own. ARPA had enlisted the Department of Agriculture to help with the ambitious effort. A few months prior, American military aircraft had sprayed the area with defoliants to dry out the vegetation. The idea was that once it was dry, it would be possible to spark a fire. But Mother Nature was not cooperating. On March 3, Military Assistance Command, Vietnam, sent fifteen B-52s loaded with incendiary bombs out to the forest, but a sudden rainstorm forced them to return. Another raid a week later was carried out successfully, though it failed to spark the anticipated forest fire. The operation, code-named Sherwood Forest, was a failure.

The agency did not give up, however. The next year, ARPA sponsored Hot Tip I and II, which involved sending some seventeen B-52 bombers from Guam to drop 172 tons of incendiary cluster bombs on another forest. This time, ARPA claimed the mission was an "outstanding operational success," meaning that the aircraft had dropped the bombs on target. But it was a "qualified technical success," meaning that no forest fire was ignited. In other words, Hot Tip, like Sherwood Forest, failed. A third attempt, in 1967, code-named Pink Rose, would also fail; rain following the bombing squelched any flames. "The country doesn't burn well," deadpanned Craig Chandler, a Forest Service employee who worked on the ARPA project.

Creating man-made forest fires evoked memories of some of the least popular aspects of World War II, like the Dresden firebombing, and it had little to do with the sorts of hearts-and-minds pacification campaign that was supposed to be part of counterinsurgency. After three failed attempts, the forest fire as a weapon project was discarded. "This was clearly one of those ideas that should have been given the very quietest funeral," an ARPA official told *Science* magazine, when the project came to light five years later. The forest fire efforts were emblematic of much of what was wrong with Project AGILE by mid-1965. Quick-reaction projects were giving way to operations, like Sherwood Forest, to support American, rather than Vietnamese, forces. It was not working.

Despite those setbacks, the proposal to seal South Vietnam's border was picked up by Seymour Deitchman, who had advised against pursuing the idea when Godel had forwarded it just a few years prior. In 1966, Deitchman was working with the JASONs, the ARPA-funded advisory group, looking for ways to cut off the Ho Chi Minh Trail. The JASONs devised a wholly unique and arguably improved border system—a virtual barrier made up of thousands of air-dropped ground sensors linked to a computer system that would cue strike aircraft to a suspected infiltration. (Prior to the JASON proposal, ideas for a virtual barrier had either relied on mines and other antipersonnel weapons or presumed that someone on the ground would have to manually relay information about an incursion.) Unlike Godel's proposal, which had floundered, the JASON proposal went all the way up to Secretary of Defense Robert McNamara, who approved the electronic barrier project. It was not to be an ARPA project, however. McNamara instead assigned it to a secret Pentagon organization obtusely named the Defense Communications Planning Group. The electronic barrier became part of a larger highly secret project initially code-named Practice Nine, and later Dye Marker and Igloo White. "ARPA was cut out of the loop," explained Stephen Lukasik, who became director a few years later.

ARPA was relegated to a bit player, providing some of the sensors used by the air force, but had no real role in the overall design of the project. James Tegnelia, who worked on the barrier for the army in Vietnam, said that ARPA's contributions were highly classified and mostly involved providing hardware for the barrier. "A lot of what we called 'dirty tricks,' all classified programs, were done by ARPA," said Tegnelia, who later served as an acting director of ARPA in the 1980s. "Silent pistols and chemical darts that would poison you, those kinds of things, strange stuff."

Soon, military aircraft were dropping strings of sensors across the jungles, which sent information to a computer command center located in Nakhon Phanom in Thailand. The sensors, a combination of sound and seismic detectors, were designed to pick up Vietcong supply trucks, while the computers would calculate the position of the target, which would be relayed to the aircraft, allowing them to strike within minutes. Computers were still in their infancy, and computer-automated killing entirely new, so the system was, at least for the time, like something from science fiction. By linking sensors and aircraft to computers in real time, the JASON proposal transformed Godel's idea of a barrier into something far more technologically advanced: the world's first electronic battlefield. What the barrier failed to do, by all accounts, was to have any appreciable effect on the course of the war. In his memoir, Godel wrote that the ARPA barrier idea was resurrected "in the desperation of a failed Secretary of Defense" when it was already too late to do any good. "It was," he wrote, "a good idea gone sour, and it never worked despite all manner of high tech communications gear, patrolling, and elaborate airborne surveillance techniques."

The barrier was technically brilliant, but its implementation was flawed: the air force saw it as an extension of its strategic air campaign to destroy the Vietcong, rather than as a tool for stopping infiltration. Soldiers and press alike mocked it, calling it the "McNamara Line," after the French Maginot Line, and the name stuck. "The barrier proved to be worthless," *The New York Times* reported in McNamara's obituary.

Deitchman, who had revived Godel's border-sealing project and transformed it into an electronic fence, took a different view of ARPA's efforts. He later acknowledged the barrier was a strategic failure but argued it was a technological success. It failed to slow the Vietcong's progress, but it demonstrated for the first time a way to automate the "sensor-to-shooter process." In other words, it sped up the process of killing. The failed barrier became, as Deitchman wrote many years later, the first articulation of "network centric warfare," a term that would gain currency in the Pentagon in the early years of the twenty-first century. Deitchman, at least back in 1966, believed that science could alter the course of the war in Vietnam, and he was about to take Godel's place as the head of AGILE.

CHAPTER 10

Blame It on the Sorcerers

Do you see anything on this card that reminds you of a penis?"

Walter Slote, a New York psychotherapist, handed the ink-blot-stained piece of paper to the Vietcong fighter.

"No, sir," the fighter replied curtly.

"How about this top part?" Slote asked.

"No."

Slote persevered: "Do you see anything here on this card that reminds you of a woman's vagina?"

"No," he replied.

Neither man was in a particularly good mood. Slote was frustrated because he had been fruitlessly going through these cards, part of the classic Rorschach test, and the Vietcong fighter was unhappy because he was sitting in a Saigon prison staring at ink blots, rather than planting bombs and killing Americans.

An American firm, called Simulmatics Corporation, had sent Slote to Vietnam in 1966 to help the Pentagon understand the growing insurgency. Slote believed the Rorschach test, popular at the time among psychotherapists to diagnose personality traits, could be used to understand the reasons behind growing resentment of the United States and the South Vietnamese government. But so far, the Rorschach ink blotches had failed to yield great insights into the psyche of the Vietcong fighter.

Slote asked the Vietcong fighter to go through all the cards and find *anything* that reminded him of a person. Nothing. What about anything sexual? Nothing, again. Slote seemed puzzled that an imprisoned Vietcong fighter being interviewed by a man hired by the Defense Department to quiz him about penises and vaginas would be so reticent. Slote finally asked the fighter to find a picture that he liked or disliked, but the imprisoned man, who had once led a sabotage squad, was reluctant to even touch the cards. "I do not understand these pictures, so I do not know which one I like, which one I dislike," the sullen fighter replied.

Slote ended up spending seven weeks in Vietnam, during which time he collected data on exactly four Vietnamese: a prominent French-educated writer, a student activist in hiding, a senior Buddhist monk, and the Vietcong insurgent. All four harbored anti-American feelings and were critical of the government of South Vietnam, but Slote found the Vietcong fighter particularly frustrating. Even the antigovernment Buddhist monk Slote interviewed was more cooperative. "You know, I've never seen one, except on a child," the monk replied in astonishment, when Slote asked him if a particular ink blot resembled a vagina.

"The Viet Cong member was a thoroughly deadened man. Unless directly addressed, he stared into space, his expression was stony and flattened, he never reached out nor truly responded," Slote later wrote in his report. "The only time he came alive was when he was telling of his exploits. His eyes would brighten and he held himself with greater dignity, but as soon as this passed, he would lapse back into a lethargic dulled apathy—a pattern that I am convinced was lifelong and not precipitated by imprisonment."

The New York psychotherapist was not interested in the nuances of Vietnamese politics; he quizzed the men about their parents, their dreams, and their sex lives, or lack thereof. Slote decided, after interviews with his four informants, that the problem with the Vietnamese people was not a thousand years of foreign domination, including French colonialism, Chinese imperialism, and now finally American intervention. Instead, the root of the problem, he believed, was their troubled family structure. "It is my strong impression that the triad of sibling rivalry, deflected parental hostility and unresolved dependency needs constitutes the central psychological core of anti-Americanism in Vietnam," he concluded.

Slote's presence as a Pentagon-funded researcher in Vietnam might have sounded ludicrous, but it was part of a much broader trend spearheaded by ARPA to study the roots of insurgency. Defense officials realized the growing insurgency was a phenomenon bullets and bombs alone could not stop, and increasingly they turned to researchers from the "softer" sciences, including anthropologists, political scientists, psychologists, and in this case even a psychotherapist. The chief proponent of this new line of work was Seymour Deitchman, an emerging member of Secretary of Defense McNamara's technocratic elite and a longtime nemesis of William Godel's. Deitchman was convinced it was an engineer's slide rule, not a soldier's intuition, that would determine who won or lost battles. More important, he believed that people could be studied and their actions predicted the way engineers measure and track the flight of a ballistic missile. Vietnam was about to become the test bed for a new science of human behavior.

In 1966, on the last day of the summer meeting of the JASON advisers in California, the same meeting where the electronic barrier was born, Deitchman was asked to help fix ARPA's troubled Vietnam program. John S. Foster Jr., who had recently taken over from Harold Brown as the Pentagon's director of defense research and engineering, pulled Deitchman aside and told him that he had a new job for him. Foster, like Brown, was a physicist, and though he supported the work being done in Vietnam, he felt the ARPA program needed some technical oversight. Foster "started to twist my arm about taking over Project AGILE in ARPA," Deitchman recalled. Foster wanted someone like Deitchman, an engineer, to bring more science into the program. Just a year and a half younger than William Godel, the diminutive, pipe-smoking Deitchman had a path that ran parallel to that of the former marine. Both Godel and Deitchman fashioned careers as counterinsurgency experts, albeit from different perspectives.

Deitchman had been applying operations research to aviation and defense issues as an analyst at the Institute for Defense Analyses. There, he befriended Jesse Orlansky, a psychologist who was interested in the growing insurgency in Vietnam. Intrigued by the idea of blending social science with the hard sciences, Deitchman began using operations research to analyze questions of "limited warfare," the Pentagon jargon at the time for insurgencies. In 1962, he published "A Lanchester Model of Guerrilla Warfare" in the journal *Operations Research,* which took mathematical formulas popular with modeling U.S.-Soviet conventional engagements and applied them to insurgent warfare.

With his growing expertise in guerrilla warfare and operations research, Deitchman was recruited to work in the Pentagon with Harold Brown, one of McNamara's whiz kids. In 1964, Deitchman became Brown's special assistant for counterinsurgency, a job that made him responsible for research and development programs in Southeast Asia. Much of what he did was patrol for dumb ideas being proposed in the hallways of the Pentagon. There were more than a few, he recalled, like an air force proposal to create an "artificial moon" over Vietnam—a satellite with a giant dish that would illuminate the Mekong delta region, allowing the air force to use starlight scopes at night.

The military services were grasping at technological solutions, yet Deitchman understood that the problem the military increasingly faced was one that defied modern weaponry. In the mid-1960s, an increasing number of Vietnamese were turning against both the South Vietnamese government and the American military. The Vietcong attacks in South Vietnam began to climb in number and scale, highlighted in February 1965 by a dramatic Vietcong infiltration of the Pleiku airbase—home to American military advisers. The attack resulted in the deaths of nine Americans and more than one hundred wounded. Villagers in South Vietnam were increasingly aiding and joining the Vietcong, a phenomenon that mystified officials back in Washington who believed that only coercion could account for this dramatic shift in support. The South Vietnamese military was losing ground to the Vietcong not for lack of better weapons or technology but because something was driving the peasants to support the enemy. It was a question of psychology, and psychology was not a problem that one could bomb into submission. But it could be studied.

ARPA had been moving into the social sciences since the early 1960s, first under J. C. R. Licklider, who had been brought in to start a behavioral science program. Licklider's behavioral science program was small in scale but continued to grow after he left and eventually became part of Project AGILE. Under the direction of Lee Huff, a political scientist, ARPA expanded even more into the social sciences, contracting with think tanks like the Rand Corporation to perform fieldwork in Vietnam. A description of AGILE in 1966 summarized it as seeking to understand the "close interrelation between the technological, behavioral, and environmental factors" involved in insurgency. AGILE would no longer be a narrow technical program to help indigenous forces; it would be directed at creating the "total solution of the problem of counterinsurgent operations."

Defense contractors also saw an opportunity to profit from this new direction. By the mid-1960s, ARPA was being flooded by proposals from companies, universities, and independent researchers, all suggesting ways to help understand why increasing numbers of Vietnamese—and which Vietnamese—were siding with the Vietcong, rather than embracing American forces. The military-industrial complex offered up its own bizarre mix of solutions to this problem, often with technologies to fix what was, essentially, a very human problem. General Electric in August 1965, for example, wrote to ARPA suggesting the company be given a "continuing open ended type contract" that would allow it to apply its technology to counterinsurgency. Its first proposal was a "mass polygraph for internal village security." The concept was a sort of modern version of witch dunking reimagined as science fiction.

"Consider the following scenario as being typical of the type of situation and method of operation," the General Electric sales manager began. "A high security Central Government anti-terror police contingent arrived by helicopter at a village suspected of being under covert Viet Cong pressure or subject to territorial activities. The villagers are assembled by their local chief such that each villager can see every other villager. Each individual is connected to the new type of mass polygraph which measures galvanic skin response (GSR) and heart beat of all the villagers simultaneously." From there, the scenario got still weirder. A suspected supporter of the Vietcong would be hauled up before the assembled villagers, who are all hooked up to the polygraph. The machine would record a "group" response, alleviating the fear that any single villager was the informant. "The process can be repeated to test as many members of the village as desired," the sales manager explained.

ARPA managed to stay out of the lie detector business, but when Deitchman arrived in November 1966 to take over Project AGILE, the program he inherited was in disarray, "a collection of $25 to $30 million worth of odd projects." He began to comb through the programs, weeding out what he deemed to be shoddy work. Godel's former cadre viewed Deitchman with a mix of suspicion and horror. AGILE under Godel had been run like an intelligence operation, employing local contacts, like a Thai prostitute who was coordinating the river surveillance project. The prostitute spoke Thai, Lao, Vietnamese, and English and knew all of the key people in the region, according to Warren Stark, the AGILE program manager. "Sy Deitchman was aghast that he would have someone like that on the payroll," Stark said.

Deitchman, however, was resolved to clean up AGILE's problems and bring in grounded science. One of his first decisions was to get rid of Herman Kahn, the portly nuclear war theorist. Kahn had left Rand to found the Hudson Institute, and now he was gallivanting around Vietnam with ARPA funding giving entertaining slide presentations filled with grandiose, if perhaps functionally useless, ideas. That included, in one report, a proposal to build a moat around Saigon to protect it from the Vietcong. Kahn's anti-infiltration moat was widely derided by the press and lawmakers. The moat became so famous that when the army general Creighton Abrams complained about the "fucking geography" of Vietnam, someone jokingly suggested that they could move some hills. Abrams's response, greeted with raucous laughter, was: "Well, you could get Herman Kahn to work on that."

Kahn and his colorful slide presentations were starting to cause political problems for ARPA, and Deitchman was determined to cut ties with him. "What is Herman Kahn doing for you out there?" Deitchman asked one senior official at Military Assistance Command, Vietnam.

"Well, he comes out here and gives us some absolutely fascinating briefings that get us thinking and we enjoy it," the official replied.

"Is that worth a quarter of a million dollars of Uncle Sam's money to you?" countered Deitchman.

"No, I guess I couldn't say that," the official said.

And with that, Deitchman got rid of Kahn, and when Kahn threatened to complain to McNamara, Deitchman called him on his bluff. "Go ahead," Deitchman said.

Kahn's flashy lectures underscored a larger problem facing Pentagon-funded social science work: there was a tendency to pay for studies that supported what the military wanted to hear, rather than what it needed to hear. For example, ARPA had been paying for Rand since the early 1960s to conduct social science work in Vietnam. That support included funding for Gerald Hickey, the Rand-employed anthropologist who had questioned the strategic hamlet program. Though highly regarded in the Defense Department, when Hickey's work ran counter to Pentagon policy, as it often did, officials simply ignored him.

One of Rand's most significant ARPA-funded projects, and the one that would become its most famous wartime social science work, was the Viet Cong Motivation and Morale Project, which sought to understand support for the the communist insurgency. Two Rand analysts, Joe Zasloff and John Donnell, were sent over to supervise the interviews, which were conducted with captured Vietcong, as well as those who had surrendered under amnesty. The initial analysis of the interviews was rather dismal for the prospects of U.S. intervention in Vietnam. The study produced insights that did not mesh with the official line, because it showed that the Vietnamese were joining the Vietcong out of genuine political conviction, rather than being forced to join, as many government and military officials back in Washington maintained. Joining the Vietcong "had to do with anger at an exploitative government, and a sense of nationalism," and the ability of the communists to manipulate those feelings, according to David Morell, who oversaw the study for ARPA. The response from Washington to those conclusions, however, was "not just negative; it was shock."

In 1964, Rand embarked on a second prisoner of war study, this time sending over Leon Gouré, a noted Rand Sovietologist and political hard-liner who had been accused of hyping the Soviet Union's civil defense programs. According to his colleagues, he went to Vietnam already sure that air strikes were the solution to counterinsurgency, and that was the answer he gave to the Pentagon. "The new theme, as expounded by Gouré in his proposal, was finding and exploiting enemy vulnerabilities to the impact of military operations," wrote Mai Elliott in her authoritative account of Rand's involvement in the Vietnam War.

Not surprisingly, the results of the new study were quite different from the first. "When the Air Force is paying the bill, the answer is always bombing," Gouré reportedly said. Even Gouré's colleagues at Rand did not trust his findings; they believed he was cherry-picking information and interviews. But Gouré's new slant on the Rand work soon got him the ear of Secretary of Defense McNamara, who upped his budget from $100,000 to $1 million. By January 1966, McNamara was briefing President Johnson about Rand's work, which not surprisingly confirmed that strategic bombing was working.

When Deitchman arrived at ARPA, he took one look at Gouré's work and realized the Rand analyst was simply repeating what Pentagon officials wanted to hear: that dropping bombs on insurgents was working, even though it was not. "I got hold of McNamara's military assistant and turned off the prisoner of war interviews because that clearly was being slanted," Deitchman recalled. "McNamara was getting a view of the war that was distorted from it."

Deitchman believed the Rand work, and much of the social science research being conducted in Vietnam, lacked scientific rigor. His view was that social science, if treated more like the "hard" physical sciences, could be used for prediction. He wanted data, not entertaining lectures or self-reinforcing studies. Yet he found that growing opposition to American involvement in Vietnam, particularly on college campuses, was making it almost impossible for ARPA to find good academics to work for the Defense Department on Southeast Asia issues.

The year before, the debate over military support for universities exploded in the midst of campus unrest when it was revealed that the army, through the Special Operations Research Office at American University, was funding civilian researchers to study insurgency in Chile. Their connection to a military-funded project had not been disclosed. With student radicalism and opposition to the Vietnam War at a near fever pitch, Project Camelot, as the ill-fated work was called, triggered an avalanche of criticism in Latin America and the United States. Professors who took Defense Department funding were accused of promoting an imperialist agenda.

A tailor-made solution was offered by a company called Simulmatics Corporation, which promised to hire academics in their time off to work under contract to the Pentagon. Its co-founder Ithiel de Sola Pool enjoyed a reputation as a brilliant and politically savvy professor of political science at the Massachusetts Institute of Technology. He had close ties to senior national security figures, including Robert Komer, the former CIA official running the pacification program in Vietnam. Deitchman believed that Pool "could easily attract other well known scholars; among them were many who were experts on Vietnam, had been there before, spoke the language, and knew many of the key Vietnamese figures who could grant 'access' for research." Deitchman figured that Simulmatics would allow ARPA to tap academic expertise, without actually having to work directly with universities. "As far as we in ARPA were concerned, a group like this had impeccable credentials and helped avoid many problems," Deitchman later recounted in his memoir, *The Best-Laid Schemes.*

In 1966, ARPA gave Simulmatics a wide-ranging contract for social science work in Vietnam, and soon the first research teams started showing up in Saigon. It was to be one of the most disastrous contracts ever let under AGILE.

Simulmatics was created as a way of making money off human capriciousness, be it in elections or war. The company's origins dated back to 1958, when William McPhee, a professor at Columbia University, developed a novel theory to predict television-viewing habits. McPhee pitched his work to Edward Greenfield, a New York businessman, who in turn introduced McPhee to Ithiel Pool. Greenfield and Pool liked the idea but thought elections were more promising as a business model, and they were right. Simulmatics shot to fame during the 1960 election of John F. Kennedy, when, in a series of reports to the Democratic National Committee, it correctly predicted voter habits. "This is the A-bomb of the social sciences," Harold Lasswell, a Yale professor, said at the time, comparing Simulmatics' work to a demonstration of the first nuclear chain reaction. Buoyed by that much-hyped success, Simulmatics began selling the services of what *Harper's Magazine* labeled a "people machine" to government and private clients.

Simulmatics was selling exactly what Deitchman and proponents of military-funded social science seemed to want. Simulmatics had a "people machine" it wanted to test, and ARPA had people it wanted to test it on. Pool originally suggested that ARPA hire the company to run experiments covering such areas as "intelligence and population control" in a "laboratory province" in Thailand. "The basic idea as I see it is that a limited area of Thailand be used as a site for a major field test of security programs that the government of that country can undertake with American assistance," Pool wrote. "Thailand is a country in which in certain areas the security problems are sufficiently real so that a field test is possible, as at the same time the government is sufficiently alert and active in meeting the threat so that sensible programs can be tried."

While ARPA did not end up hiring Simulmatics in Thailand, it did give the company a wide-ranging contract for Vietnam, where the agency wanted a quick-reaction force that could deploy social scientists to answer questions or provide analysis on specific problems.

The Walter Slote study on the Vietnamese psyche, which had drawn inferences on an entire nation based on the dreams and sex lives of four men, was only the start. One of Simulmatics' key employees was Joseph Hoc, a Vietnam-born Catholic priest who had been teaching at Boston College. Hoc worked on a report for "testing psychological warfare weapons" under contract to ARPA. "My research work has been carried out with the purpose of providing to the advanced research projects agency a possible means to predict and even to control human events in Viet Nam," he wrote.

Hoc's ideas were almost as nutty as Slote's, though potentially more harmful. "It is possible to control the Vietnamese hamlet people and manipulate them through informal means of communications and to persuade them to react to a given situation in a desired manner," Hoc wrote. Standard interview techniques used by social scientists are "not adequate," he wrote. Hoc proposed to have Simulmatics pay villagers to spread false rumors in a hamlet, and then secretly record people's reaction, or what he called "techniques of human manipulation."

Several of those techniques were tried out with ARPA funding. Under Hoc's guidance, Simulmatics ran a study testing "psychological weapons" in hamlets, some controlled by Vietcong forces, and others loyal to the South Vietnamese government. The "weapons" included an American-style chain letter that would be spread in hamlets to trick the Vietcong into rallying. Villagers balked at distributing the letter, believing it to be a Vietcong trick.

Simulmatics also sought to use Vietnamese faith in "prophecy and the power of holy men" by publishing and distributing five thousand copies of a booklet that prophesied the Vietcong defeat. By unfortunate coincidence the distribution took place just as the Tet Offensive was starting, an event the prophecy did not foresee. Several of the projects were, even Simulmatics admitted, a total failure, such as using folksingers to spread pro-government messages and creating cartoons with political messages. Most dismal, perhaps, was a "sorcerer's project," which had enlisted Vietnamese sorcerers—essentially local magicians—to sway villagers against the Vietcong. It failed because, as Hoc put it without a hint of irony, "the sorcerers did not say what they were supposed to say."

Not surprisingly, Garry Quinn, an ARPA official, criticized Hoc's report: the "variables were contaminated," "error sources were not systematically investigated," and "rules of inference were violated." Hoc did not seem concerned by the criticism, reporting that Quinn, who kept a picture in his office of Snoopy's famous admonition, "Curse You, Red Baron," was just in a bad mood. "I don't think that we should take it seriously," Hoc said, proposing to continue the psychological warfare study. He was shocked when ARPA refused.

To head the Simulmatics office in Saigon, Pool hired Alfred de Grazia, a political scientist. During World War II, de Grazia had worked on propaganda and psychological operations; his academic career took a more unusual turn. In the early 1960s, de Grazia had sided with Immanuel Velikovsky, a best-selling author whose revisions of world history based on ancient myths had sent the scientific community into an uproar (among other ideas, Velikovsky claimed that Mars had left its orbit around 750 B.C.E. and almost collided with the earth). Regardless of the merits of Velikovsky's theories, the selection of de Grazia, who had alienated much of the academic community, was an odd choice for a program designed to recruit academics. The highly qualified social scientists Pool promised to send to Vietnam never materialized. Deitchman and other officials complained repeatedly that Simulmatics was sending to Vietnam unqualified, inexperienced people who seemed more intent on flouting military contracting regulations than conducting rigorous research. When told that spouses could not accompany researchers to Vietnam, one Simulmatics employee promptly put his wife on payroll as a researcher. In a letter, Deitchman railed against the "rank amateurishness" of Simulmatics employees.

Many projects failed from sheer incompetence. Military experts dismissed as amateurish the results of a study on the Vietcong amnesty program, known as *Chiêu Hồi*. A study on Vietnamese TV-viewing habits, run by a nurse, was conducted as if "someone had taken a book of rules about scientific methodology, then systematically violated each one." Simulmatics sent what one ARPA official in Vietnam called "Briefcase Directors," who traveled on short stints to the region while on break from university, bringing students with no regional or subject expertise to conduct fieldwork. Simulmatics let go the entire ARPA-trained Vietnamese interviewing team. Outraged, the fired employees staged a "Hate the U.S." banquet and protested to the American ambassador.

Other complaints were more serious; at one point, an ARPA program manager back in Washington expressed concerns over reports that Simulmatics employees were "running around Vietnam with pistols, rifles and even automatic weapons." The director of ARPA's field unit in Vietnam confirmed that Simulmatics employees had asked for M16s and .38-caliber guns, but wrote that the request was denied. Perhaps the Simulmatics employees were just bragging about carrying weapons, the field director suggested. "Wonder what brand of pot they're smoking?" he wrote.

De Grazia's relationship with ARPA was soon no better than with academia. He began to send a series of increasingly irate letters back to Washington. The laundry list of complaints included everything from the military's not providing transportation for Simulmatics employees (ARPA pointed out that Vietnam was, after all, a war zone) to broken copy machines. "If ARPA is to perform its vital job well, it has to watch the higher priorities of the struggle and avoid the chicken shit," de Grazia wrote in one rant. When ARPA complained that Simulmatics' reports were of poor quality and submitted without basic copyediting, de Grazia responded by pointing out the bad grammar used on a sign in the ARPA urinal: "Time after time I must stand facing it and read 'If you want them to work, do not put cigarette butts in the urinals.'" Pool later conceded that de Grazia did not "pan out."

By 1967, relations between Simulmatics and ARPA had reached a breaking point. Someone at ARPA drafted an entire list of Simulmatics' failures, ranging from unqualified personnel to general incompetence. Pool called it "a malicious set of baseless stories." In December, W. G. McMillan, the science adviser to the U.S. Military Assistance Command, Vietnam, wrote to Deitchman with a simple message: terminate the Simulmatics contract. "This Corporation has been operating in the Republic of Vietnam for approximately eighteen months attempting to perform social science research under ARPA contracts," he wrote. "Review of these efforts makes it clear that Simulmatics has failed to meet contractual requirements."

In interviews and in his book, Deitchman said simply that the Simulmatics contract did not work out for administrative reasons. His correspondence in the archives tells a much franker story. "One aspect of it may be that their severest critics are looking for sound scientific effort backed up by numbers and rigorous methodology," he wrote. "Frankly, I don't think we'll ever get that from Simulmatics."

A succession of ARPA and Pentagon officials demanded repeatedly that the contract be canceled for gross incompetence. The Simulmatics "scientific approach" to counterinsurgency had not succeeded, and though he knew there would be blowback, Deitchman finally ended the contract in early 1968. Edward Greenfield, the head of the company, showed up in Deitchman's office, saying he would complain to McNamara; it was no idle threat given the company's close ties to senior administration officials. Deitchman's reaction was identical to what he told Herman Kahn: go ahead.

Almost bankrupt, Simulmatics made one last desperate attempt to get Pentagon money: the company in 1968 turned to William Godel, who was out of prison. Godel, whose post-ARPA career had included gunrunning in Southeast Asia, still had contacts with top Asian officials, particularly in Thailand. He worked out a deal with the head of the Thai air force to request that ARPA fund a program to study regional police and security forces. Simulmatics would run the contract. "Greenfield told me Godel would likely be associated with Simulmatics through a third-company arrangement," Deitchman wrote to a colleague.

Deitchman knew he was in a bind. If he refused a senior Thai official, it would put ARPA's work in Thailand in jeopardy. If ARPA agreed, it would mean putting Godel back on the agency's payroll. In a memoir covering his years at ARPA, Deitchman recalled the Simulmatics episode only briefly, as one of generally good work but administrative missteps. ARPA, he wrote, was forced to terminate the contract for "bureaucratic reasons."

His official correspondence, much of it classified at the time, told a different story. The files include dozens of lengthy memos chronicling the Simulmatics disaster. The return of Godel was apparently the final straw. "I ain't having any," Deitchman wrote. "I view it as an affront to the integrity of the ARPA program." Deitchman swore that Simulmatics would not see another penny from ARPA, regardless of the political repercussions. He ended with a request that was never fulfilled, because his original letter was lying in the National Archives more than forty years later: "Burn this after reading."

From start to finish, the Simulmatics experiment in Vietnam lasted barely eighteen months. Its work was a disaster, leaving a trail of incompetence, shoddy research, and politically disastrous missteps. If there was a lesson to be learned from Simulmatics, or ARPA's attempt to solve the problem of insurgency using social science, it was that trying to study and influence human behavior is infinitely more difficult than collecting data on the flight of a ballistic missile. Deitchman, in acknowledging the failure, called it a form of the Heisenberg principle: "The fact and means of measurement and observation affect and change the phenomena being observed and all the participants." In other words, humans will realize they are being observed, and they will adapt their behavior.

Nothing illustrated this truism better than Deitchman's proposal to lighten the ninety-pound load that soldiers carried into the field. Much of that weight was due to inefficiency, Deitchman decided, because soldiers were carrying redundant items. If combat patrols operated as a "system," rather than a collection of individual soldiers, then they could divide up the equipment more efficiently. One soldier, for example, could carry communications equipment, and another, extra ammunition. At first, the idea seemed to work, but then an interesting thing happened: soldiers began to use the extra room to load up on cans of Coke, and their packs got right back up to ninety pounds. "As we thought about that, we said, 'Well, you know, whatever satisfies them,'" Deitchman recalled. "'If they are getting nourishment out of that, maybe that's okay.'"

For all of the criticism that Deitchman directed at Godel's approach to counterinsurgency—and much of the criticism was valid—Deitchman's attempt to transform social science into a hard science was also a failure. ARPA's counterinsurgency work, though largely unsuccessful, rolled forward out of classic bureaucratic inertia. It had become central to ARPA—its third-largest program after missile defense and nuclear test detection—so to admit failure would have been to resign from a core mission. But with the Vietnam War ramping up, Congress was growing weary of ARPA's meddling in world affairs and its sponsorship of social science research, which seemed to be a strange preoccupation for a military agency. In addition to its Vietnam work, ARPA was studying childhood nutrition in the Middle East, and it was measuring the shape and size of Iranian soldiers' heads and feet to design better uniforms. For critics of the Vietnam War, the ARPA programs were more grist for the mill.

"How does this get to be your responsibility?" demanded Glenard Lipscomb, a Republican representative from California, in one hearing reviewing AGILE's expansion. "Why is this not the responsibility of the State Department, or the regular military departments?"

Because no one else was doing it, argued Herzfeld, an answer that did little to assuage the congressional doubters. In fact, ARPA was expanding AGILE still further around the world. "We are planning to concentrate more on the non-southeast Asia part of AGILE," Herzfeld told Congress, when asked for the agency's plans in 1967.

"How many countries would you expect to go into?" Lipscomb demanded. "How large an operation is Project AGILE to become? There seems to be no end of it."

Herzfeld argued that ARPA's expansion was justified, because its work was going well. "I think to some extent, Mr. Chairman, we are breaking ground here for a new way of looking at insurgency, how to stop insurgency while it is small," Herzfeld told lawmakers. "This is absolutely a major military problem for the United States and it is largely unsolved. We were not able to stop insurgency while it was small in Vietnam. It got very big. It is not insurgency now but a war."

That counterinsurgency strategies in Vietnam had utterly failed to stop the escalation into a full-scale war did not enter into the discussion. When another lawmaker asked Herzfeld if he saw an end to the war in Vietnam, and if so, when and how, Herzfeld responded with enthusiasm. "I am quite convinced, if one compares what is going on now with what was going on three or four years ago, that on the military side we are now really winning," he said. "On the civilian side I think we have stopped the decay and we are gradually pulling up. This coupled with the winning on the military side, makes me quite convinced that we will win."

The lawmaker then pointed out that the French had been in Vietnam for a decade, with more than half a million men, and yet "they didn't win."

"Yes," Herzfeld agreed, "but they didn't do nearly as well the things we are doing."

Herzfeld was wrong. In the predawn hours of January 31, 1968, North Vietnamese and Vietcong forces carried out a simultaneous wave of attacks, bringing the war out of the jungles and into South Vietnam's cities. Timed to coincide with the lunar New Year, the Tet Offensive redefined the nature of the conflict. What had been guerrilla warfare was clearly veering toward conventional war, and much of the work that ARPA had been sponsoring seemed almost beside the point. Saigon's once bustling expatriate nightlife wound down, and ARPA decided to arm its civilian employees.

In the end, it was Congress that put an end to ARPA's support for the social sciences. The growing unpopularity of the war in Vietnam led lawmakers to question more of ARPA's work, particularly AGILE. In 1969, the Democratic senator Mike Mansfield, an ardent critic of the Vietnam War, pushed through what became known as the Mansfield Amendment, which prohibited the Defense Department from funding research that "lacked a direct or apparent relationship to a specific military function." The amendment struck at the heart of

ARPA's social science funding, ending much of the agency's work in Southeast Asia and the Middle East. With AGILE drawing to a close, Deitchman went back to the Institute for Defense Analyses.

Over the next few years, almost all of the governments that ARPA worked with on counterinsurgency collapsed. Only Thailand managed to avoid a total political implosion: a 1973 revolution toppled the military-led government, but the country never fell to a communist insurgency. In 1974, a communist junta took power in Ethiopia. The following year, South Vietnam fell to a conventional invasion by the North, and the country was united under a communist government; sectarian divisions in Lebanon ignited a civil war that would go on for fifteen years. In Iran, the shah's repression, corruption, and reliance on foreign benefactors fueled a slow-burn implosion. In 1979, the regime finally collapsed. In its disastrous wake, Ruhollah Khomeini, better known in the West as the Ayatollah Khomeini, created one of the most enduring anti-American regimes in the world.

Vietnam presaged those counterinsurgency disasters. Deitchman recalled during one trip there taking a break from endless military briefings to visit a Taoist temple with John Boles, a brigadier general in charge of the Joint Research and Test Activity at Military Assistance Command, Vietnam. They traveled together through the heat and humidity on a cyclo, a three-wheeled bicycle taxi common in Vietnam. Inside the incense-filled temple, they spotted an elderly fortune-teller. To Deitchman's amusement, the general declared that he wanted his fortune read. The fortune-teller obliged, relating a series of seemingly prescient statements, like Boles's recent promotion and an upcoming visit with his family. "As to the reason you are here," the fortune-teller told the two Americans, "it will be like scissors cutting water."

Deitchman at the time thought the simile was wonderful but wrong. He believed that with science the Pentagon could succeed in Vietnam. If he thought it was ironic to be told by a fortune-teller that the American mission would fail—considering that ARPA had been paying for a variety of magicians, sorcerers, and holy men to spread rumors of a Vietcong defeat—he did not mention it. Years later, however, he acknowledged the fortune-teller's prescience. "How right he was," Deitchman wrote, the month before he died. ARPA's work in Vietnam—like much of what the American government did there—accomplished almost nothing.

AGILE became about helping American conventional forces fight abroad, the very thing that the original counterinsurgents had wanted to avoid. AGILE—and counterinsurgency in Vietnam more generally—failed in large part because it was supporting a government that was incapable of providing the security the population wanted, and no amount of American forces could change that. ARPA could not change that equation. Technocrats like Deitchman and Herzfeld had deeply believed that science could solve almost any problem related to warfare, even human problems. They were ultimately proved wrong, a lesson that Deitchman took to heart. Herzfeld, however, defended the counterinsurgency work to the end as part of his enduring philosophy about ARPA's role in coming up with solutions to big problems. "AGILE was an abysmal failure; a glorious failure," Herzfeld later recounted without a hint of irony. "When we fail, we fail big."

Yet of all the failings tied to ARPA's global experiment in counterinsurgency, perhaps the most troubling was the arrogance involved in treating nations as living test beds. It was an arrogance born, in many cases, from smart and well-meaning scientists. It was an arrogance that pervaded the agency in the mid- and late 1960s. Whether it was insurgent warfare or nuclear warfare, ARPA was going to the edge of science and policy, drawing itself into some of the biggest, most controversial, and most secretive projects of the Cold War. And like AGILE, it did not always end well.

CHAPTER 11

Monkey Business

On October 22, 1964, half a mile underground in a salt mine in Mississippi, the United States set off a nuclear device about a third as powerful as the bomb that leveled Hiroshima. Above where the nuclear device was detonated, someone had placed a Confederate battle flag next to a sign that read, "The South Will Rise Again."

It was an unexpectedly apt description for what happened next: shock waves from the nuclear explosion, which scientists had expected to be contained inside the underground cavity, sent tremors through the earth. In the nearby town of Baxterville, chimneys and shelves of the modest homes collapsed, plaster cracked, and residents returned to houses that looked as if they had been ransacked. It was what Charles Bates, chief of ARPA's Vela Uniform program, described as a "bad roll."

The only atomic weapons tests conducted east of the Mississippi in the United States were carried out under the auspices of ARPA's nuclear detection project. Carrying out a nuclear test, even so close to a populated area, was "easy in those days," Bates explained in an interview four decades later. ARPA and the Atomic Energy Commission consulted with Senator John Stennis from Mississippi, who sat on the Senate Armed Services Committee. Then Stennis spoke to the governor, who spoke to the sheriff, the local judge, and the local newspaper editor. Soon, everyone was on board, and 150 local residents were evacuated, whether they liked it or not. It was, after all, the Cold War, and the nation believed the American military was involved in a life-or-death battle to avoid nuclear Armageddon. "These were poor people. We put them on government per diem," Bates said. "Even babies were on per diem for two days. They got to go to Hattiesburg and stayed in a hotel that they couldn't afford on their own." Adults got $10 per day, and the babies and children $5.

The test, called the "Salmon shot," pushed the walls of the mine out, like a "spoon into Jell-O," as Charles Herzfeld later told Congress. Unlike Jell-O, however, the walls did not bounce back. Instead, what remained after the explosion was a spherical cavity some hundred feet in diameter filled with melted salt and poisonous gas. It took the government two years of pumping in fresh air to remove the gas and lower the interior temperature to a still scorching three hundred degrees Fahrenheit. Then, on December 3, 1966, ARPA set off another nuclear explosion in the same salt mine. This test, called Sterling event, involved a much smaller 380-ton nuclear explosive. Both shots were part of Project Dribble, which was designed to see whether the Soviets might be able to hide a nuclear test by conducting it inside an underground cavity, essentially muffling the signal by reducing the amplitude of the seismic waves—a process called decoupling.

That ARPA could set off a nuclear weapon in a salt mine in Mississippi reflected the ambition, power, and reach the agency had by the mid-1960s, and not just in nuclear test detection. Its missile defense work was also expanding globally: on a tiny sliver of land in the South Pacific, known as Kwajalein Atoll and Roi-Namur, ARPA paid to build radar that looked like giant golf balls to track warheads launched over the Pacific. By the mid-1960s, ARPA's nuclear test detection work, such as that conducted in Mississippi, had produced a variety of technologies, particularly sensors, and agency officials were eager to find ways to expand that work to new areas of national security.

In 1965, ARPA established the Advanced Sensors Office, specifically to market technology—derived from its work in nuclear test detection—to the CIA and intelligence community. The office's first director was Sam Koslov, a physicist who came to ARPA after helping the CIA and the air force develop sensor payloads for balloons that would be flown over the Soviet Union to detect nuclear tests. The new office was, as Stephen Lukasik, who later became ARPA director, described it, the agency's "first attempt to get in bed" with the intelligence community.

ARPA's relationship to the intelligence community had always been fraught. The agency was established to provide technology for the military, not the intelligence community, even though there was natural overlap. In the days of the Corona spy satellite, ARPA had been an unwanted interloper from the CIA's perspective. Similarly, ARPA's involvement in nuclear test detection intruded on what the intelligence community believed was its territory. And, of course, there was William Godel and his work in Vietnam, which had also been viewed with suspicion by the CIA. Yet some ARPA officials looked to the intelligence community as a way to expand the agency's influence: in essence, spies were just an additional customer, or so went the theory.

In his memoir, Charles Herzfeld mentions almost in passing setting up the innocuous-sounding Advanced Sensors Office, which he describes as dedicated to work on "some special projects." In congressional testimony, the office's work was described obliquely as supporting "research in such fields as acoustics, electromagnetics, optics, biology and chemistry that have important applications to new and advanced sensor concepts and hardware." Individual projects were rarely discussed, and when they were, many details were deleted from the public congressional record.

The office was the most "sparsely described" of ARPA divisions, and its activities were often hidden even from senior officials, according to the agency's history. "Its operation, greatly complicated by its relationship to intelligence applications, was questioned, from the beginning," the history states. The office was, simply put, ARPA's conduit to the spy world. Wrapped in a blanket of secrecy, it existed for barely seven years, and its tenure was often stormy, with a director who ran roughshod over his bosses. The Advanced Sensors Office was established in 1965 with a modest budget of just under $5 million, and its first program was Project Pandora, a top secret research program into mind control.

In 1965, medical workers began showing up at the American embassy in Moscow, drawing blood from the employees inside. The American diplomats were told that doctors were looking for possible exposure to a new type of virus, something not unexpected in a country known for its frigid winters.

It was all a lie. The Moscow Viral Study, as it was called, was the cover story for the American government's top secret investigation into the effects of microwave radiation on humans. The Soviets, it turned out, were bombarding the embassy in Moscow with low-level microwaves. The "Moscow Signal," as officials in Washington called the radiation, was too low to do any obvious harm to the people in the building. At five microwatts per square centimeter, the signal was well below the threshold needed to heat things, as a microwave oven does. Yet it was also a hundred times more powerful than the Soviets' maximum exposure standards, which were much more stringent than those of the United States. That was cause for alarm.

The intelligence community was worried that the Soviets knew something about non-ionizing radiation that the United States did not. With research into the effects of low-level radiation still in its infancy, one of the first theories forwarded by the CIA was that the Soviets were trying to influence the behavior or mental state of American diplomats, or even control their minds. The United States wanted to figure out what was going on without tipping off the Soviets that they knew about the irradiation, and so the diplomats working in the embassy—and being exposed daily to the radiation—were kept in the dark. The State Department was responsible for looking at biological changes associated with microwaves, and the newly created Advanced Sensors Office at ARPA, headed by Sam Koslov, was assigned to look at the possible behavioral effects of microwaves.

The new ARPA office was only up and running for a few months when Koslov took another position in the Pentagon. That left his deputy, Richard Cesaro, in charge. Cesaro was one of ARPA's earliest employees, hired back in 1958, and also perhaps its most notorious, with a reputation for being creative, aggressive, and obnoxiously rude. No one quite knew what Cesaro was doing, and that was the way he liked it. Some officials, like Robert Frosch, who served as the deputy director of ARPA for several years in the 1960s, thought Cesaro might even be a resident spook. "He did his best to act like one, always leaving the suggestion that he knew things you didn't know drifting in the air behind him," he said.

Cesaro was no intelligence operative, but he would surely have enjoyed the idea that senior ARPA officials thought he was. In the early days of ARPA, Cesaro had been an ardent space enthusiast, promoting the agency's role in rocket programs. In the late 1950s, he worked with Godel when ARPA was in charge of the Corona spy satellite program. When ARPA lost its space work, Cesaro kept his ties to the intelligence world and eventually found a new home at ARPA with the Advanced Sensors Office. He was described in one official government history as "a master at maneuvers in government decision-making processes, a technical gadfly, and an aggressive proponent for using advanced technology." Colleagues at ARPA also remembered Cesaro, who stood just over five feet tall and wore elevator shoes, as an unrepentant bully known for belittling colleagues to their faces. He reveled in his access to classified projects, and by 1965, with Godel gone, Cesaro was the senior-most ARPA official with connections to the spy community. As director of the Advanced Sensors Office, he enjoyed the prestige and independence that came with running highly classified programs, often on behalf of the intelligence community. He traveled around the world, working on secret projects that even his bosses at ARPA often could not ask him about. The Advanced Sensors Office quickly followed the tradition of spooks, according to Lukasik, which dictated, "Tell your nominal boss as little as possible about what you are doing."

In the past, the intelligence community had treated ARPA as either an unwanted interloper or, during better times, a convenient cover for classified projects. ARPA in its early years was often used as a "cutout," meaning an intermediary for classified funds, a legacy of Godel's intelligence world connections. On the books, the program would belong to ARPA, but in reality it was a facade, and even the director might have only basic knowledge of what the project involved. "We were used as a cutout for some very important classified things. The cutouts were generally there if it went wrong. It was all a guise of ARPA being very intimately involved where in reality we weren't," said Kent Kresa, a former senior ARPA official. "There was no money for ARPA. There wasn't anything. We would have to show up so it looked like we were involved."

One example of a cutout, though former ARPA officials have never confirmed it, was the National Security Agency's top secret spying facility in Australia, known as Pine Gap. In the early 1960s, Godel had gone to Australia to negotiate what Lee Huff described simply as a facility "to use for some of the space work." Soon, American engineers started showing up to survey land in a valley located in central Australia, and within a few years giant golf-ball-looking radomes popped up along with security fences and more than a dozen buildings. The only public information released either by the Australian or by the American government about the facility was that it would be operated by ARPA.

In fact, ARPA had almost nothing to do with Pine Gap's eventual operation, short of the occasional official's visiting to give the ARPA imprimatur. The facility was a ground station for signals intelligence satellites operated by the NSA. "When I visited a foreign country not to be named, I was there as an announced 'company man,'" recalled Lukasik, the former ARPA director, who even forty years later declined to specify it was Pine Gap, or in Australia. "I was the cover for their station, the owner of what the sign at the gate called ARPA Joint XXX Space Defense facility."

Yet the Moscow Signal investigation was a rare opportunity for ARPA to work directly with the intelligence community. In October 1965, Cesaro addressed a secret memo to ARPA's director, Charles Herzfeld, explaining the justification for this new research effort. The White House had charged the State Department, the CIA, and the Pentagon to investigate the microwave assault in secret. The State Department was the lead on the program, code-named TUMS, and ARPA's responsibility, Cesaro explained, "is to initiate a selective portion of the overall program concerned with one of the potential threats, that of radiation effects on man." Thus was born ARPA Program Plan 562, better known by its code name, Project Pandora, an exploration of the behavioral effects of microwaves and one of the more bizarre episodes in the history of Cold War science.

With the passage of time, the government's concerns about microwave-induced mind control might sound like something born of the worst sort of Cold War paranoia—the sort of thing easily parodied as a tin-foil-hat conspiracy—but set in the landscape of the 1960s, it seemed a plausible concern. The discovery of the Moscow Signal came amid a flurry of American and Soviet research reports on the possible biological effects of low-level microwave radiation. Anecdotal reports of fatigue and confusion fueled theories that microwaves could be used as a weapon for behavior modification, or even mind control. One theory that officials floated was the Soviet Union might be using microwaves to influence the behavior of embassy workers, perhaps to induce clerks to make mistakes on encrypted messages, allowing Soviet cryptographers to crack American codes. In fact, ARPA-sponsored translations of Russian-language research at the time indicated that the neurological effects of microwaves fascinated the Soviets, which American officials took as possible evidence that the Moscow Signal was some sort of weapon.

ARPA's role in Pandora immediately evoked concerns among the few Pentagon scientists who were cleared to review the program. Bruno Augenstein, a German-born physicist who worked for the Defense Department, sent a top secret memo to Harold Brown and Gene Fubini, two of the Pentagon's top technology officials, to let them know that ARPA was evaluating proposals looking at the neurological effects of microwaves. In his note, Augenstein alluded to "past unsavory history of experiments of this kind in this country, which has made a number of people rather leery of further experiments in this field," likely a reference to the CIA's infamous MKULTRA mind control experiments begun in the 1950s, in which agency officials tested the effects of LSD as a possible mind control agent on humans. Augenstein wrote that there did appear "to be some internal resistance in ARPA to the suggestion that ARPA proceed with these experiments, probably because there is a feeling that at one time it certainly attracted a number of crackpots."

If the ARPA program was supposed to avoid the mistakes of prior scandals in human experimentation, then Cesaro was an inauspicious choice to lead Pandora. A propulsion expert, he had no apparent expertise in the biological sciences, but he relished running a top secret project that had high-level attention from the White House and the CIA. He embraced the assignment with an enthusiasm that might have been admirable, had it not been quite so morbid. It soon became clear Cesaro's primary interest was pushing forward with actual microwave weapons, rather than understanding the underlying biology.

To see if the Moscow Signal really affected human behavior, ARPA first started by testing microwave radiation on monkeys. Because Pandora was top secret, the primary research had to be run at government laboratories rather than at universities. The air force was assigned to provide electromagnetic equipment needed to generate the microwaves, and the Walter Reed Army Institute of Research was responsible for selecting the monkeys and running the experiments. The initial tests were designed to see how primates performed work-related tasks when exposed to radiation that matched the Moscow Signal, which was beaming every day at the men and women inside the American embassy in Moscow.

The test protocol involved training the monkeys to press certain levers in response to signals. If the monkeys pressed the lever correctly, they would receive a reward of food, "much as embassy employees might be rewarded with a dry martini at the end of the day," wrote the columnist Jack Anderson. Researchers would then measure whether the monkeys performed worse when subjected to the Moscow Signal, compared with when there was no radiation. By December 1965, shortly after the lab work had started, Cesaro was already enthusing over the results. The normal process for accepting any new, significant scientific phenomenon would have involved submission of the results for peer review, publication in a respected journal, and eventually replication by an independent group. Pandora, on the other hand, operated in the world of classified science, where results were conveyed not by the researchers conducting the experiments but by the manager in charge, in this case Cesaro. In December 1966, Cesaro reported that the first monkey involved in the tests had demonstrated "two repetitive, complete slowdowns and stoppages" as a result of exposure to the Moscow Signal. "There is no question that penetration of the central nervous system has been achieved, either directly or indirectly into that portion of the brain concerned with the changes in the work functions and the effects observed," Cesaro wrote.

The radiation results were so convincing to him that he recommended the Pentagon immediately start to investigate "potential weapon applications." He initiated a new phase of Pandora intended to move toward human testing, taking the ARPA program dangerously close to the very work that Augenstein, the Pentagon scientist, had warned of. Cesaro also wanted to make Pandora even more secretive than it had been previously. "The extremely sensitive nature of the results obtained to date, and their impact on National Security, has resulted in establishing a special access category for all data results and analysis, under codename Bizarre," he wrote. Bizarre, as it turns out, was an appropriate name for the project, because at this point the number of monkeys involved in the testing stood at one.

Initially, the Pandora scientific review committee seemed to go along with Cesaro's enthusiastic proposal to move directly to human testing. The committee even suggested recruiting human subjects from Fort Detrick, Maryland, home to the army's biological research program (the conscripts assigned to Fort Detrick have been a continuous source of human subjects for Defense Department research for decades; subjects there have been exposed to everything from yellow fever to hallucinogenic drugs). In minutes from a May 12, 1969, meeting to discuss human testing, the Pandora scientific committee discussed plans to move forward with eight human subjects. The human subjects would be exposed to the Moscow Signal and then given a full battery of medical and psychological tests.

The committee was aware of the potential for a conflict of interest involved in classified human testing; the idea of informed consent becomes hazy when the subjects are not even aware of the true purpose of the test. To address this problematic issue, the committee recommended having medical personnel on hand to assure the "medical well being" of the subjects. Yet even those medical personnel would not be told the reason for the testing and would instead be given a cover story. Humanely, at least, the committee did recommend "gonadal protection be provided" to the male test subjects.

Fortunately for the would-be recruits and their gonads, the human tests were never pursued. The committee's views on Pandora quickly began to change as they reviewed the actual data, which eventually included more primates and additional testing. The scientific committee's minutes, declassified and released years later, demonstrate increasing doubts about the testing protocol, in particular the lack of controls used in testing the monkeys. Among the concerns was that there was never a solid baseline established to compare how the monkeys' performance allegedly degraded after exposure to radiation, the members noted. In other words, it was never established how well monkeys performed the tasks during a test period when not exposed to any periodic bouts of radiation.

While Pandora never progressed to testing on humans, it did look at the effects of occupational radiation exposure on humans. One experimental protocol, called Big Boy, examined sailors on the USS *Saratoga*, comparing those who worked above deck, and were exposed to radiation from the radar, to those who worked below deck (the sailors were not told that they were part of a human radiation study; an unspecified cover story was used). The conclusion was that there were no psychological or physical effects as a result of exposure to low-level microwave radiation.

In 1968, Joseph Sharp, the lead Pandora researcher at Walter Reed, left the program. Major James McIlwain, a medical doctor who had been drafted into the army, was selected to replace him. It took almost a year before McIlwain was cleared for Project Pandora, but once that happened, he got to work on a rigorous review of the data, poring over the computer printouts detailing each animal's behavior. Within a year, McIlwain completed the statistical analysis, and what he found was not encouraging for the prospects of microwave mind control weapons. The basic question, he recalled in an interview years later, was whether it was more likely that the animal would stop working when the radiation was on compared with when it was off. "The answer to that was no," he said. The Pandora scientific review committee agreed, concluding, "If there is an effect of the signal utilized to date on behavior and/or biological functions, it is too subtle or insignificant to be evident." In other words, microwaves could not be used for mind control.

By 1969, Stephen Lukasik, then the deputy director of ARPA, had some serious doubts about Cesaro, whom he regarded as a serial liar. The impresario of ARPA's black programs acted as if he reported to no one, alluding to orders from high-level intelligence agencies but refusing to provide any specific information. "He was all over the place, cloaked in special access programs," Lukasik said, a reference to highly classified national security programs.

Pandora, the mind control project, was particularly worrisome. At that point, the research had been going on for almost five years, and millions of dollars had been spent for construction of a new microwave laboratory. Lukasik asked Sam Koslov, the original director of ARPA's Advanced Sensors Office, to review the Pandora file and let him know what he thought. Koslov was an old hand at intelligence projects and less likely to be snowed by claims of secrecy and overwrought concerns about the potential for Soviet exotic weapons. Koslov, then at Rand, reviewed the materials and discussed the results with McIlwain, at Walter Reed, and reported back to Lukasik in November 1969.

Like other review committee members, Koslov criticized the original experiments for having almost no baseline and noted the experimental procedure changed over time. Also, if the question was whether a modulated microwave beam, such as the Moscow Signal, was causing deleterious effects, why was it never measured against a continuous wave? he asked. Simply zapping monkeys with the Moscow Signal was an entirely wrong approach, if the goal was to understand whether the effects were associated with a specific signal. "One should start with an examination of various basic wave forms and then the combinations resulting in possible intermodulations and demodulations by biological tissue," Koslov wrote.

Koslov also rightfully questioned whether the entire program truly needed to be secret. One could much better run a more open program that looked at the health effects of microwaves generally and then have a secret program looking at technology or weapons, if it was warranted, he argued. "In brief, I am forced to conclude that the data do not present any evidence of a behavioral change due to the presence of the special signal within the limits of any reasonable scientific criteria," Koslov wrote to Lukasik. In 1969, ARPA ended its support for Pandora, and the remaining work was transferred to Walter Reed. At the time it was closed down, the program had spent upwards of $5 million—a sizable amount for a biological sciences program at the time and a good portion of the Advanced Sensors Office's budget.

Even before Pandora drew to a close, Cesaro found another area where he could market ARPA's secret sensor technology: Vietnam. Instead of zapping people with microwaves, however, Cesaro was trying to use ARPA's sensors to hunt them down and kill them.

In 1967, the American military in Vietnam was facing a politically delicate but militarily challenging situation: the North Vietnamese military was increasing its attacks across the demilitarized zone, but the United States was restricted by its rules of engagement, which required the military to identify what it was shooting at. "You can't just hit a target because it moves, you have to know what it is," Eberhardt Rechtin, ARPA's director at the time, explained to lawmakers. "That means, in effect, that you have to be looking at it."

Cesaro, once again, had a solution, or many solutions, in fact. By the late 1960s, he had managed to expand his classified work from microwave research in the United States to a series of controversial Vietnam War projects. All involved using the sensor technology that ARPA had developed from the nuclear world and applying it to identifying targets. Cesaro was not always practical, but he was creative when it came to foreseeing the potential applications of technology. His most ambitious—and eventually most high-profile—plan was an audacious effort to arm surveillance drones in order to kill "high priority targets" in North Vietnam. Thirty-five years before the CIA used an armed Predator in Afghanistan, ARPA was preparing to arm a strange-looking drone that looked as if it were made out of spare parts in someone's garage.

When Constantine "Jack" Pappas, a young naval officer, showed up for a scheduled appointment at ARPA in 1967 to talk about ARPA's drone plan, he got a quick lesson in why so many people disliked Richard Cesaro. After being kept waiting more than an hour for Cesaro, who was in his office, Pappas finally lost his temper at the secretary, threatening to "blow the door down" if Cesaro did not come out to see him. When Cesaro emerged, however, Pappas's view soon changed. "Cesaro was a prickly guy with a big ego," Pappas said, but he was genuinely interested in new technologies.

Pappas had been assigned to the QH-50 Dash, short for Drone Anti-Submarine Helicopter, an unmanned aircraft armed with a nuclear weapon. The QH-50 was a coaxial helicopter, which means that it used two main rotors that spun in opposite directions, and had no tail rotor. The navy originally bought the drone helicopter with plans to use it in antisubmarine warfare, because its compact size made it ideal for operations off a small ship deck. The QH-50 would hunt for Soviet submarines; when it found them, it would drop a nuclear depth charge on them. The oddly shaped drone was innovative but prone to crashing, at least according to its critics. However, the QH-50 was secretly drafted into war in Vietnam by the U.S. Navy, which loaded the drone with sensors for reconnaissance. The sensor-laden version of the QH-50 was nicknamed Snoopy.

In the fall of 1967, the air force and ARPA launched a project code-named Blow Hole, to find a way to deal with the North Vietnamese incursions across the demilitarized zone. The goal was to have a technology that could be delivered in forty-five days. ARPA's plan was to take the helicopter drone, arm it, and send it flying over the DMZ in search of Vietcong, who were launching artillery at American forces. Cesaro initiated two drone projects in early 1968: a QH-50 loaded with a television camera and electronics, which was called Nite Panther; and a second, armed version, which was called Nite Gazelle. Over the course of the next four years, ARPA would experiment with putting guns, grenade launchers, bombs, and missiles on the QH-50. In one of the first demonstrations of "precision targeting," the Nite Gazelle was equipped with a laser that designated targets, which could then be destroyed by weapons fired from air force or navy aircraft.

In 1969, Cesaro expanded his killer drone work with a project called Egyptian Goose, which would place a radar on a "leftover" balloon from World War II (the name was derived from a parallel Israeli project that would use the balloons to spy on Egypt). Another balloon, called Grandview, would relay television surveillance video from the battlefield in Vietnam, identifying targets so that the armed drone could destroy them. Neither balloon made it to Vietnam, but the project was later credited with leading to the U.S. government's use of tethered balloons for surveillance, which by the early twenty-first century ranged from protecting the American border with Mexico to military bases in Afghanistan.

The QH-50, on the other hand, was sent to Vietnam, even though, as a later ARPA history noted, the aircraft had a "checkered" history, an understated way of referring to the drone's multiple crashes. Reliability problems plagued the QH-50. Trying to get the weapons to accurately hit anything also proved futile. Shortly after the program started, the QH-50 equipped with a 7.62 mm mini-gun and gravity bombs was tested at the Naval Air Station in Patuxent River, Maryland. The tests, a later report notes, were "unsuccessful." While some of ARPA's QH-50s made it to Vietnam for testing, it appears none of them were ever used in combat. Stephen Lukasik says be believed all the armed QH-50s ended up crashing. Though Nite Gazelle was never used operationally as a weapon, it demonstrated the technological feasibility of a drone that could both spot and kill an enemy.

Cesaro's aggressive pushing of his office's technology brought him into conflict with almost everyone, including Seymour Deitchman, who was in charge of AGILE at the time. Deitchman regarded Cesaro as an aggravation who pushed technologies without regard to their operational utility, like advanced night-vision devices that required an unwieldy battery pack. Deitchman recalled clashing with Cesaro over a project named Dancing Bells, which was supposed to spot Vietcong hiding in the underbrush. The idea was to equip helicopters with sensors that used constantly shifting frequencies to detect possible human movement under jungle foliage. Deitchman decided the vibration would make the images unusable and canceled the project over the objections of Cesaro.

Cesaro's office also paid for the MIT Lincoln Laboratory to build radar that could see through the jungle, protecting military outposts from surprise Vietcong attacks. The Camp Sentinel Radar went to Vietnam in 1968 at Lai Khe. When it was mounted on top of a tall tower, its electromagnetic energy could penetrate the dense foliage to pick up any potential Vietcong lurking in the underbrush. Though the six prototypes of the radar sent to Vietnam were hailed as a technological success, Fred Wikner, a physicist who served as science adviser to General Creighton Abrams, said the only Sentinel radar he ever saw in Vietnam was knocked out during a typhoon and took two years to repair.

The legacy of ARPA's most secretive and controversial office is mixed. By the end of the 1960s, the intelligence community concluded that the Soviets were using the pulsed radiation to activate listening bugs concealed in the embassy's walls, and not to control diplomats' minds. Yet concerns about the Moscow Signal lingered even after the scientific testing ended, though mind control was generally ruled out. A State Department doctor in charge of the blood tests, Cecil Jacobson, asserted that there had been some chromosomal changes, but none of the scientific reviews of his work seemed to back his view. Jacobson achieved infamy in later years, not for the Moscow Signal, but for fraud related to his fertility work. Among other misdeeds, he was sent to prison for impregnating possibly dozens of unsuspecting patients with his own sperm, rather than that of screened anonymous donors as they were expecting.

Richard Cesaro never attained that level of personal notoriety, but he asserted, even after he retired, that the Moscow Signal remained an open question. "I look at it as still a major, serious, unsettled threat to the security of the United States," he said, when interviewed about it nearly two decades later. "If you really make the breakthrough, you've got something better than any bomb ever built, because when you finally come down the line you're talking about controlling people's minds."

Perhaps, but Pandora resonated for years as the secrecy surrounding the project generated public paranoia and distrust of government research on radiation safety. Project Pandora was often cited as proof that the government knew more about the health effects of electromagnetic radiation than it was letting on. The government did finally inform embassy personnel in the 1970s about the microwave radiation, prompting, not surprisingly, a slew of lawsuits. In the end, the government found that the best method for dealing with the incessant Moscow Signal was to build an aluminum screen to shield the building from microwaves. "The lesson learned is to treat your people as if they have some intelligence," said Koslov, reflecting on the controversy.

The armed drone program also crashed and burned. In 1972, Lukasik transferred Nite Gazelle to the military services, which discontinued the project, though the QH-50 was used in tests for a number of years. Lukasik called Nite Gazelle a "showy stunt" that failed. The legacy of ARPA's work on armed drones would not become clear until thirty years later, when, in the weeks following September 11, 2001, the air force took a QH-50 and put a Hellfire missile on it. The test was a failure. "The bird tumbled out of the sky," recalled Peter Papadakos, the son of the man who designed the QH-50. It did not matter, because at that point the air force and the CIA had a new armed drone, called Predator, based on yet another ARPA project. Afghanistan, and not Vietnam, would ultimately be the testing ground for armed drones.

With seemingly little to show from its Vietnam work, let alone Pandora, ARPA was losing support in the Defense Department. Vietnam began to take over the Pentagon figuratively and literally. ARPA's offices were moved out of the Pentagon to make way for analysts working on the Vietnam War, and the agency moved into leased office space on Wilson Boulevard in Rosslyn, Virginia. It was a clear demotion for ARPA and distanced the agency even further from the Pentagon's senior leadership. Herzfeld, the director, was appalled, calling the move the loss of a "great gift" for the Defense Department. "I argued very hard against moving ARPA out and lost, just plain lost," Herzfeld recalled.

By the fall of 1967, Herzfeld found himself pushed out of ARPA. Domestic opposition to an escalating conventional war was bringing the agency's projects under even greater scrutiny, and ARPA was facing new challenges to its existence.

CHAPTER 12

Bury It

On July 16, 1969, at 9:32 a.m., Apollo 11 lifted off from Florida on its way to fulfilling President Kennedy's vision of putting a man on the moon by the end of the decade. The rocket that launched the Apollo mission was the Saturn V, a descendant of the Saturn rocket that ARPA's first director had fought for over "dying and bleeding bodies," as William Godel had put it. The rocket that led to man's first steps on the moon was credited to Wernher von Braun and his team of rocket scientists at NASA. The critical contribution that ARPA made to that mission had been long forgotten. By the time NASA reached the moon, ARPA was stuck in Vietnam.

Earlier that year, when Richard M. Nixon was sworn in as president, troop levels in Vietnam peaked at more than half a million. That same year, the journalist Seymour Hersh published shocking details of a massacre conducted by American troops in My Lai, a village in South Vietnam. My Lai was just the tip of the iceberg: the American public was being flooded with graphic pictures and reporting depicting the scale of civilian suffering and death. The war was reaching new heights of unpopularity, and ARPA was under attack.

Cyrus Vance, when he was the deputy secretary of defense, had even advocated disbanding the agency. Congress also started to wonder why the Pentagon needed ARPA at all. "Would it be desirable to abolish ARPA as such and consolidate this work elsewhere?" asked Representative George Mahon, the Texas lawmaker and frequent ARPA critic, during one contentious hearing.

By the early 1970s, Secretary of Defense Melvin Laird announced a new policy called Vietnamization, which entailed shifting responsibility to the South Vietnamese government. Henry Kissinger, the national security adviser, was in secret peace talks with the North Vietnamese government. The American involvement in Vietnam was coming to an end, and so was ARPA's work there. The Vietnam War, and ARPA's involvement in it, had made the agency a punching bag on Capitol Hill. "Congress hated AGILE," Stephen Lukasik said.

Lukasik, who had been serving as the deputy director, was appointed director in 1970, just as relations between the White House and the Pentagon were about to go nuclear. In June 1971, *The New York Times* began publishing excerpts culled from a top secret Pentagon study that revealed the missteps and deceptions that had allowed the Vietnam War to escalate. *United States–Vietnam Relations, 1945–1967: Study Prepared by the Department of Defense,* better known as the *Pentagon Papers,* had been commissioned several years earlier by the then defense secretary, Robert McNamara, to review the history of the conflict. It did not take long before a Pentagon military analyst named Daniel Ellsberg was identified as the leaker.

Relations between the White House and the Pentagon were strained even before the leak. Nixon was suspicious of the Defense Department's civilian leadership, and Henry Kissinger did not want to share power with the Pentagon. Together, the two men made a habit of bypassing the Office of the Secretary of Defense, going straight to military commanders. The explosive leak of the *Pentagon Papers* cemented Nixon's distrust.

In 1972, Nixon slashed the number of personnel from the Office of the Secretary of Defense, a move meant to reduce the power of the Pentagon's civilian leadership. To protect its bureaucracy, the Pentagon began designating offices as field agencies to immunize them from reductions. On March 23 of that year, ARPA officially became the *Defense* Advanced Research Projects Agency, which meant that ARPA would hence be DARPA. On its own, the new name had no significance for the agency's direction, but it was a symbolic defeat. Lukasik despised the name change, insisting that the acronym remain "ARPA," which it did, until the end of his directorship. "This was not a minor point of civil disobedience," Lukasik said. "I said freely that DARPA sounded like a dog food."

The renaming of the agency was the culmination of several years of decline. In September 1967, Lukasik, then the deputy director, had been summoned along with the agency's acting director to the Pentagon, where John S. Foster Jr. informed them that ARPA's missile defense work—the agency's second-largest program—was being transferred to the army. By the early 1970s, there was talk of taking away ARPA's nuclear test detection work. It stayed, but it was growing smaller every year as arms control faded into the background of national policy. The agency's budget was also declining, from around $300 million in the mid-1960s to just over $200 million by the early 1970s.

The agency Lukasik took over in the early 1970s was very different from the one that had existed just a decade prior. In ARPA's turbulent, and still quite brief, existence, it had transformed itself from a space agency into an agency that specialized in nuclear test detection, missile defense, and, somewhat incongruently, counterinsurgency. Missile defense was gone, arms control had taken a backseat to deterrence, and counterinsurgency was ending, too. Now, as Lukasik described the situation, "the flow of 'Presidential assignments' had dried up and we had to figure out what to do absent guidance from the highest levels of government."

Pushed out of the Pentagon and Vietnam, ARPA was in search of a mission, and there was no clear road map. One of the earliest justifications for the agency had been to "avoid technological surprise," meaning that its mission was to prevent another Sputnik, or an unexpected technological advance. That phrase actually did little to guide the research agency, in Lukasik's view, because it could encompass anything. "It did not serve as a useful planning concept," Lukasik said. "There is too much technology, too many possible wars."

Lukasik's predecessor, Eberhardt Rechtin, had focused on quietly getting rid of projects that were subject to congressional scrutiny, like a $1 million "mechanical elephant." Operated by a man sitting inside the machine using hydroelectric controls, the four-legged "Cybernetic Anthropomorphous Machine" was designed to navigate the jungles of Vietnam, hauling equipment for soldiers. Rechtin called it a "damned fool" project bound to land ARPA in hot water.

There were, by the late 1960s, quite a few "damned fool" projects, or at least projects that were not going to change the course of the war and risked making the agency a target of ridicule. ARPA, for example, was paying for an army project to develop a jet belt, a wearable device that would allow soldiers to fly around the battlefield. Developed by the Bell Aerosystems Company, the jet belt had been under development for several years and finally went into flight tests in the late 1960s. The innovative but unwieldy contraption required the pilot to wear something akin to a fiberglass corset, to which the engine and the flight control system would be attached. The jet belt, the company claimed, would "open the door to a new type of counter-guerilla operations." The as-yet-undeveloped mini-rocket system would allow soldiers to fire weapons while flying over the battlefield, according to Bell. The problem with the jet belt, however, was limitations in the technology, as well as the concept of operations. Even though the company had moved from a pressurized hydrogen peroxide rocket (which allowed just seconds of flight) to a mini turbojet that ran on kerosene, the system could still only carry enough fuel to allow the soldier to travel for a few minutes—not enough to really be useful in battle. ARPA eventually stopped funding for it, and the jet belt never reached Vietnam, though a modified version of the engine ended up being used later in air force cruise missiles.

Other Vietnam technology projects were more successful but also died with the end of the war. ARPA, for example, funded a silent helicopter for the CIA based on the Hughes OH-6A light helicopter. Two silent helicopters were eventually built and used to tap phone lines in North Vietnam. The aircraft were retired after the war, however. "The agency got rid of it because they thought they had no more use for it," James Glerum, a CIA official at the time they were fielded, later told *Air & Space Magazine.*

A similar fate befell ARPA's silent aircraft, called the QT-2, which was designed for "covert air operations." The silent aircraft was based on the Schweizer SGS 2-32 sailplane, converted into a powered aircraft using a car engine. The idea was that such a plane, which would have removable wings, could be easily transported where needed, assembled, and then flown on reconnaissance missions with a two-man crew. The concept, widely regarded as innovative, did lead to the eventual deployment in 1970 of the YO-3 "Quiet Star" aircraft, but the planes were mothballed after the Vietnam War. Counterinsurgency, in all its multitude of incarnations—technological and analytical—was losing currency.

When Lukasik was formally appointed director, he recognized that something needed to be done with the agency, which was in turmoil. Lukasik had inherited a few career personnel, like Richard Cesaro, who felt as if they answered to no one, even the director. In 1971, on Lukasik's first official day as director, he fired Cesaro, for "general dishonesty." Getting rid of Cesaro was just the first step: the real problem was Vietnam, and the Project AGILE field offices. The AGILE name was radioactive on Capitol Hill, where lawmakers associated it with the disastrous war in Vietnam. "AGILE, counterinsurgency, was an embarrassment," Lukasik said.

ARPA in the 1970s was still a young agency. It had been around long enough to have a reputation but not a legacy. For ARPA to survive, Lukasik needed to find new areas to pursue and redefine the agency's role in research and defense strategy. First, he had to kill AGILE.

In Washington, the key to killing something controversial is to never admit you actually killed it. To do that, you first change the name. Perhaps a year later you change the name again, to confuse those who might be tracking it. Then you kill it. By that point, most people will have forgotten about it. Faced with an impending congressional showdown over AGILE, Lukasik sat down with his deputy Don Cotter, a longtime government scientist known for his frank advice. "Look, the Nixon Doctrine says we have got to strengthen our allies so that they can take care of themselves," Cotter told Lukasik.

Cotter suggested renaming AGILE to sound like something that would fit under the Nixon Doctrine. Lukasik agreed and coined a new term for AGILE, the wonderfully bland-sounding "Overseas Defense Research." The move was, he later admitted, an effort just to "bury it someplace" deep inside ARPA's bureaucracy. The program would no longer have anything to do with battling guerrillas. "We shifted from little guys with insurgency problems to bigger guys facing the Soviet

Union," Lukasik said, and in the process performed some alchemy. "I've now moved counterinsurgency from shit to gold."

On the day after Christmas 1972, Lukasik sent a letter to Secretary of Defense Laird titled "Taking Stock," which laid out the agency's work over the previous four years and his views for the future. The agency's work abroad was couched carefully in the language of the Nixon Doctrine and Vietnamization. ARPA was no longer helping governments fight insurgency; it was bringing Western technocracy to their defense bureaucracies, teaching them how to buy weapons. This was not entirely new, of course. One of the earliest ARPA projects in the Middle East was an effort to bring Secretary of Defense McNamara's passion for cost analysis to Iran. The U.S. Army in 1964 had helped the Iranian military establish a professional research and evaluation group to train a cadre of military officers to evaluate weapons based on cost and performance. It sounded simple, but it was something new to the Iranians, and "the organization foundered." ARPA was asked to step in and help.

ARPA did not have much better luck than the army. The agency tried to teach mid-level Iranian military officers how to test and evaluate weapons through a failed effort called the Combat Research and Evaluation Center, or CREC. By 1969, the height of the center's activities, the best example ARPA could come up with in a report was an evaluation of "field bakery equipment" to determine whether troops could make their own bread in the field. The same report expressed "serious doubts about the validity of the program," which had been around for five years. "Iran's military leaders neither assign CREC qualified and motivated officers nor give it real work to do, especially because they themselves are not technically oriented enough to recognize the potential value of CREC," the report griped. The center was "disappointing," an official later acknowledged, and ARPA soon ended support for it. The only activity that met with any interest from the shah was an "anthropometric survey," which was being used to help design military uniforms. That work pleased the shah, who was himself known for donning extremely elaborate uniforms.

In 1970, ARPA proposed a new approach in Iran, which it called "high level systems analysis," advising senior ministry officials on how to buy military weapons. The idea was to teach the staff the basics of systems analysis, which meant, for example, not just comparing the cost of an American missile with that of a British one but calculating the actual "kill per engagement" to come up with a total cost comparison. The bottom line was how much did killing the enemy cost you? If it takes three British missiles to destroy a tank, as opposed to one American missile, then even if the British missile was half the price, it ended up being more expensive. The shah liked the idea of professional bean counters, or at least well enough to approve the ARPA project.

ARPA recruited Joseph Large, a Rand analyst, to run the program, but Large quickly saw its futility. "The reason is simple," Large wrote to Alex Tachmindji, the deputy director of ARPA. "His Imperial Majesty makes many of the little decisions and all of the big ones." Large also expressed concerns about ARPA's plans to send a second analyst who had a reputation as bright and hardworking but also "pompous." The new analyst was Anthony Cordesman, a young man frequently described as a sidekick of "Blowtorch Bob" Komer, the head of the pacification campaign in Vietnam. Komer was known, as his *New York Times* obituary put it, for his "near-religious faith in the power of facts and statistics." Cordesman, who would go on to become one of Washington's most influential national security analysts, was Komer's disciple. He was also, like Komer, divisive. "The trouble was that Tony Cordesman was a pain in the ass and he got everybody mad," Lukasik said.

When Cordesman arrived in Iran in 1972, he was put to work with the vice-minister of war, Hassan Toufanian, who was ostensibly in charge of Iran's weapons procurements. Yet the only person really in charge of weapons buying was the shah. "Toufanian was the top clerk, if you will, bringing things to the shah," said Henry Precht, a diplomat in the American embassy in Tehran at the time. "The shah made all the decisions."

Not only did the shah make all the decisions, but also the decisions he made were often influenced by corruption. Who cared about the comparative costs of tank killing when the deciding factor was a bribe paid to the shah's middleman? Cordesman wanted to study Iran's plans to buy hovercraft, but the shah's nephew was in charge of the purchase and thus it "would not be politic," Colonel Harold Kinne, an ARPA official, wrote. Toufanian told Kinne he simply did not have officers to spare for Cordesman and Large and in any case the Iranians "did their own analysis in their heads."

For the next few months, ARPA and embassy officials sparred over whether to keep the systems analysis work going. Some thought the ARPA men were doing good work, even if they were not teaching the Iranians much about systems analysis. The assignment helped "mask the analysis work now being done by Cordesman and Large," one official wrote. What other analysis was being masked is unknown, and even fifty years later Cordesman declined to discuss any details of his work for ARPA, claiming that he believed it to still be classified.

One thing is clear: ARPA's attempt to educate Iranian officials on weapons analysis failed to sway Iran's ruler. When the shah came to the United States in 1973 to shop for a new fighter aircraft, Malcolm Currie, the director of defense research and engineering, was tapped to play host, taking him to Andrews Air Force Base for a private demonstration of American airpower. The air force went first, flying its new F-15 through a seemingly impressive series of air maneuvers. When the navy demonstrated its F-14, the pilot put on an even better show: he did a roll, circled around, and made a loop right over the shah's head. After the performance, the shah said to Currie, "You know, I've always viewed Iran as kind of an island in the middle of an ocean."

Currie marveled at the shah's nonsensical statement. The F-14 was a fighter built for aircraft carriers, something that Iran did not possess. It was more expensive and had a shorter range than the F-15, but it was clear the Iranian leader's mind was made up. Iran became the only foreign country to ever buy the F-14, a decision based on a thirty-minute air show. The results of Cordesman's work, at least as measured by its impact on the shah's spending spree, were also clear. As Precht, the American diplomat, summed it up, "He failed."

By 1973, ARPA's days operating abroad were numbered. AGILE's field offices were an unending and unwanted source of troubles: In 1970, two ARPA officials, James Woods, a researcher, and Robert Schwartz, the head of the Thailand office, were held for two weeks in

Jordan after their commercial plane from Germany was hijacked and diverted to the Middle East. Fearful of the hijackers' finding his Pentagon documents, Schwartz attempted to flush the papers down the toilet and, in desperation, even swallowed some of them. Both men were eventually released unharmed, though Woods, the ARPA researcher, later groused that the government denied him per diem for the days he was held, because the hijackers had provided him with room and board.

The American embassy in Iran was also eager to show ARPA the door. Cordesman was brilliant but arrogant, producing thick analytical reports faster than most people could read them. "He projected an air of superiority over lesser beings, particularly Iranians who felt insecure to begin with," recalled Precht. The ARPA office was becoming a liability; visitors to the embassy suspected it was somehow involved in intelligence operations, and the embassy worried about a congressional investigation. "This would be unfortunate for there are several classified, highly important operations which could be compromised by a general investigation," Precht noted in a confidential memorandum to senior embassy officials.

Was the Middle East office a worthwhile endeavor, or a fool's errand, operating in a part of the world where a science and technology agency had no business being? Charles Herzfeld's argument to Congress in the late 1960s had been that ARPA was doing valuable work that no one else was doing, and sometimes he was right. In 1971, the Beirut office was asked to look into the little-known threat of "improvised explosive devices," the technical name for homemade bombs. More than forty years before IEDs entered the popular lexicon and became the leading killer of American and coalition troops in Iraq and Afghanistan, ARPA hired a contractor to study them and prepare a comprehensive report, which concluded that "there are limits to the benefits one can expect to derive from improved tooling and procedures, and technological development efforts devoted to exceeding those limits are unwarranted." In other words, the report found there are ultimately a limited number of ways to detect and destroy crude bombs; no magic bullet exists.

It is unclear if the report had any resonance at the time, but it accurately predicted the outcome of what happened decades later, when, faced with a flood of IEDs in Iraq, the Pentagon invested in a new agency tasked to do precisely what the ARPA-sponsored report had recommended against: develop technologies to detect and destroy these bombs. After spending nearly $20 billion, the director of that agency, called the Joint Improvised Explosive Device Defeat Organization, admitted in 2010 that the best method the Pentagon had for detecting bombs was still a dog. The 1971 ARPA report, in the meantime, sat in a box in the National Archives and Records Administration in College Park, Maryland.

Those moments of neglected prescience aside, ARPA's accomplishments in the region were limited. The problem in Iran was not whether the military forces could bake their own bread but whether a corrupt monarchy that used torture and fear to maintain power was worth supporting at all, though that was a question beyond ARPA's purview. Systems analysis—or the idea of looking at all the parts and components of a problem—just was not feasible in a monarchy dominated by nepotism and corruption. ARPA's work in Iran "was a neat idea," Lukasik later said, "except the only organization that can take a systems view about a country is the country's government." In 1974, Lukasik canceled the systems analysis project, and the office in Iran was soon shut down completely, as were the rest of ARPA's field offices.

Even with the closure of the field offices, there was one more move needed to distance ARPA from the fiasco of Vietnam: a second name change. Lukasik took what was left of AGILE, or Overseas Defense Research, and packaged it in a new division called the Tactical Technology Office. And then, for good measure, he swept into the office two more problem children: the Advanced Engineering Office and the Advanced Sensors Office.

The Advanced Engineering Office was doing some good work, but many of its projects, like a flippable barge that would serve as a mid-ocean military base and a ten-ton air cushion vehicle designed to battle the Soviets in the Arctic region, made little sense to Lukasik. More problematic was that its head, a Chinese-American scientist, had landed himself in hot water for speaking to Chinese nationals in Canada, raising security concerns. And then there was the Advanced Sensors Office, the spooky outfit headed by Richard Cesaro, whom

Lukasik had fired. Those offices got wrapped up into the new office, with AGILE as its core. "I took a really bad dish and I poured in some crap and some shit, and it became the Tactical Technology Office," Lukasik joked.

ARPA already had a Strategic Technology Office for nuclear warfare, investigating lasers and antisubmarine warfare technology, and now it had the Tactical Technology Office for conventional warfare, creating weapons, like drones, sensors, and bombs. It looked nice and neat, at least on paper, but it needed a larger purpose. The new Tactical Technology Office was populated by novelties that had grown up in Vietnam. Those technologies, from drones to sensors, had not always proved useful for counterinsurgency, but perhaps they might be better suited for a conventional battlefield. The problem was figuring out how to persuade the military services to use them. Or even more important, why they needed them.

Most advances look obvious in hindsight, but it can often take many frustrating years to understand what new technology can do. When the inventor Nikola Tesla first publicly demonstrated remote control in 1898, operating a small boat using radio signals, it caught people's attention, but it did not instantly spark a revolution. Whether it is trying to make people see how you could use networked computers or getting the military to understand why it might want bombs that can be guided to a specific location, one often needs to make a dramatic presentation to convince people, but even that may not be enough.

For seven years, the air force and the navy tried in vain to destroy Dragon's Jaw bridge, a critical North Vietnamese supply line, dropping bomb after bomb, in hundreds of sorties. One pilot described 250-pound warheads "bouncing off" the bridge. Even direct hits did little more than minor damage to the bridge, which the air force historian Richard Hallion called "a notorious graveyard for dozens of strike aircraft and airmen." It was not until 1972, when the air force struck the bridge with a barrage of newly developed laser-guided bombs, that it was finally destroyed. The era of precision warfare had begun, though in Vietnam it was too late to make any difference in the outcome of the war.

Fred Wikner, a physicist who worked for the Pentagon, was interested in precision weapons and in ARPA. He had seen ARPA's Southeast Asia work firsthand, in 1969, when he served as the army general Creighton Abrams's science adviser. One of the problems Wikner saw in Vietnam was that the Pentagon, including ARPA, was employing high-tech solutions to what were often low-tech problems. Dragon's Jaw bridge was actually the exception; most of the problems the military faced in Vietnam did not call for technological novelty. Wikner angered military officers because he would tell them they did not understand the science, and then he angered scientists, because he told them they did not understand war. Military officers took to calling Wikner the FSA, short for Fucking Science Adviser, a riff on Foreign Service Officer, or FSO, the formal name for diplomats (another group often derided by the military). If FSA was meant as an insult, Wikner wore it as a badge of honor.

By the time Wikner arrived in Vietnam, the majority of American casualties there were from mines and booby traps, the types of weapons later referred to as improvised explosive devices, or IEDs, in the wars in Iraq and Afghanistan. Nearly impossible to spot in dense jungle foliage, an invisible wire stretched across a jungle trail would trigger the explosive. Wikner, recalling his Boy Scouts training, devised a novel fix to the trip-wire problem. He went out on patrol with marines, showing them how a six-foot stick could be used in the jungle to feel for a possible trip wire without actually setting off the explosive. "That's the greatest accomplishment of my life," the physicist later recalled.

When Wikner returned to the Pentagon in 1970, he was appointed to head the newly created Office of Net Technical Assessment, an in-house think tank for the Pentagon. Wikner's office was supposed to look at factors that might affect the future strategic balance between the United States and the Soviet Union, and come up with possible solutions. The Vietnam War was drawing to a close, and the Pentagon's attention was back in Europe, where military doctrine called for using tactical nuclear weapons in case of a Soviet invasion. If the bang was big enough, precision would not matter. The Soviets, however, had a different strategy. With the United States focused on Southeast Asia, the Soviet Union had increased its conventional forces, introduced new weapons and technology, and modernized its military doctrine. The Soviets believed they could win a war in Europe "with or without nuclear weapons," wrote the army general Donn Starry, who served in the 1970s as the commander of the Fifth Corps in Germany. "Their preferred solution: without." American troops deployed in Europe, on the other hand, felt like "no more than speed bumps for Soviet forces en route to the Rhine and beyond," Starry wrote. Wikner summed up the army's attitude for facing off against the Soviets in Europe: "Nuke 'em till they glow."

In 1973, James Schlesinger became the defense secretary, and he turned his attention to NATO, which had been relying on American nuclear weapons as a deterrent to a Soviet conventional attack in Europe. The Soviets outnumbered NATO conventional forces, but Schlesinger felt the nuclear deterrent had become an excuse for NATO not to modernize. In July of that year, he summoned Wikner and Lukasik's deputy, Cotter, who was leaving ARPA to become a special assistant on nuclear policy. The defense secretary had an assignment for them. "What we want is to have a viable conventional response to a Soviet invasion of Western Europe that doesn't rely totally on nuclear weapons," he said.

Wikner realized that ARPA and its high-tech gadgetry from Vietnam might now have a role to play in military strategy. AGILE had experimented with just about every imaginable newfangled battlefield technology, from armed drones and tethered balloons to laser-guided rockets and advanced radar. The newly created Tactical Technology Office inherited a wealth of technologies that had been percolating over the course of the Vietnam War and were directly relevant to precision warfare. "Steve, we have to do something about hitting things," Wikner told Lukasik.

In good Washington fashion, Wikner proposed to Lukasik that ARPA fund a study that would bring together nuclear theory and conventional weapons. Wikner, who knew Washington's national security bureaucracy, told Lukasik that ARPA would co-sponsor the work with the Defense Nuclear Agency so that the study could look at both conventional and nuclear forces, something ARPA could not do on its own. More critically, Wikner advised Lukasik to name the study something so bureaucratically obscure that no one would notice it.

"If we had called this Project Smart Kill, it would have been dead in the water," Lukasik joked. "I'm not sure whether Congress would have killed it, or the services would have killed it, or someone up in the Office of the Secretary of Defense would have killed it."

In 1973, Lukasik signed off on the Long Range Research and Development Planning Program, or LRRDPP, an unpronounceable acronym tailor made not to appear in a *Washington Post* headline or be spotted by an eagle-eyed congressional staffer. Left unsaid was that it was a study meant to transform the technology of counterinsurgency into weapons for the modern conventional battlefield.

In Washington, studies of conventional and nuclear warfare line bookshelves, fill cardboard storage boxes, and more often than most people would care to admit are eventually fed into shredders. Making a study influential in a city overwhelmed by wonk reports requires choosing the right person. That was why ARPA's director tapped Albert Wohlstetter, one of Rand's most influential nuclear theorists, to head the study. If Herman Kahn was the court jester of the nuclear world, Wohlstetter was its cardinal whose advice carried real weight with policy makers. Wohlstetter had been a mentor to Kahn, who often repackaged Wohlstetter's ideas for his own popular writing. If Kahn could spin up a lecture hall with briefings that tallied civilian deaths like a football score, Wohlstetter by comparison "wrote as colorlessly about nuclear catastrophe as is humanly possible," one historian noted.

Wohlstetter's writing might have been dry, but he was much more effective than Kahn at influencing politicians, a skill that eventually made him a confidant of Ronald Reagan's and later a favorite of the neoconservatives. An unusual figure in his own right, Wohlstetter had honed his anti-Stalinist thinking in the 1920s as a Trotskyite before transforming himself into a darling of the conservatives at Rand, where he joined a critical set of rational thinkers who helped formulate Cold War nuclear strategy. Notoriously pompous but politically adroit, Wohlstetter knew how to synthesize technical information into jargon-free language that would appeal to policy makers. Kahn popularized the notion of second strike, the ability to survive an initial nuclear attack and retaliate, but it was Wohlstetter who first described it in morbid detail in a 1958 article, "The Delicate Balance of Terror." The doctrine of mutually assured destruction, in which both sides could end the world, was not nearly enough to prevent Armageddon, he argued. "To deter an attack means being able to strike back in spite of it," Wohlstetter wrote.

The LRRDPP was approaching a different aspect of deterrence: the ability to respond to a Soviet conventional attack without relying on nuclear weapons. To do that, however, required developing plausible scenarios of a Soviet attack. Most scenarios, according to Lukasik, "are one paragraph of bullshit," but Wohlstetter and his colleagues bandied about a series of not-quite-apocalyptic possibilities, such as a Soviet invasion of Scandinavia through Norway and Finland or a surprise attack on Iran. Then they looked at what the American response to those attacks might be if the military had all the high-tech weaponry that ARPA had been developing, largely for Vietnam. This was dangerous ground for ARPA to tread; nuclear strategy was well outside the agency's purview. "We put together what Albert called contingencies—nothing wrong with that, this is just long-range planning contingencies, right?" Lukasik said.

The real focus of the study was the high-value Cold War real estate known as the Fulda Gap, which extended from the East German border to Frankfurt, West Germany, and was pinned as the likely invasion route for Soviet conventional forces. The lowlands there were perfect terrain for Soviet tanks to barrel across on their way west. The Soviet Union enjoyed overwhelming conventional superiority, and American policy at the time was to threaten to use tactical nuclear weapons, a sort of take-no-prisoners approach to the European battlefield. This dismal scenario had driven development of some of the seemingly more Strangelovian nuclear weapons, like the Davy Crockett, a tactical nuclear recoilless gun.

Lukasik's theory was that ARPA for years had been funding technologies, many of them intended for Vietnam, that could, in theory, transform this battlefield, undoing the Soviet advantage. ARPA had developed drones that could operate autonomously, computer systems that could calculate targets, and precision weaponry that could destroy those targets. And none of it was being used in any meaningful way by the military. It was well outside the bounds of a "technology agency" like ARPA to tell the services how they should be fighting in Europe, or anywhere else. Even in Vietnam, ARPA had faced tough opposition from the military services, particularly the army, which did not want eggheads telling commanders how they should fight their wars. That was where the LRRDPP study came in. Ostensibly, it was just to demonstrate how some of the new technologies could be used in hypothetical war scenarios. "Look, we have something to accomplish, i.e., hit targets, and we have this load of technology around—space technology, airborne technology, infrared technology, radar stuff, we have computers, we have all this sort of thing. What do we do? How do we tie it together to solve the problem?" Lukasik explained.

In fact, the study was a blueprint for an entirely new type of war. Stripped of its wonkish language and pared down to its essentials, the study's conclusions advocated using very precise conventional weapons in battle, in place of tactical nuclear weapons. Conventional weapons with "near zero miss may be technically feasible and militarily effective. If so, such nonnuclear weapons, under a wide range of circumstances, might satisfy the current United States and Allied damage requirements that now require the use of nuclear weapons," a study report stated.

The conclusion reflected Rand's long-standing opposition to the doctrine of mutually assured destruction. Conveniently, it also provided a case for ARPA's new generation of weaponry. Of course, the idea was still theoretical, because ARPA had not yet developed all these weapons, just the underlying technology. The specific ARPA proposal that followed was something like a cross between the *Star Wars* Death Star and *Terminator*'s Skynet. It was a weapon, or actually several weapons, that could collect and integrate data from various radar, crunch the numbers using a computer-driven targeting system, select targets, and then send a drone-packed mother ship over Soviet lines. Once there, it would release its kamikaze drones, called guided submunitions, which would hunt down and destroy Soviet targets.

In other words, the ARPA concept was not a single weapon but many weapons operating together, or what would later in Pentagon parlance be called a "system of systems." The eventual name for this program underscored its ultimate purpose: Assault Breaker, a weapon system for defeating the Soviets in the Fulda Gap, without having to rely on a nuclear assault. To the cynically minded, ARPA had just done a classic end run around the defense planning process. It had funded a study of defense intellectuals to tell the Pentagon that it needed a new generation of weapons that could be developed at ARPA, and specifically at its new division, the Tactical Technology Office, whose sensors, drones, and bombs had been developed for Southeast Asia.

Wohlstetter and his acolytes eventually got what they had wanted out of the ARPA study: a new way of fighting on the European battlefield. By 1982, the army had adopted this strategy, which relied on new technology, much of it out of ARPA. Some of the credit, according to Wikner, goes to Starry, the four-star general in the army who developed the new military doctrine. "ARPA gets 60 to 70 percent credit for the technologies that were used to implement it," Wikner added.

AGILE was buried, and from its grave arose the Tactical Technology Office, whose weapons enabled a fundamental shift in warfare. It was Vietnam, not outer space, that would prove formative for the modern agency. Over the next three decades, the projects in that office would give rise to precision weapons, drones, and stealth aircraft—the modern tools of warfare, or what some would later call a "revolution in military affairs." In a few short years, Lukasik had overhauled ARPA, creating the foundations for weapons that would in the coming decades transform the battlefield. "It all came from counterinsurgency," Lukasik said.

The backlash against the Vietnam War drove ARPA's weapons out of the jungle and onto the modern battlefield. Yet that same antiwar sentiment would have an altogether different effect on ARPA's seminal work in computers. The growing counterculture in the United States would transform J. C. R. Licklider's man-machine symbiosis into something far more ambitious and controversial: machines controlled directly by the human brain.

CHAPTER 13

The Bunny, the Witch, and the War Room

On the morning of January 28, 1969, a red and blue flag emblazoned with a yellow star was hoisted over the Stanford Post Office in Palo Alto, California. Students for a Democratic Society used the Vietcong flag to protest the university's ties to the Pentagon and its ongoing research in Southeast Asia. In particular, the students were concerned about classified military work at the Stanford Research Institute, an offshoot of the university created after World War II. The institute was supposed to take on research that fit with the university's overall mission, but that work had increasingly evolved into Defense Department contracting. ARPA, one of the Stanford Research Institute's major funders, particularly for computer science, was also paying its researchers to conduct studies in Southeast Asia.

Over the course of the year, the protests at Stanford grew, much as they did at campuses across the United States, with some erupting into violence, like the bombing of the University of Wisconsin's Sterling Hall. The attack, which was directed at the university's ties to the Pentagon, ended up killing a researcher with no connection to the military-funded work. Stanford never experienced that level of violence, but in April 1969 several hundred students took over the university's Applied Electronics Laboratory, bringing work there to a temporary halt. That month, Stanford's board of trustees voted to end the university's classified research and cut off its historic relationship with the Stanford Research Institute. Just months later, ARPA-funded computer science work there would make history.

At 10:30 p.m., on October 29, 1969, a one-word message arrived at a computer console at the Stanford Research Institute. "Lo," read the message. That was the entire content of the first transmission sent across the ARPANET. Charley Kline, a student programmer working for Professor Leonard Kleinrock at the University of California, Los Angeles, sent the message to Bill Duvall, a computer programmer at the Stanford Research Institute, and it was supposed to be "login," but the system crashed before it could be transmitted in its entirety, sending just the first two letters.

At that point, the ARPANET consisted of just four sites, or "nodes": UCLA, the Stanford Research Institute, the University of California, Santa Barbara, and the University of Utah. ARPA had funded those four sites to each receive an Interface Message Processor, which broke up data into small chunks, a method known as packet switching. Even with its first brief transmission, the ARPANET already contained most of the underpinning of the modern Internet. It was still basic computer science work, with no direct military mission, and the technical details often befuddled ARPA's directors. Without true believers, the ARPANET could have easily been killed, either by protesters worried that it was a Pentagon project to conduct nuclear warfare or by lawmakers convinced it was not doing enough for the Pentagon.

The ARPANET continued through those years largely unscathed, because ARPA officials believed in the vision of man-computer symbiosis that had been laid out by J. C. R. Licklider and worked hard to protect it. Stephen Lukasik, like Charles Herzfeld before him, understood the broader importance of the project and struck a delicate balance of justifying its work to Congress as relevant to the Pentagon while publicly downplaying any potential military role. "ARPA's computing program continued to lead its charmed life, rather like a person sleepwalking through a battlefield without getting a scratch," M. Mitchell Waldrop wrote in his history of the ARPANET.

Not all of ARPA's computer science work was so fortunate. At the University of Illinois, the ARPA-funded Illiac IV computer became the subject of outrage for students who believed it would be used to help the war in Vietnam. Illiac IV was focused on demonstrating a massively parallel processing computer, not performing calculations related to military operations in Southeast Asia, but its public image became enmeshed in the war. In May 1970, students organized a "Smash Illiac" protest day on the campus quad, with speakers from the Black Panther Party and the Chicago 15. A poster advertising the protest featured a cartoon image of the Illiac IV with a screen showing "kill die factor." Concerned about the ability to protect the expensive supercomputer, ARPA ended up moving it to a NASA facility in California.

Meanwhile, Congress was chipping away at ARPA. The Mansfield Amendment of 1969 had ended much of the Defense Department's basic research, cutting off ARPA's social science work and forcing the Behavioral Sciences Office to refocus its work (to avoid controversy, Lukasik later renamed it the Human Resources Research Office, causing many people to mistake it for a personnel office). The new congressional legislation also affected the hard sciences, forcing ARPA to hand off its interdisciplinary materials laboratories, which had provided long-term funding for universities and had been a fundamental part of ARPA's basic research portfolio, to the National Science Foundation. Everything ARPA did now had to have a military justification.

Congress was on the lookout for any programs that did not seem appropriate for ARPA to pursue. "What exotic studies are being made under the behavioral sciences?" a lawmaker asked Lukasik in one hearing. "Are we still studying the mating of bugs and the behavior of monkeys?"

"Nothing like that is going on in the program," Lukasik answered gravely.

Lukasik found himself literally patrolling DARPA's halls for potential problems. One day, he was walking by the office of Al Blue, who headed the Information Processing Techniques Office, and dropped in to say hello. On Blue's desk was a computer science report by MIT. Lukasik felt his throat seize up when he spotted the title, "Computer Assisted Choreography." He could almost imagine the next congressional hearing, where lawmakers would grill him about why ARPA was researching dance. "Al, I understand what this is about. This is a good idea, but, Jesus Christ, don't give me any more reports called Computer Assisted Choreography," Lukasik said. "Change the name [to] Man-Machine Coordination; it's just as good."

Whether it was congressional attacks or student protests, ARPA could not escape the Vietnam War and the rising counterculture enveloping the country. Yet it also was about to be influenced by those events in unexpected ways. Whether it was Timothy Leary's lectures on the benefits of psychedelic drugs or newfound interest in Eastern mysticism, even the Pentagon was not impervious to the cultural anarchy of the late 1960s. Unconstrained by conventional wisdom but bound by a belief in rigorous science, ARPA was about to create a new field of research, transforming J. C. R. Licklider's notion of man-computer symbiosis into technology that would allow people to control computers with nothing more than their thoughts. Its origins would weave together the Stanford Research Institute, the ARPANET, and a growing fascination with the powers of the human mind.

The Stanford Research Institute in the early 1970s harbored a dark secret that would have shocked even the student protesters outraged by its military research. Among its many classified research projects was a contract supported by the CIA's Office of Technical Service, a division headed by Sidney Gottlieb, perhaps the most notorious scientist ever to work for the spy agency. The secret program was testing different forms of parapsychology, such as whether humans had the ability to use their minds to visualize or even influence remote objects. Believing the work was showing promise, Gottlieb one day invited ARPA's director, Stephen Lukasik, over to his CIA office to discuss it.

The CIA's Office of Technical Service was housed in a low-slung office building on the grounds of the U.S. Navy Bureau of Medicine and Surgery, across from the State Department's headquarters in Foggy Bottom. The buildings were the old headquarters of the CIA, before it moved to Langley, Virginia, but in the early 1970s the facility continued to house a few CIA activities. Gottlieb, a chemist by training, was both an unconventional thinker and an unwavering patriot who believed his work served the good of the nation. Born with a clubfoot that kept him out of military service, and afflicted with a stutter that he channeled into the study of speech pathology, Gottlieb was known for his iron determination. "Friends and enemies alike say

Mr. Gottlieb was a kind of genius, striving to explore the frontiers of the human mind for his country," read a *New York Times* obituary of Gottlieb, "while searching for religious and spiritual meaning in his life." In the end, however, Gottlieb would be remembered most for what looked like a willful contempt of common decency.

As the head of the Office of Technical Service, Gottlieb led a wing of the CIA whose failed innovations to assassinate the Cuban leader, Fidel Castro, included poison pens and exploding seashells. He also worked on one of the agency's most notorious projects, the use of LSD as a mind control drug. Under Gottlieb's supervision, LSD beginning in the 1960s was tested on unwitting human guinea pigs including, among other unfortunate victims, the mentally ill, prostitutes, and even one unsuspecting army scientist who committed suicide. When the program was first exposed in 1975 by the Rockefeller Commission, and then detailed by the congressional Church Committee, Gottlieb's public legacy as some sort of mad scientist was all but assured.

Lukasik had a more charitable view of Gottlieb and his CIA colleagues. He viewed them as having ARPA's penchant for creativity and freethinking but without the checks that public oversight provides. "They are, and I'll say this in the most positive way, truly wonderful people," he recounted. "They didn't so much care about the laws, but in terms of creative people who are just told they don't have to worry about anything, and just create. They were good people."

The day Lukasik went to visit Gottlieb, the CIA scientist was in fine form. He had taken over the Foggy Bottom office that once belonged to the CIA director Allen Dulles. It was an unusually long office, perhaps fifteen feet wide and twice as long. There was a curtain across the length of one of the long walls, and Gottlieb told Lukasik, "Steve, let me show you what outsiders don't see." Gottlieb rather dramatically yanked back the curtain revealing a wall-size map of the world flecked with dots. Gottlieb said, "These are the 146 locations of our acoustic taps."

Lukasik knew that Gottlieb was just showboating, but he indulged him, waiting for the main point. What Gottlieb really wanted to discuss was bunny rabbits and nuclear Armageddon. In the early 1970s, the Soviet Union and the United States were locked in a cat-and-mouse game with nuclear submarines. Submarines equipped with nuclear missiles were difficult to spot when prowling the deep seas, making them a potent weapon in the nuclear balance of power with the Soviet Union. The submarine's chief vulnerability was its need to communicate. In the early 1970s, there was no good way to communicate with submarines deep underwater to let them know, for example, that they needed to launch their missiles because nuclear Armageddon was under way. The solution usually involved coming to the surface, which would make them vulnerable to detection and attack.

That was where Gottlieb's new pet project came into play. In 1970, the best-selling book *Psychic Discoveries Behind the Iron Curtain* described the Soviet Union's and other Eastern bloc countries' enthusiasm for psychic phenomena of all sorts. "Major impetus behind the Soviet drive to harness ESP was said to come from the Soviet Military and the Soviet Secret Police," the book's co-authors asserted. The book detailed dozens of investigations into psychic phenomena conducted behind the Iron Curtain, ranging from Kirlian photography, which sought to capture the "aura" of living things, to telepathic projection of emotions. The idea that the Soviets were investing money in parapsychology quickly became a self-reinforcing justification for the Americans to do the same.

According to *Psychic Discoveries,* one theory of parapsychology the Soviets were testing involved a projected emotional link between a newborn and its mother, which allowed the mother to "sense" her offspring's death even over long distances. Because actually killing a newborn human child was not really an option, they resorted to experimenting with baby rabbits and their mothers. The experiment was as ghastly as it sounded: a baby rabbit would be killed out of sight and sound of its mother, while scientists in a separate lab room observed the mother for a reaction.

The Soviets claimed it worked and it could be used for communicating with submarines, even if they never quite laid out the protocol for how this would be done. Presumably, a mother rabbit would be kept aboard the submarine, with a submariner assigned to monitor it for signs of distress. The idea was not that an overly excited mother rabbit would prompt a nuclear exchange, but such a signal could be used, as Lukasik put it, as a "bell ringer for Soviet boomers." It would be a message for them to come above water and get a more detailed message, such as an order to launch their nuclear missiles. The very absurdity of the scenario did not dissuade Gottlieb. The CIA had begun funding experiments at the Stanford Research Institute to conduct a "quiet, low-profile classified investigation" into parapsychology. Gottlieb was interested in having ARPA look at the work and possibly support it with its own funds. "I thought this was a lot of bullshit," Lukasik admitted.

Though they sounded dubious, the purported Soviet experiments claimed applications in antisubmarine warfare, an area that ARPA was pursuing at the time. Perhaps more important, the late 1960s and early 1970s had sparked widespread interest in parapsychology, even among some members of Congress, who were pressuring agencies, like ARPA, to support it. Lukasik figured that at the least the agency could make a good faith effort to see if there was anything worth funding. He turned to Austin Kibler, an air force colonel in charge of the Behavioral Sciences Office. "Follow this stuff," Lukasik told him. "Show me anything that works. I don't care what."

The reaction back at ARPA was not enthusiastic. "Everyone pretty much felt that it was just a big pile of crap," said Robert Young, who later took over as head of the Behavioral Sciences Office, "but we had been given the responsibility from those on high to deal with it. And we did." The scientist selected to lead the parapsychology investigation was ARPA's resident expert in counterculture, George Lawrence. The thirty-nine-year-old scientist cut a distinctive figure at ARPA. He favored bell-bottoms and wide-collared shirts over suits and pencil holders and brought his son to skateboard in the hallways of the Pentagon. Pictures from his work trips usually included him lying half-naked by a swimming pool, or holding a pitcher of beer, rather than examining pieces of missile guidance systems.

ARPA in popular culture may be portrayed as the lair of proverbial mad scientists, but the truth was that, socially speaking, even in the 1960s and 1970s the agency was almost as straitlaced as any other part of the Pentagon. It was home to intellectual freethinkers, but they were largely drawn from the hard science faculties of universities, the defense industry, and the military, not exactly the hotbeds of 1960s counterculture that were dropping acid and embracing mysticism.

Lawrence was an exception. He had adopted at least the trappings of a bohemian. While the era of free love was alien to most of those who worked in the Defense Department, the newly divorced Lawrence embraced it, showcasing a revolving door of girlfriends. Lawrence, at least by ARPA standards, was "out there," not just in his clothing and his freewheeling lifestyle, but also in his choice of research, which drew heavily on popular cultural notions of "mind over body."

He was not your typical antiestablishment hippie, however. In the late 1960s, when many men were looking for a way to get out of going to Vietnam, Lawrence was looking for a way to get in. He was offered a job at the Walter Reed Army Institute of Research, which, like ARPA, was interested in counterinsurgency and was hiring psychologists to go to Vietnam to study the effects of stress on Special Forces advisers. Lawrence, then a PhD psychologist at Albert Einstein College of Medicine, was offered a government position for about $12,000 a year. The money was not great, but it sounded like an adventure. Just as he was packing his bags, the counterinsurgency campaign in Southeast Asia turned into a conventional war, and Walter Reed closed down its research program in Vietnam. A colleague of Lawrence's told him about a possible job at ARPA.

In the fall of 1968, Lawrence joined ARPA as deputy director for the Behavioral Sciences Office. For Lawrence, the ARPA job was a dream come true: he could propose almost any sort of research program, write up a brief proposal with some nominal military justification, and get it funded. He also had a blanket travel authorization that allowed him to plan trips "to such places at such times in such frequency as may be necessary in the performance of your official duties either within or outside the continental limits of the United States and return to Washington, D.C."

Lawrence had a penchant for research that captured the cultural zeitgeist of the late 1960s, when explorations of mind-body interactions and consciousness research combined science and spiritualism. He was also, like J. C. R. Licklider, part of a small but growing number of psychologists fascinated by computers, particularly human interaction with computers. Lawrence was initially interested in research helping people deal with stress and pain. His first major program, started in 1970, was in biofeedback, a relatively new area of investigation that involved training people to control physiological functions, such as breathing and heart rate, by providing subjects with real-time information from sensors.

In the 1960s, biofeedback still had the stigma of New Age mysticism, because the idea was that a person could essentially will his or her way to a different physical state. It melded biology and Eastern philosophy and evoked comparisons to Timothy Leary's promotion of LSD. It was also starting to attract interest from scientists, who thought biofeedback might allow people in a stressful situation to slow their heart rates or lower their blood pressure purely through mental concentration. Researchers like Joe Kamiya at the University of California, San Francisco, studied the brain's alpha and theta waves to see if subjects could alter their state of consciousness with the help of real-time electronic monitoring. "The image of electronic equipment guiding human beings to a greater awareness and control over their own physiology and consciousness appealed to both white-coated experimental scientists and the white-robed gurus of the higher-consciousness movement," the psychologist Donald Moss wrote in a retrospective of the field.

The justification for ARPA's interest in this field was to help troops in combat; biofeedback could in theory allow soldiers to shoot more accurately, or even to slow their bleeding after being shot by allowing them to control their own heart rate. Researchers hypothesized that pilots of damaged aircraft could be taught to lower their own heart rate and blood pressure, allowing them to carry out emergency procedures without panicking. There was little in the way of documented experiments, however, and Lawrence considered biofeedback an area ripe for examination. His ARPA program was the first systematic exploration of the field, bringing the scientific method to an area that had previously been dominated by anecdotes. Among the people he recruited to work on the project was a future director of ARPA, Craig Fields, then a young Harvard professor and a colleague of Licklider's. He submitted a proposal on "autonomic conditioning under computer control," which had people monitor their heart rate with EKG feedback. Soon, Harvard students were running up and down stairs with cardiovascular electrodes attached to their chests. Lawrence brought hippie counterculture to scientists, and scientific rigor to hippies.

When Lawrence tried to take the work from the lab to the battlefield, he ran into resistance. Lawrence wanted the researchers to travel to Vietnam to test biofeedback on Special Forces in the field, but the response was overwhelmingly negative. No one wanted to go. "Someone wrote to a congressman and said I was trying to coerce university professors into going into the jungles in Vietnam," he said. "It was asinine. It was completely misperceived. I thought they would look at it as an interesting adventure the way that I did."

It probably did not matter in the end, because Lawrence concluded that the more ambitious applications for biofeedback, such as soldiers' ability to slow down their heart rate enough to prevent them from bleeding out, were probably too ambitious. "Biofeedback offers much less powerful and robust self-control over certain internal physiological events than many researchers anticipated on the strength of early anecdotal evidence," a report concluded. "Clearly, from the work which has taken place in many laboratories, it is extremely difficult, if not impossible, to train subjects to regulate nervous system events to a level contrary to the best interests of their own physiology." On the flip side, Lawrence wrote, at least no one was going to die by consciously willing their heart rate to stop, a concern expressed at one point by some researchers.

While biofeedback was not necessarily successful, it cemented Lawrence's reputation at ARPA as the go-to guy for counterculture ideas, particularly mind-over-matter research. So it was not a total surprise when Lawrence was assigned to look at the CIA's parapsychology research to see if it was something that ARPA should fund.

By today's standards, the idea of ARPA, a technical agency, investigating spoon bending and ESP might sound outlandish, but even some of the conservative elements of the Defense Department and intelligence community were being swept up in popular enthusiasm for psychic investigations. Best-selling books like *The Secret Life of Plants* combined botany with New Age ideas to argue that plants were sentient beings, while *The Tao of Physics* merged quantum theory with mysticism. A cover of *Time* magazine was dedicated to the nation's "booming interest" in psychics. Interest in ESP and psychic phenomena went lockstep with popularization of counterculture in mainstream America.

When ARPA launched its investigation into parapsychology, it was hard to tell how seriously anyone, even Lawrence, really took the investigation. At least at face value, Lawrence embraced the assignment. He played around with Kirlian photography to see if it could really capture "auras," attended a parapsychology conference in Scotland, and traveled around the country meeting witches, psychics, and other purveyors of the paranormal. He liked the witches most of all. But Lawrence's most famous psychic investigation—and one that ended up attracting national attention when it was picked up by the press—involved a trip he made in December 1972 to the Stanford Research Institute, where the physicists Russell Targ and Hal Puthoff were being funded by Gottlieb's CIA office to investigate psychic phenomena. Lawrence's boss, Austin Kibler, had made an initial trip out and was apparently impressed at least with the seriousness of the institute, which received ARPA funding for a variety of projects.

At the time Lawrence visited, the Stanford Research Institute work was focused largely on testing the skills of Uri Geller, a charismatic Israeli entertainer turned paranormalist. Geller's most well-known spectacle was bending spoons, purportedly with his mind. He also claimed a host of other psychic abilities, such as thought projection and "remote viewing," the term given to the ability to describe objects in far-off, or at least unseen, locations. The last ability was of particular interest to the national security community, because it would in theory enable spying on foreign bases and technology.

According to Puthoff, one of the CIA officials in charge of the Stanford Research Institute contract suggested that Lawrence visit in the hopes that ARPA might agree to fund the institute's parapsychology work. Puthoff and Targ, who were eager to get mainstream recognition, agreed to host Lawrence for an informal demonstration but told him he could not observe their controlled experiments. "At the time we were concerned that it might be a setup in which Lawrence and Geller could have set up a collaboration, i.e., could have colluded in advance," Puthoff said. So deep was this paranoia that after each day of testing he and Targ would check the ceiling tiles in the lab to look for possible listening bugs or hidden cameras.

Lawrence invited two other scientists to accompany him: Ray Hyman, an amateur magician and university psychologist, and Robert Van de Castle, a professor of sleep studies who believed in psychic premonitions, including his own. Van de Castle, whom Lawrence knew from graduate school, studied the ability of people to predict the future and receive thoughts while dreaming. "He and Hyman and I made this trip to the Stanford Research Institute where Geller was going to convince me that his stuff was valid and I was going to pump a lot of money into it," Lawrence said.

The trip got off to an inauspicious start: Van de Castle and Lawrence met up the evening before in San Francisco and went out for Chinese food. Lawrence was drinking heavily and playing the part of the raconteur, telling Van de Castle that what he really wanted was to find a female psychic he could have sex with and test her powers. Van de Castle, a proponent of parapsychology, was upset by Lawrence's cavalier attitude. The next morning did not start off much better: Hyman and Van de Castle showed up at the institute to meet with Puthoff, Targ, and Geller. Lawrence swaggered in late, looking a mess, according to Van de Castle. Lawrence fell back into a chair, put his feet up on the conference table, looked around the room, and pronounced, "Okay, show me a fucking miracle."

And so the day started with one hungover military scientist, one amateur magician turned psychologist, a professor who studied psychic dreams, two seemingly credulous physicists, and Uri Geller, the would-be psychic superweapon. It went downhill from there. Geller began his repertoire demonstrating his ability to mind read numbers. The Israeli performer dramatically covered his eyes with his hand and had Lawrence write down a number on a piece of paper. Hyman, sitting to the side, later recalled that he could clearly see that Geller was peeking, watching the motion of Lawrence's hand write out the number ten. Geller also wrote down ten.

In another demonstration, described later in a letter by Hyman to the head of the Stanford Research Institute, Geller wanted to show his psychic ability to receive someone's thoughts, so he took Van de Castle aside into a separate room. Geller asked Van de Castle to choose a cartoon from a magazine and draw it by hand, because magazine pictures were harder to "receive." Both pictures—the original and the hand-drawn duplicate—were placed in separate envelopes. Van de Castle placed the envelope with the original picture in his breast pocket and the one with the hand-drawn image under his elbow. Geller then instructed the professor to close his eyes, and stood directly behind him—close enough to touch him—and prepared to receive Van de Castle's thoughts. Geller soon emerged triumphant: he had drawn a stick figure facsimile of the image, a feat no one observed because only Van de Castle was in the room and he had his eyes closed the entire time.

Hyman was perplexed: What were the conditions of the experiment? Why did no one observe Geller drawing the image to ensure Van de Castle had not coached him? The answers were evasive at best. And so it went with the rest of the demonstrations. Either Geller, on close examination by outsiders, could not or would not perform, or when he did seem to get results, there was little credible examination. "Targ and Puthoff, from the way I have encountered them by day in their laboratory, seem to emerge as bumbling idiots rather than as respected, accomplished physicists," Hyman wrote.

None of the demonstrations involved any scientific controls. Puthoff countered that the experiments were not supposed to be controlled—it was only a demonstration. Even so, Hyman asked why the scientific duo did not search Geller prior to an experiment where it was claimed he could erase a single frame from a film using only his mind? Perhaps Geller had used a device to alter the film. The answer, Puthoff told Hyman, was that such a device sounded improbable. Hyman was astounded that Targ and Puthoff found it more believable that Geller had psychic powers that allowed him to erase film than that he possessed a device capable of doing the same thing. Hyman believed Geller's work had all the classic hallmarks of a trained magician: befriend, distract, and dazzle.

If Hyman was doubtful of Geller's psychic capabilities, Lawrence was outraged. In one demonstration, Geller moved a compass needle by five degrees. Lawrence, after stomping his foot to imitate what he believed Geller had done, moved the needle forty-five degrees. And so it went for most of the visit. By the time Lawrence and his crew left, it was clear that Puthoff and Targ were not going to get ARPA funding for psychics.

On one of the West Coast trips to investigate the paranormal, Lawrence was invited to a party in the hills above Westwood, on Sunset Boulevard. The woman hosting the party had been a private donor to psychic work. At one point Lawrence ended up seated next to Ed Mitchell, an astronaut who had come to believe in his own telepathic powers. A wealthy woman at the party described a recent trip to India, where a mystic had produced a ring, seemingly out of thin air, and gave it to her as a parting gift. She turned to the astronaut and asked, "Colonel Mitchell, what do you think? I never really understood it. Did he materialize it out of the elements, or do you think he teleported it from somewhere else?"

What a lunatic question, Lawrence thought. But Mitchell replied, "I think it was probably teleported," and then provided an explanation for how that might happen. The woman turned to Lawrence.

"Dr. Lawrence, what do you think? Do you think it was teleported here?"

Lawrence found himself starting to answer the question with what might be a scientific explanation. "Then I thought, I've been sucked into this just sitting here listening to it, and I'm taking it seriously. What the hell is the matter with me?"

If Lawrence at some point had taken parapsychology seriously, that time was long gone. "The whole thing," he concluded, "was just garbage."

Though Uri Geller's demonstration in California had failed to impress Lawrence, the idea of reading people's minds captured his imagination. That same year he visited the Stanford Research Institute, Lawrence launched a different sort of mind-reading project: instead of relying on the paranormal, researchers would use measurable brain signals to control a computer. From his exploration of parapsychology, Lawrence found science.

The brain-driven computer dreamed up by Lawrence drew heavily on, and referenced, Licklider's idea of "man-computer symbiosis." If Licklider's vision of a man-computer symbiosis was futuristic, Lawrence's program, which he named biocybernetics, was outright audacious. In biocybernetics, the machine would not just be a part of man's decision-making process through inputs provided by a keyboard or joystick, as envisioned by Licklider; it would interact directly with the human mind, using sensors that monitor brain activity. Lawrence funded researchers looking at neural signals like the P300, a brain wave that can be detected by electroencephalography and that occurs approximately three hundred milliseconds after the brain recognizes an object. In practical terms, such work might free a quadriplegic to think messages or words or to control a machine by wearing an EEG cap, or what ARPA officials called "the intelligent yarmulke."

ARPA under the auspices of biocybernetics funded a raft of researchers tapping brain signals, such as Jacques Vidal, a UCLA researcher who coined the term "brain-computer interface" for the work. "Can these observable electrical brain signals be put to work as carriers of information in man-computer communication or for the purpose of controlling such external apparatus as prosthetic devices or spaceships?" Vidal wrote in a seminal paper in 1973. Within a few years, Vidal's research yielded promising results: in one experiment, test subjects were able to move an electronic object through a maze on a computer screen just by thinking about it.

Those were fantastic times, according to Emanuel Donchin, a professor at the University of Illinois who was funded by Lawrence. It was not that ARPA was the only agency funding such work, but it was at the time the most important for developing the field. Donchin, who was looking at cortical slow waves, described how ARPA was able to turn cartwheels around other government funding agencies. Donchin recalled sitting in on a National Institutes of Health study group to review a $5,000 grant proposal for Eric Kandel, a neuroscientist who would later go on to win the Nobel Prize. The scientists were debating the funding for hours. At one point, Donchin excused himself to make a brief phone call to Lawrence about his ARPA grant. Donchin needed $15,000 for some equipment, and Lawrence's response was "You got it." When Donchin returned to the room, the scientists were still debating the $5,000 for Kandel. That, according to Donchin, best illustrated the difference between ARPA and other science agencies.

On the other hand, ARPA programs, like biocybernetics, were often outrageously optimistic on military applications. "Soon, for example, a computer monitoring electrical brain activity of an aircraft pilot (or any other individual engaged in a task where continuous vigilance is required) should be able to determine whether a warning signal not only had been seen but that the pilot understood its significance and intended to respond appropriately," one early program description read. Lawrence knew full well that such applications were years away. "I made it up," he recalled of some of the more fantastical applications.

Like Licklider, Lawrence was interested in fundamentally transforming how people interacted with machines, which would eventually have far-reaching applications. But the challenge of biocybernetics was weighing the fantastical applications it offered—brain-driven computers and mind-controlled aircraft—with the reality that such work was decades away. For example, according to a 1975 summary, ARPA hoped to achieve a capability to translate an eight-word vocabulary based on EEG signals. "The discipline of biocybernetics is essentially being created by ARPA," the program summary at the time stated.

Lawrence's brain-driven computers and soldiers tapping into the body's autonomic functions were on the edge, but so was killing bunny rabbits to communicate with submarines or funding an Israeli magician to remotely view Soviet bases. ARPA was a place in the early 1970s that tolerated and even encouraged exploring such outlandish ideas, but unlike some other agencies it required good science.

In one final meeting to discuss ARPA's possible funding of parapsychology, Lawrence sat with Lukasik, the ARPA director, and CIA officials who had been funding the work. At the end, one of the CIA officials turned to Lawrence and said, "They certainly haven't been wasting our money. Dr. Lawrence, what do you think about all this?"

At that point, Lawrence's investigation of psychic phenomena had introduced him to a colorful array of mystics and frauds. "You have been wasting your money," he exploded in frustration. "Every damn dime of this is nonsense."

There was dead silence. Lukasik quickly changed the subject, and as Lawrence recalled, no one ever asked him to look at parapsychology again. Nor did ARPA ever fund a psychic program. "I worked so long, and so hard, and dealt with so many fools and charlatans," Lawrence later recalled. "There is no question in my mind that all of it is bunk."

Geller's advocates, who believed the magician could help the United States spot Soviet submarines, looked on Lawrence's role with great disappointment, but Lawrence helped save ARPA from the embarrassment that befell the intelligence community when it was revealed the nation's spies had spent tens of millions of dollars on psychics. And for those who questioned whether ARPA's open-ended investigation of parapsychology was a good idea at all, the reality is that the same attitude that allowed Lawrence to meet witches and psychics also enabled him to pursue the brain-driven computer, which in the 1970s sounded like pure fantasy. The intelligence community's support of psychics continued through 1995, producing claims of successful results but little in the way of evidence that was ever accepted by the scientific community. Biocybernetics, on the other hand, blossomed.

It was an audacious idea in the early 1970s, when the ability to read brain signals was crude at best. By 2013, however, biocybernetics had spawned an entire industry of brain-computer interface devices used in everything from commercial video games and car sensors to tools that allow "locked in" patients, those with no way to communicate with the external world, to type messages and control external devices. Applications that were once decades away are now being built, and Lawrence's "made up" vision is becoming a reality. As for parapsychology, Lawrence joked years later that maybe he should not have been so forthright with his criticism, instead playing it out even longer. "At the very least," he said, "I could have met some more witches."

Those days were coming to a close anyway. Over in the Pentagon, Malcolm Currie, ARPA's overseer, was getting increasingly annoyed. He could not understand why a Pentagon agency would be involved in things that did not have an immediate military application. ARPA's independence from the Pentagon meant that it did not typically need permission from higher-ups to fund individual projects, but now that autonomy appeared almost foolhardy, at least to Currie. News of the spoon-bending investigation was one of the final straws. Currie decided that ARPA was "wandering off the territory." It was time, Currie decided, for some changes at the agency.

ARPA, in the view of a new crop of defense officials, was meant to be a corporate laboratory of the Pentagon—a place that built weapons—not a think tank or a scientist's playground, let alone a place that tried to address strategic-level problems, as past directors had done. The new ARPA—now to be called DARPA—would build technology that could find enemies and kill them.

PART II

SERVANTS OF WAR

CHAPTER 14

Invisible War

"Oh my God, he's dead," Alan Brown thought as he watched the chief test pilot for Lockheed's Skunk Works division float down to earth. At least the pilot had ejected, but his head was slumped awkwardly to one side, and Brown thought the force of the ejection killed him.

It had not, but the pilot was in bad shape. He had broken his collarbone and was knocked unconscious, which meant he was in danger of suffocating when he landed in a sandy area of the test range. It was fortunate that a chase helicopter with medics was already in the air, because when they got to the pilot, he was already turning blue, his mouth and nose filled with sand. The helicopter rushed him to Southern Nevada Memorial Hospital. Then things got complicated.

Neither the aircraft nor the test site where it was being flown existed, at least officially. It was May 4, 1978, and the pilot, Bill Park, was flying a top secret aircraft known by its code name, Have Blue. He had ejected after the experimental aircraft's undercarriage was damaged, making a safe landing impossible. The flight tests for Have Blue took place at Area 51, sometimes called Groom Lake, a classified site in the Nevada desert that has long resided uncomfortably between lore and secrecy. The secret test area was established in the mid-1950s as a place to test the U-2 aircraft for the CIA away from prying eyes. Kelly Johnson, the head of the secretive Lockheed Skunk Works division that built the U-2, looked at a number of areas across the United States and settled on a barren area of Nevada pocked with dry lake beds. They named it Paradise Ranch; Johnson later called the name a "dirty trick" to help attract people to what was an otherwise inhospitable area. For decades its existence was not acknowledged by the government, even as conspiracy theorists, press, and tourists made pilgrimages to the edge of the restricted zone, well aware that secret work was going on inside.

That contradiction did not necessarily matter; the Pentagon could live with the public knowing that it was a top secret testing area, as long as no one knew what precisely was being tested. The colorful mix of fact and fiction that linked the area to everything from antigravity research to aliens ended up helping the military shield real technology. From the Pentagon's perspective, it was better for people to believe it housed little green men than secret spy aircraft. Military officials often contributed to the false aviation rumors, feeding cover stories to the press. Even if the military did not acknowledge it, most people knew that part of southern Nevada, just a couple hours' drive from Las Vegas, was used for testing secret aircraft. Yet the day of Park's crash, official secrecy almost turned tragic.

There was a cover story for an accident. After all, crashes of experimental aircraft were hardly unexpected. But in the chaos that ensued after the crash, and in the rush to get Park to the hospital, the cover story went out the window and was replaced with an impromptu and rather unbelievable fabrication. "He was on a scaffold and he fell off," blurted out one of the air force security officials who accompanied Park.

When the hospital staff saw the outline of flight goggles around Park's face, they grew suspicious. Park's helmet had been ripped from his face when he was ejected, and his skin was bright red from windburn, except where his goggles had been. Again, the explanation was ludicrous. "Oh, that was his gas mask," the official said.

The scaffolding story failed miserably. Hospital workers started asking more questions, and when Park regained consciousness, he only added to the mystery. The injured pilot refused to say what happened, providing staff with just his name and his employer's name, Lockheed. He gave his address as "General Delivery, Las Vegas." Someone eventually alerted the media of a suspected aircraft crash. Military officials at Nellis Air Force Base denied any knowledge of a crash.

The Pentagon a week later finally confirmed that a plane had crashed and that a pilot suffered minor injuries. (In fact, the crash ended Park's career as a test pilot; he would reveal years later in an interview that his heart had stopped beating by the time paramedics reached him.) The press quickly followed up on news of a secret aircraft crash, even if some of the details were wrong. One press account reported that the pilot was testing a classified aircraft called the "TR-1," a high-altitude aircraft meant to "perform missions along a country's border without entering its airspace." That report was wrong, either an attempt at deliberate misinformation or confusion with another project that was in fact named the TR-1, a version of the U-2 spy plane.

The specialized aviation press quickly forwarded a different theory. *Flight International* reported that the crashed plane was a "stealth" aircraft project led by Kelly Johnson, Lockheed's famed aircraft designer, who had been developing secret planes for the CIA for years. This aircraft, in fact, was sponsored by the agency now known as DARPA, whose future hinged on the program's success. At $100 million, it was one of the largest aircraft programs the agency had ever sponsored, and it was a huge risk for an agency that some critics were lobbying to close down. In the late 1970s, around the time of the secret aircraft's development, a letter was circulating calling for the abolishment of the agency, according to James Tegnelia, the former acting director of DARPA in the 1980s. The military chiefs wanted "to disestablish DARPA, because it wasn't paying off for the services," he recalled.

Designed to give the United States a strategic edge in combating Soviet conventional superiority, the stealth aircraft grew out of research that emerged in the waning days of the Vietnam War and matured in DARPA's newly formed Tactical Technology Office. Born of Stephen Lukasik's vision to carve out a strategic role for the agency, the secret aircraft had the potential to save DARPA. If it failed, however, it might put the agency out of business permanently.

The stealth aircraft's journey began in 1974 during a chance meeting in Lukasik's office with a man obsessed by an invisible rabbit. Charles "Chuck" Myers, the Defense Department's director for air warfare, had been making the rounds in the Pentagon, pitching anyone who would listen to his idea for a new type of aircraft. Myers was part of a self-described subversive group of military experts known as the fighter mafia, who battled the prevailing air force preference for technologically complex fighters. The fighter mafia had successfully lobbied for the development of a maneuverable, lightweight fighter that eventually became the F-16 Fighting Falcon.

Now Myers, who had flown combat missions in both World War II and the Korean War, was on a one-man mission to sell his concept for a small, radar-evading aircraft inspired by an invisible rabbit named Harvey. In the 1950 James Stewart movie of the same name, Harvey was a six-foot-tall "pooka," a mythical creature only the protagonist could see. Myers's pooka was much like Stewart's—an invisible plane, or a "stealth aircraft." By 1974, Myers had pitched Harvey to anyone and everyone who would listen. Like James Stewart's Harvey, no one believed in an invisible aircraft except Myers. His wife had even made him a model of "Harvey," which consisted of a stuffed Easter Bunny, to which she added a top hat, a plastic cocktail olive sword, and a band around its hat that read, "Walk Stealthily and Carry a Sharp Stick."

Myers was concerned about the escalating threat from ground radar and surface-to-air missiles. Soviet-made surface-to-air missiles had proved deadly to American pilots in Vietnam; even without a direct confrontation with the Soviet Union, the Vietnam War had demonstrated that Soviet technology was advancing at a fast pace. The United States had put its money into an expensive war, and the Soviet Union in the meantime had put its money into technology, like surface-to-air missiles. The lessons of Vietnam were reinforced in 1973, when Israeli pilots during the Yom Kippur War faced a barrage of these deadly Soviet-supplied missiles. Israel reportedly lost about a third of its combat aircraft. For a growing circle of military experts, the lesson was that even advanced aircraft were becoming increasingly vulnerable to air defense systems. "Signature is the most important feature of a combat system," Myers wrote, regardless of whether it is an aircraft or an infantry soldier. "When fighting in the jungle if your canteen sloshes and you smell like Burma Shave, survival is unlikely."

No one in the Pentagon was interested in Myers's invisible aircraft until a chance meeting with DARPA officials changed everything.

Myers was invited to attend a meeting to review the programs of the Tactical Technology Office, the division formed from the remnants of ARPA's Vietnam work. Robert Moore, then the deputy head of the office, presented the DARPA projects, which did little to excite Myers. After the meeting, Myers cornered Moore and asked him if DARPA might have funds to pay for a study on the "invisible aircraft." He handed Moore a copy of a white paper he had written, titled simply "Harvey."

Moore was intrigued but unconvinced. He was getting regular briefings from intelligence officials, who were warning that the Soviets had developed incredibly sophisticated air defense systems. The primary concern was the Fulda Gap scenario, where American pilots were trained to evade Soviet radar with what Moore described as "drastic operational techniques and countermeasures," a somewhat understated way of describing how pilots would have to fly just one hundred to two hundred feet off the ground—the equivalent of buzzing Yankee Stadium—and then "pop up and roll over." Not only was this a complex technique; it also required half a dozen support aircraft to relay target information and jam the enemy radar.

Though Moore was not crazy about the Harvey metaphor, he understood the potential of a "stealthy" aircraft. DARPA had been quietly working on its own stealth experiments with remotely piloted vehicles, or drones. DARPA's Vietnam War–era drone work had not led to any operational weapons, but it had sparked interest in developing lower-cost mini-drones. At Wright-Patterson Air Force Base in Ohio, where the air force conducts advanced aircraft research, two engineers, Allen Atkins and Ken Perko, had begun work with DARPA funding on a mini-drone that was launched inside a capsule of a surface-to-air missile and could travel seven hundred miles downrange. Once it reached its destination, the drone would emerge from the capsule like a butterfly from a cocoon, extend its wings, and then fly around the Fulda Gap looking for targets. When it spotted a Soviet tank or surface-to-air missile, it would relay the information to an F-4 combat aircraft, which could then fly over and destroy the target.

The mini-drone was not invisible to radar, but it was small enough that it would be harder to detect and harder to hit, particularly for a surface-to-air missile. Yet even a small pilotless aircraft shows up on radar, and the latest Soviet radar-guided anti-aircraft weapons could take it down. Perko and Atkins proposed to DARPA another mini-drone, one specifically designed to reduce what is known as radar cross section, or how visible something is to radar. DARPA agreed to fund the project, which used small drones made by McDonnell Douglas.

McDonnell Douglas ended up building half a dozen of these radar-evading drones, called the Mark V, and testing them at Eglin Air Force Base in Florida against a host of Soviet weapons, part of a secret arsenal of equipment that the American military had acquired through black market channels. Just as DARPA had hoped, radar could not lock on and track the little drones. By today's standards, the mini-drone's radar signature was large, but it performed better than anyone had expected. It flew undetected past every air defense system, except the Soviet ZSU-23-4 Shilka, an advanced radar-guided anti-aircraft weapon. Even then, it was only spotted when flying directly overhead. By the fall of 1974, Perko and Atkins knew they were onto something. The drone was stealthier than they had predicted. The equations everyone had been using to predict an aircraft's radar cross section were not based on the "real world," Atkins said. In the real world, there are insects, clouds, and birds.

Moore told Myers a bit about the DARPA work, but Myers was unimpressed by the concept of mini-drones. Both men seemed to agree that at least exploring the idea of a stealth aircraft made sense. Myers wanted $2 million for his Harvey study, a sum that Moore told him DARPA did not have. Moore was intrigued by Myers's idea, however, so he asked Perko, who by then had been recruited into DARPA, to survey the military aircraft companies to see which ones might have experience with reducing radar cross section. While Perko was on his fishing expedition, a memorandum landed on Moore's desk from Malcolm Currie, the director of defense research and engineering. The Pentagon's chief technologist had some money available, but was lacking in new ideas. Moore realized that Harvey, a stealth fighter that could survive close air support against Warsaw Pact forces in the Fulda Gap, might fit the bill. Instead of Harvey, Moore changed its name to the "the high stealth aircraft," and DARPA requested proposals from five companies: Northrop, McDonnell Douglas, General Dynamics, Fairchild, and Grumman. Then the stealth aircraft almost died.

RIGHT Before joining ARPA in 1958, William H. Godel built his reputation as a legendary intelligence operative. He started his career as a marine in World War II, and was later sent to Europe undercover—posing as a German veteran—to recruit foreign scientists to work with the Pentagon. Godel became one of ARPA's most influential early employees, pushing the agency into an ambitious counterinsurgency program in Southeast Asia.

ABOVE, RIGHT Roy Johnson, a vice president from General Electric, was unknown in Washington, D.C., when he was appointed as the founding director of ARPA, the nation's first space agency. A fierce advocate of ARPA's role in manned space missions, he clashed with the White House and the president's science advisers. He resigned after less than two years, to pursue a career as an artist.

LEFT Herbert York, a nuclear physicist and the agency's first chief scientist, resented having Roy Johnson, a businessman, in charge of ARPA. York aspired to be the military's space czar, and got his wish in December 1958, when he was appointed to be the Pentagon's director of defense research and engineering, a position above ARPA. He then stripped the young agency of its space work.

RIGHT Herbert York and Roy Johnson addressing a Senate committee. Though they presented a united front in public, the two men clashed over the future of ARPA. Most significantly, York ended ARPA's work on the Saturn rocket, giving the project to NASA. Disappointed, Johnson resigned from ARPA. In 1969, a Saturn rocket launched Apollo 11, the first manned mission to the moon.

LEFT Nicholas Christofilos, pictured here in March 1958, draws a radiation belt around the earth. The "mad Greek," as he was known, proposed to fill the belt with charged particles to create an impenetrable shield against ballistic missiles. His idea led to Operation Argus, one of ARPA's first projects, which involved setting off nuclear explosions in the upper atmosphere. The missile shield did not work, though Christofilos's theory helped confirm the existence of what became known as the Van Allen radiation belt.

CLOCKWISE FROM RIGHT In 1961, South Vietnam's president, Ngo Dinh Diem, personally approved ARPA's work and the founding of the Combat Development and Test Center. In the first picture, Colonel Bui Quang Trach (Army of the Republic of Vietnam), director of the ARPA-sponsored center, escorts President Diem for a visit to see the agency's projects. In the second, Diem examines ARPA's chemical defoliation spray system. And in the third, Colonel Vito Pedone, head of ARPA's field unit, discusses the military dog program, which involved sending trained canines to help South Vietnamese soldiers track Vietcong insurgents hiding in the jungle. Diem remained an enthusiastic supporter of ARPA's work until his death in 1963.

RIGHT J. C. R. Licklider, who came
o ARPA in 1962 to manage research
n command and control, paved the
vay for the Internet and personal
omputing. Pictured here using a
ight pen to interact with a computer,
ie articulated a vision for "man-
omputer symbiosis" that would
'think as no human brain has ever
hought and process data in a way
not approached by the information-
handling machines we know today."

RIGHT Situated at a former French
barracks, the ARPA compound
n Saigon became the center of
the agency's wartime work. The
compound was home to the Combat
Development and Test Center,
run jointly with Vietnam's armed
forces, and ARPA's Research and
Development Field Unit. For more
than a decade, ARPA coordinated
everything from chemical defoliation
to psychological operations from the
compound.

LEFT ARPA director Stephen Lukasik visits the agency's field unit in Vietnam. In front of him is Lieutenant Colonel Bien (full name unknown), acting director of the Combat Development and Test Center. Following him is Colonel Ephraim M. Gershater, the head of ARPA's Research and Development Field Unit. By the time of Lukasik's visit in 1971, the Vietnam War was sparking nationwide protests in America, and ARPA was heavily criticized for its involvement. Lukasik eventually closed the field offices, shut down the counterinsurgency program, and transformed the research into the Tactical Technology Office, which developed many of the weapons, such as drones, associated with modern warfare.

ABOVE, LEFT AND RIGHT George Lawrence helped bring the counterculture to ARPA in the 1960s. He was hired by the agency because of his background in psychology and computers. Yet it was his interest in unconventional science that led him to create the field of biocybernetics, which sought to control computers with the human mind. He also became the agency's point person for parapsychology investigations, popular at the time with parts of the intelligence community.

LEFT AND ABOVE During the Vietnam War, ARPA funded Bell Aerosystems to build an Individual Mobility System, better known as the Jet Belt. In the version pictured here, the propulsion system, which uses a small bypass gas turbine engine, was attached to a fiberglass corset worn by the soldier. The company touted the Jet Belt for counterinsurgency operations, claiming it could be used to deploy special operations forces to track down insurgent fighters. ARPA eventually canceled the project, but the unique engine was later incorporated into cruise missiles.

LEFT In 1964, ARPA sponsored a nuclear test, called the Salmon Event, which involved the denotation of a 5.3-kiloton device in the Tatum Salt Dome in Mississippi. The test, one of two nuclear detonations conducted under Project Dribble, was to help determine if the Soviet Union could conceal their nuclear tests by conducting them in underground cavities. The work was part of Project Vela, the agency's nuclear test detection program, which was credited with paving the way in 1963 for the Limited Test Ban Treaty.

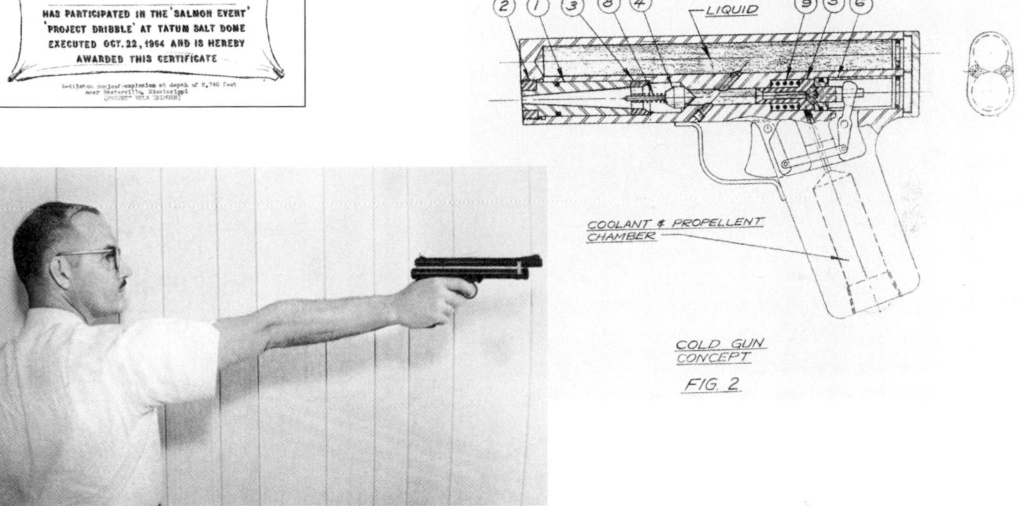

ABOVE, LEFT AND RIGHT Following the assassination of John F. Kennedy, ARPA was assigned a classified program, code-named Star, to protect the president. The work drew on ARPA's experience with counterinsurgency in Vietnam, leveraging research on armor, small arms, and threat assessment. As part of Star, the agency sponsored an investigation of various nonlethal weapons for the Secret Service, including a gas-propelled impact projectile, pictured left in testing, and, right, a cold liquid weapon, which officials dubbed the "squirt gun."

ABOVE Physicist John S. Foster, center, the Pentagon's director of defense research and engineering, speaks with Colonel Trach, second from left. Foster supported ARPA's Vietnam program, AGILE, but felt it needed more science. In 1966, he brought in Seymour Deitchman, far right, to take over the agency's counterinsurgency work. ARPA director Charles Herzfeld, third from right, also pushed to expand the agency's counterinsurgency work around the world.

LEFT Deitchman, an engineer, wanted ARPA to help the military understand the social factors feeding the rising insurgency in Vietnam, but he was appalled by much of the work being sponsored by AGILE. One of his first decisions was to cancel a contract given to Herman Kahn, the famed futurist and nuclear theorist, who proposed, among other absurdities, building a moat around Saigon. The political fallout from firing Kahn, who was close to defense secretary Robert McNamara, was depicted in a cartoon given to Deitchman as a farewell present by his colleagues in 1968.

RIGHT In the mid-1960s, the insurgency in Thailand was concentrated in the northeast, and communist insurgents frequently used the Mekong River to cross the border with Laos. The ARPA-sponsored Mekong River surveillance system, which provided boats and other gear to the Thai military, was often regarded as one of the few bright spots of ARPA's work in Southeast Asia. The radar platform shown here, together with a few modest boats, was jokingly called "Deitchman's Navy," because of Seymour Deitchman's key role in the project.

LEFT Throughout the war, ARPA used Thailand as a laboratory for technology that could be deployed in Vietnam. The ARPA-sponsored Combat Development and Test Center in Thailand, like its counterpart in Vietnam, developed a wide range of equipment for use in counterinsurgency operations. Pictured here is an ARPA-developed "Psychological Operations in Box," a sunlight projector that could be used to show propaganda films in villages susceptible to communist sympathies. The rugged equipment, which used sunlight rather than an electric light to project the film, was designed for austere conditions and typical of ARPA's approach to jungle warfare.

ABOVE In 1963, Warren Star (center), a young Harvard MBA, was put to work on ARPA's programs in Southeast Asia and eventually became a senior official in charge of a global counterinsurgency program. He is pictured here with Richard D. Holbrook (left), the director of ARPA's research in Thailand, and James Woods (right), an anthropologist employed by the agency to conduct fieldwork in Southeast Asia.

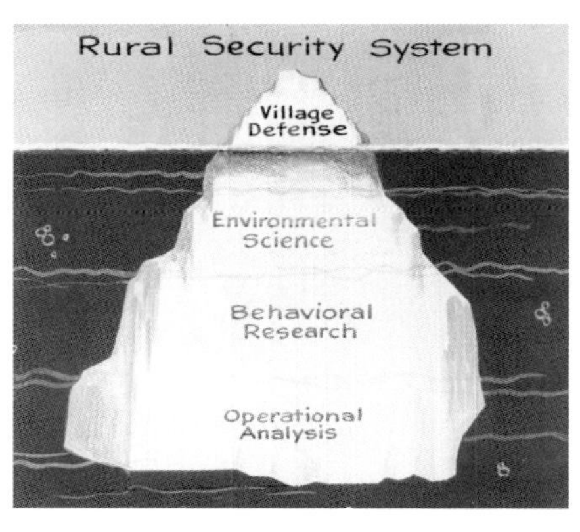

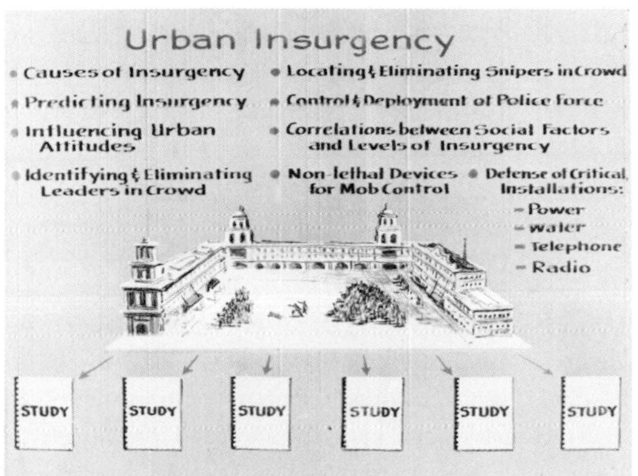

CLOCKWISE FROM ABOVE, LEFT By the mid-1960s, Project AGILE had become a worldwide scientific program to address global counterinsurgency. These briefing slides helped describe the new, holistic ARPA approach, which looked at everything from behavioral research to assassination of "leaders in a crowd." The first slide depicts ARPA's work in developing environmental science for warfare; the second shows how various areas of research, including behavioral science and operations research, formed the scientific underpinning of village defense in Vietnam. The final slide describes the types of research and studies ARPA was pursuing in order to combat insurgency in urban areas, which became a major concern as the Vietnam War progressed.

ABOVE In 1967, ARPA sponsored Lockheed to develop the QT-2 aircraft, or "quiet airplane," for use in Vietnam, based on a modified Schweizer SGS 2-32 sailplane. The aircraft—essentially a powered glider—used a muffled Volkswagen engine so that it could fly at night without being detected, to spot Vietcong laying booby traps. The modified aircraft was painted beige to mask its visibility, and in domestic testing it was given the fictitious company name San Jose Geophysical Inc. to hide its military connections. Though sometimes touted as the first example of a "stealth aircraft," the QT-2 was not designed to evade radar.

ABOVE One of ARPA's first experiments with radar-evading stealth involved a small drone called the Mark V. The agency contracted with McDonnell Douglas to build half a dozen of the unmanned aerial drones. The completed drone—held here by Allen Atkins, who went on to lead ARPA's Aerospace Technology Office—demonstrated that it might be possible to build an aircraft that could evade detection by radar, and gave Pentagon officials the confidence to move forward with a prototype stealth aircraft.

In 1975, as the idea of designing stealth aircraft was being discussed in DARPA, Lee Huff, a former DARPA employee, sat down with William Godel, the agency's onetime deputy director. Huff had been hired to write a history of the agency and interviewed his onetime mentor and boss, who, after serving time in prison, made a career in private business. Reflecting on his time at DARPA, Godel said that program managers were a "dime a dozen," but true innovators were rare. Asked what DARPA should be working on at the present time, Godel suggested an "unmanned non-detectable bomber."

Godel had no way of knowing that just two years earlier DARPA had started a secret study looking at ways to use technology to help overcome the Soviet conventional advantage, but he was clearly aware of the strategic debates at the time. President Nixon's pursuit of détente with the Soviet Union and associated negotiations for nuclear reductions had also brought renewed debate over the Warsaw Pact's conventional advantage in Europe.

Godel's prescience about the concept of a stealth aircraft was based on his understanding of the political and military situation in Europe as much as technology: he knew that penetrating Soviet air defense was a strategic imperative, just as global counterinsurgency had been in the early 1960s. Godel had always believed that DARPA's role had to be guided by strategic planning, which required understanding current problems but also looking ahead at future threats. That had been a view supported by the first three scientists to serve as the Pentagon's director of defense research and engineering. Herbert York, John S. Foster, and Harold Brown, all physicists and former heads of Livermore Laboratory, had supported an expansive role for DARPA that bridged strategy and technology.

Malcolm Currie, on the other hand, was closer to the world of engineering and defense contracting. The engineer had risen to be a vice president of research at Hughes Aircraft before moving over to the Pentagon. He viewed DARPA as the Pentagon's industrial lab and was irritated by anything he saw that smacked of policy or strategic planning. And despite his enthusiasm for the idea of a stealth aircraft, Currie was not happy with DARPA, or at least not with Lukasik, its director. Lukasik was traveling around the world discussing arms control, he was sponsoring studies on military strategy, and then there was the parapsychology investigation. Currie decided that Lukasik, by then the longest-serving director of DARPA, needed to go. The pick to replace him was George Heilmeier, an engineer from RCA, who was working for Currie in the Pentagon. Heilmeier had already distinguished himself as the father of the liquid crystal display, a technology that would eventually be used in everything from cockpit displays to home alarm clocks. Perhaps because of their mutual background in industry, Currie and Heilmeier "clicked."

The new DARPA director brought immediate changes. Heilmeier believed Secretary of Defense James Schlesinger wanted DARPA to go back to its roots, or at least what Schlesinger believed were its roots. "He wanted more technology there," Heilmeier said. "He didn't want a lot of work done in foreign policy by a technical agency." Lukasik had allowed DARPA officials unprecedented freedom. DARPA orders, the brief summaries used to authorize new programs, were unimportant to Lukasik, and he rarely if ever read them. Heilmeier, on the other hand, pored over every word of the orders, often sending them back for revisions. He also began to comb through the budget, weeding out programs he felt had limited relevance to the military.

J. C. R. Licklider, whose vision of an "Intergalactic Computer Network" had been enthusiastically embraced by the likes of Herzfeld and Lukasik, suddenly found himself in conflict with DARPA's new director. The soft-spoken Licklider had returned to DARPA in 1974 but found himself facing a director who did not share his broad vision of computer science. "When I looked at the so-called proposals I thought, 'Wait a second; there's nothing here,'" Heilmeier said. "It was just, 'Give us the money and we will go do good things.'"

Heilmeier wanted to know how research would help a tank operator. He wanted a way of translating Morse code. He wanted a way of working with a pilot. Computer scientists protested that creating artificial intelligence was not like building an aircraft that was set to fly on a specific day, but in Heilmeier's view it should not be any different. Scientists were aghast that they now had to write proposals with specific goals in mind. "That's pure bullshit," he said. Licklider left DARPA for a second time, not long after Heilmeier came on board.

Heilmeier was convinced that any Pentagon support for science had to be aimed at developing a specific technology that could be used by the military. A good part of his philosophical approach to technology came from his time at RCA, where he felt the company had failed to recognize—and take advantage of—his invention, the liquid crystal display. He knew from personal experience that it was not enough to come up with an innovative technology; there needed to be a plan to take that technology to market, whether in the commercial world or with the military. He looked at his job as akin to a venture capitalist who invests in high-risk technologies but only those that meet a set of criteria. Heilmeier created a series of questions that he called the "catechism," which he made the litmus test for any program brought to his office:

> First what are you trying to do?
> How is it being done now, and what are the limitations of current practice?
> What's new in your approach?
> Why do you think it can be successful?
> If you are successful, what difference does it make?
> How much money does it need, and how long is this going to take?
> What are the midterm and final exams?

Coming from the private sector, Heilmeier had little interest in science for the sake of science. He admitted there was resistance to his view, particularly from DARPA-funded researchers who had not needed to justify their work in the past in terms of an immediate military application. "Why don't you try to perform an impossible sexual feat?" Heilmeier said was his response to them.

While a stealth aircraft might sound like an ideal program for a new DARPA director interested in advanced military technology, Heilmeier was initially skeptical when he called in Moore to review the Tactical Technology Office's portfolio. The new director was housecleaning, and he made it clear that Moore's programs, and even his job, were far from guaranteed. "He went through all my programs and after he finished all the other reviews, he called me into his office and told me that the only program he found questionable was the stealth aircraft," Moore recalled.

Heilmeier said he did not see how an aircraft that could evade Soviet radar involved any advanced technology. Prior to becoming director of DARPA, he had listened to Myers talk about the Harvey concept, which was actually very different from what DARPA was proposing. Myers was advocating for a modestly priced aircraft with a *reduced* signature that would fly in the Fulda Gap to assist other, more capable fighters. The idea of the aircraft being "invisible" was hyperbole; Harvey was not an aircraft that would be undetectable to all enemy radar, just less visible than other aircraft. Moore's idea for the high stealth aircraft, on the other hand, was for something more ambitious—an aircraft that could slip past Soviet radar completely unseen.

Moore argued to Heilmeier that the DARPA concept of stealth was something that would affect the entire design of the aircraft. It was not just about making a tweak here or there, as had been done with the small drone. DARPA was proposing a new type of aircraft that would make the radar cross section as low as the physics would allow while accepting whatever aerodynamic design that required—so long as it could still fly. This was not Harvey but a radically new aircraft. Heilmeier listened closely, but Moore was convinced the invisible aircraft was dead.

Heilmeier might have been skeptical about an invisible aircraft, but Harvey had another patron. Currie, Heilmeier's boss, had also been on the receiving end of the Harvey briefings, and he was enthusiastic about stealth. A meeting among Moore, Heilmeier, and Currie settled the matter; Currie liked it, and he was Heilmeier's superior and mentor. "We needed to penetrate enemy air defenses," Currie recalled. "If we could do that and essentially nullify the radar threat against us, then that was an obvious thing to do."

Back in the Pentagon, Myers was still preaching his Harvey concept when Russ Daniel, a Lockheed engineer at the company's Skunk Works division, dropped by his office. Lockheed had never been contacted by DARPA when Perko was canvassing companies on possible interest in a stealth aircraft, because Lockheed at that point did not build fighters. Daniel reported back to Ben Rich, the new head of Lockheed's Skunk

Works, about the missed opportunity. It turns out, around the same time, that Rich also found out about the stealth aircraft from his resident Soviet weapons expert, Warren Gilmour, who had gotten wind of the DARPA project from a friend at Tactical Air Command. "Ben, we are getting the shaft in spades," Gilmour told Rich.

What DARPA did not know was that Lockheed had been secretly working for the CIA on stealthy aircraft designs for years, first with Project Rainbow, an attempt to make the U-2 high-altitude spy plane less visible to radar. The experiment failed, killing a pilot in the process, but the company had more luck with the CIA's A-12 reconnaissance aircraft, the precursor to the air force's supersonic SR-71. The A-12 was not a true stealth vehicle, but the "cobra shaped" aircraft, as Rich called it, had been built with radar-absorbing material, a closely held secret. Even so, the improvements were modest; the radar signature was roughly that of a single-engine Piper Cub. In other words, the A-12 still looked like an aircraft when picked up by radar, albeit a small one.

Rich wanted in on the stealth competition, but Lockheed was in a bind. Kelly Johnson, the legendary leader of the Skunk Works, had just retired; he had made his reputation with projects like the U-2 and the SR-71. Rich was desperate to make his own mark. "This was exactly the kind of project I was looking for," the Skunk Works president recalled in his memoir. "But we had been overlooked by the Pentagon because we hadn't built a fighter aircraft since the Korean War and our track record as builder of low-radar-observable spy planes and drones was so secret that few in the Air Force or in upper-management positions at the Pentagon knew anything about them."

Johnson, who was still working for the Skunk Works as a consultant, persuaded the CIA to let Lockheed give DARPA details about the company's stealth work. The CIA agreed and Lockheed sent a briefing to DARPA and asked for the chance to compete for the stealth work. George Heilmeier warned Rich there were no funds left to pay Lockheed for its participation, but DARPA would at least consider the company's proposal. Rich was undeterred, and asked to brief Perko, the DARPA official in charge of stealth.

"We want to be part of this study," Rich told Perko, after giving his stealth briefing.

"We don't have any more money," Perko replied apologetically. "We've spent all our money."

Rich then made an offer DARPA officials could not refuse: Lockheed would do the study for the grand price of $1, and the Skunk Works would cover the rest of the costs. It was an expensive gamble, but Rich sensed that stealth aircraft could be the project that would make his name and keep the Skunk Works in the good graces of Lockheed's executives.

Perko agreed to the offer and, as legend had it, reached into his pocket and pulled a dollar out, and the two men shook hands on the spot. Lockheed was in the stealth competition.

The new DARPA director was also now on board, but he agreed to fund two prototypes of the stealth aircraft only if the air force would pay half the costs. A prototype aircraft would be an expensive endeavor, and even if it were successful, it would not do anything more than collect dust in an aviation museum if someone in the military was not willing to buy it. The prototype was going to cost $50 million, and Heilmeier wanted the air force to pay 49 percent of the costs, leaving DARPA with 51 percent and control over management. Considering the air force's steady opposition to Myers's efforts to market Harvey, it did not look promising.

Pilots ran the air force, and pilots want to fly combat aircraft. The idea of an aircraft that would be "invisible," at the expense of maneuverability and performance, was not immediately appealing. The air force had been dubious of stealth claims, and even when the DARPA studies showed that a stealth aircraft was possible, the air force's leadership did not see the point: Why field an aircraft that was aerodynamically unstable? Worse, a stealth aircraft would compete in the budget with the air force's top priority, the F-16 fighter. When DARPA officials first briefed Alton Slay, then a three-star air force general in charge of research, his response was not just no, but hell no. The air force was not interested in spending its money on a DARPA project. Currie, a fan of the stealth project, offered a deal. At a breakfast with General David Jones, the air force chief of staff, he laid his cards on the table: "I will support your lightweight fighter in Congress with everything I've got if you will establish stealth as a real air force program and put some money behind it and let us go to Congress with that." Jones agreed and the two men shook hands.

Shortly after the breakfast, Currie, along with DARPA's Heilmeier, Moore, and Perko, attended a meeting with General Jones and General Slay. At that point, no one except Currie and Jones knew about their agreement, and Jones was "like the inscrutable sphinx" as the DARPA officials briefed the invisible aircraft program. At the end of the briefing, Jones, sitting at the end of a long oval coffee table, announced, "The air force ought to support this program." He then turned to Slay and asked him what he thought. Slay replied, "Well, to be against that would be like being against motherhood."

Stealth had triumphed, not necessarily on a persuasive argument about the technology, but through a handshake deal. The air force kept its commitment, Slay's prior opposition to the stealth aircraft evaporated, and DARPA officials later credited him with being an avid supporter. "You have to admire a man that when he's ordered to do something, he didn't pull any punches," Heilmeier recalled. "He did it."

The idea of designing aircraft to be less vulnerable to Soviet air defense systems was not new; it was just incredibly hard. Military engineers had experimented with ways to make aircraft harder to detect since the advent of radar during World War II. The problem was an exponential one, meaning that even large reductions in radar cross section translated into quite modest advantages when it came to evading detection. For example, if American engineers wanted to reduce by half the range at which Soviet radar would pick up an incoming bomber, from twenty minutes to ten minutes, they had to reduce the radar cross section by a factor of sixteen. And even that was not nearly enough for a strategic advantage.

In most aspects of aviation, even a 10 percent improvement in performance would be fantastic, but in the world of low observables, the technical term for stealth, a 50 percent reduction does not translate into much military capability if the aircraft is, for example, trying to slip past Soviet air defense radar. Even if the radar cross section were cut in half, the aircraft would still be picked up by enemy radar in time for it to be shot down. "The difference for the Soviet guys is they no longer will have the leisure to talk about whether the Moscow Dynamos is the better football club than Kiev, or will they necessarily finish up their cup of coffee," Alan Brown, Lockheed's chief engineer at the time, joked. "But they still have all kinds of time to call up the surface-to-air missile systems and the airfields and say, 'Go and get those guys, they are ten minutes away.'"

In other words, if engineers wanted to reduce the range of detection for an aircraft by a useful amount, say by a factor of ten, they would have to reduce the radar cross section by a factor of ten thousand, something that key air force officials deemed impossible. Tweaking the designs of existing aircraft would not yield that sort of reduction; it would require a total redesign, and most designs suitable to stealth were not very suitable to flight. As Kelly Johnson, the former head of Lockheed's Skunk Works, noted early in development, the shape best suited for stealth was something that resembled a flying saucer. Short of antigravity technology, a flying saucer was not likely to make for an effective aircraft.

Lockheed might have joined the stealth contest late, but it started with a strong advantage because of its aviation work for the CIA on projects like the U-2. Its other advantage was that the Skunk Works division operated a bit like DARPA, meaning it was flexible, had minimal bureaucracy, and could quickly assemble a team of experts for a particular project. One of those experts was Denys Overholser, a young electrical engineer and mathematician. Previous attempts to design aircraft to avoid radar put aerodynamics first, while making it stealthy was an afterthought. As a mathematician rather than an aerodynamicist, Overholser took an entirely different approach to the idea of a stealth aircraft. Instead of thinking about what made for an effective aircraft, he looked primarily at ways to design something that would reflect radar. He suggested designing the aircraft in a series of flat panels that would be arranged in such a way so as to ensure the radar's energy was deflected away from the source, and thus could not be easily detected. The panels gave the aircraft a distinctive faceted design and a decidedly un-aerodynamic shape.

Overholser had settled on that design because, in 1974, he could create a computer program to calculate the return for those flat panels based on physical optics but not for something extremely complex, like curves. He also could not estimate what happened on the edges of those panels. That is, until he stumbled across an unclassified Russian scientific paper that had been translated by the U.S. Air Force Systems Command's Foreign Technology Division, which regularly combed unclassified Eastern bloc scientific literature for work of interest to the military. The translation of the Russian scientist Petr Ufimtsev's paper "Method of Edge Waves in the Physical Theory of Diffraction" had gone unnoticed for several years until Overholser realized that it could help him calculate the radar cross section for the edges of the panels. Brown, Lockheed's chief engineer, estimated that Ufimtsev's theory contributed about "30 percent" to Lockheed's stealth calculations. "It wasn't like the Russians saved us," Brown said.

The Russian's formulas might not have helped immediately, but they clearly emboldened Overholser, who argued that flat panels would allow the engineers to predict the radar cross section. The resulting design, created to test stealth and not its ability to fly, was faceted like a diamond and shaped like a sort of swept-back pyramid. "Well, that's stupid. We'll never make that fly," was the reaction from Lockheed's designers, recalled Brown. "They christened it the Hopeless Diamond."

The name Hopeless Diamond stuck, even after Lockheed's designers modified the shape to something that ended up looking a bit like a faceted dart. It was not going to win any beauty contests, but it was enough to make Lockheed, along with Northrop, a competitor in what was now called the High Stealth Experiment, a classified program. Both companies built a scaled-down model and tested each on a stationary pole. Northrop's aircraft took a roughly similar approach to Lockheed's, using a faceted flat-panel design to deflect radar. Neither company had a clear advantage, but DARPA in April 1976 picked Lockheed as the winner of the competition, a decision that might have been based as much on the Skunk Works' reputation as it was on the actual design. According to one Northrop official, "Grown men cried that day." Lockheed would get to build and fly two prototypes of the world's first stealth aircraft, now code-named Have Blue.

On December 1, 1977, Heilmeier stood in the Nevada desert for the first flight of Have Blue. DARPA wanted the program under wraps, so neither the Skunk Works plant in Palmdale, California, nor the adjacent Edwards Air Force Base would work given the level of secrecy required. Area 51 was the logical choice.

It was dawn when Have Blue rolled out of the hangar. The pilot followed the instructions: turn right, go about three-eighths of a mile until you get to the main runway, turn right again, and then go seven thousand feet and take off. There were no lights on the runway (to avoid any hint of an aircraft test), and the only illumination came from the landing lights on the wheels of the aircraft—three lights that help guide the pilot down the runway. Heilmeier stood at the end of the runway watching as engineers made last-minute checks to the aircraft, and then watched, fists clenched, as it started to roll. No amount of computer simulation can ever make up for a first flight test, when engineers sweat over all the calculations that might have gone wrong.

For Have Blue, the Hopeless Diamond, the anticipation was even more intense, because there were a lot of things that could go wrong with an aerodynamically unstable, computer-controlled aircraft. "The air gods did not like that vehicle at all," joked Atkins, the air force engineer.

As the aircraft lifted off from the end of the runway, Heilmeier reached down and gathered a few pink stones from the Nevada desert and put them in his pocket as a trophy. The Have Blue aircraft was flying, and there was a collective sense of relief and jubilation. Not just because it was invisible to radar, but because it did not drop out of the sky.

The first flight for Have Blue coincided with Heilmeier's last day as DARPA director; he had specifically selected that day to end his tenure at the agency to ensure his legacy was linked to the aircraft's success. Have Blue was also the final project for Kelly Johnson, the Lockheed designer who had pioneered a generation of secret aircraft. To celebrate, Johnson brought out a bottle of champagne that had been flown in from Europe on the SR-71 Blackbird. Both men signed the bottle, which Heilmeier took back to Washington with him. When his wife asked what it was, he replied, "It's an empty bottle of champagne."

"An empty bottle? What do you want to do with it?" she asked.

"Someday I'll be able to tell you," he told her.

She would not have to wait long. By the time Park crashed in 1978, reports began emerging in the aviation trade press about the stealth aircraft. The second prototype aircraft crashed in July 1979 (the pilot, Ken Dyson, safely ejected), but at that point the future of stealth was already secured. The air force, based on the success of the DARPA program, had started work on an operational aircraft under the code name Senior Trend, the F-117. Ironically, though the *F* designation would normally mean it was a fighter, it was really a ground-attack aircraft. The reason for the subterfuge, according to Lockheed's Brown, was that designating it a fighter would make it easier to recruit pilots; the prestige job in the air force was being a fighter pilot. "No self-respecting fighter pilot is going to fly an attack aircraft or God forbid a bomber," Brown said.

Stealth remains one of DARPA's most highly cited accomplishments. After it had proven possible to build an aircraft that could evade radar, stealth was eventually incorporated into a variety of aircraft and weapons, from bombers to helicopters, including the modified Black Hawks that were used to raid Osama bin Laden's compound in Abbottabad, Pakistan, in 2011. One of the few people to express disappointment with the stealth aircraft was Myers, the Pentagon official whose invisible rabbit inspired the original DARPA program. Even four decades later, he felt betrayed that the small, affordable fighter he wanted—with a reduced signature, but not invisible to radar—was never built. "I'm still convinced that Harvey is a good idea," he said years later. "We should try it sometime."

By 1980, keeping the existence of a stealth aircraft under wraps was all but impossible. In the midst of a reelection campaign, when President Jimmy Carter was facing questions about the cancellation of the B-1 bomber, Harold Brown, now the defense secretary, decided it was time to confirm what was already an open secret. "I am announcing today a major technological advance of great military significance," Brown said. "This so-called 'stealth' technology enables the United States to build manned and unmanned aircraft that cannot be successfully intercepted with existing air defense systems. We have demonstrated to our satisfaction that the technology works."

Have Blue did one more thing that most people never realized: it saved the agency from extinction, according to James Tegnelia, who was the deputy director when the stealth aircraft flew. "You can fail at a $10 million program," he said. "When you're putting in $100 million, you can't afford to fail." Tegnelia credited the sudden growth of "big" technology programs, like Have Blue, with protecting the agency from critics who wanted to shut it down. After the success of Have Blue, "no one questioned the value of DARPA's investment," Tegnelia said. If there is an irony to this success, it was that the stealth aircraft emerged from the broken fragments of William Godel's Project AGILE, which were dusted off and put back together as the Tactical Technology Office. Godel's early efforts were coming to fruition just in time for what would become the largest military buildup of the Cold War.

CHAPTER 15

Top Secret Flying Machines

On Christmas Day 1979, Soviet airborne forces landed in Kabul, paving the way for the invasion and occupation of Afghanistan. In Iran, a onetime base for the U.S. military—and DARPA—students loyal to the Iranian revolution were holding fifty-two American hostages, a nightmarish spectacle played out nightly on television. In April 1980, a daring rescue attempt of those hostages authorized by President Jimmy Carter ended in an embarrassing failure. After a series of mishaps that forced the military to abort the mission, a departing helicopter collided with a parked C-130 aircraft in the desert of Iran, killing eight American servicemen.

The Carter administration's clumsy attempts to bolster its image of military prowess, such as publicizing development of the stealth aircraft, were ineffective. Approaching an election year, the country was facing a trifecta of inflation, unemployment, and recession. The price of oil peaked in December 1979 at more than $100 a barrel. Abroad, the United States was doing little better, as one by one, from Iran to Nicaragua, countries that had been bulwarks of American support fell to insurgent movements. The Soviet Union's influence, on the other hand, seemed to be expanding, from Cuba to Afghanistan.

Facing allegations of military weakness and economic decline, a charismatic former actor and governor of California stepped onto the political scene, promising to reinvigorate America through military strength. The United States is "already in an arms race, but only the

Soviets are racing," Ronald Reagan told an audience of veterans, in one of his seminal campaign speeches. "They are outspending us in the military field by 50 percent and more than double, sometimes triple, on their strategic forces." Reagan promised to reverse that trend, reinvigorate the military, and America. It was a message that resonated with voters; Reagan carried forty-four states in a landslide election.

Shortly after Reagan's victory, Richard DeLauer, the new administration's pick for undersecretary of defense for research and engineering, called his friend Bob Cooper, a defense scientist, with big news: the White House was going to double the Pentagon's budget over the next five years. "Sure, Dick, yeah, I know, I've heard that story before," Cooper replied.

DeLauer insisted it was true. Reagan was adamant that the United States should increase its defense spending to send a strong signal to the Soviet Union. DeLauer wanted Cooper, a former football player known for his forceful personality, to return to the Pentagon and take over two jobs. Cooper would work directly for DeLauer, as an assistant secretary of defense, while also heading DARPA, which, following the success of the stealth aircraft, was regarded as a hotbed of innovation, a sort of corporate lab that could quickly push out military technologies. And with Reagan in the White House, there was going to be a huge demand for new military weapons. Cooper agreed to take the job.

When he arrived at the Pentagon in 1981, he was appointed to sit on the Defense Resources Board, which made funding decisions on major weapons. The powerful board consisted of the secretary of defense, the deputy secretary, the military chiefs, and other senior military and civilian officials. Essentially, his position on the board created a direct bridge from DARPA to the Defense Department's senior leadership. For the first time since the mid-1960s, when DARPA was exiled from the Pentagon, the agency was thrust back into the epicenter of the country's military decision making. With Cooper at the helm, DARPA would no longer be a stand-alone research agency, a few metro stops away from the Pentagon. Instead, its director would have a full-fledged voice on the Pentagon body that made decisions about what weapons to develop and buy. And military technology was about to become the focus of national policy under Reagan. According to Cooper, his job was to "give an enema to the pent up technologies" at DARPA.

Shortly after Cooper arrived, Secretary of Defense Caspar Weinberger held a meeting for his staff and other senior Pentagon officials. Weinberger stood up and repeated the claim that Reagan would double the defense budget. At this point, Cooper interjected, "Cap, I've heard that song before, and I don't think the public is going to support doubling the defense budget in the next three years." Weinberger, however, was adamant. "We are going to spend them into the dirt," he said. Yet even Weinberger was shocked by what happened next.

On March 23, 1983, President Ronald Reagan addressed the American people directly, warning of a grave threat of war and nuclear annihilation, balanced with a Hollywood message of hope. "The solution is well within our grasp," Reagan assured the nation.

That solution, it turned out, was one of the most expensive and technologically foolhardy projects ever undertaken by the Pentagon: a space-based missile shield to protect the United States and its allies from a Soviet nuclear attack. "I am directing a comprehensive and intensive effort to define a long-term research and development program to begin to achieve our ultimate goal of eliminating the threat posed by strategic nuclear missiles," Reagan said, much to the surprise of the Pentagon's leading missile defense experts, who had been telling the president for the past year that such technology was currently impossible. Reagan's dream was soon derisively labeled "Star Wars," a name that would stick with the program for life, and death.

Not since the days of DARPA's 1950s-era Project Argus, the nuclear-powered astrodome, had the government pursued a scheme to create a missile shield that could protect the entire country. Exotic proposals like Project Argus never got beyond the conceptual stage. Now Reagan was proposing to build an equally ambitious defensive system. Despite a few exploratory efforts, most of the missile defense work over the prior two decades had focused on ground-based inceptors that might take out a few incoming missiles, and even those systems had not gone very far. None of that concerned Reagan, however. "What other President, after all, could persuade the country of something that did not, and could not for the foreseeable future, exist?" Frances FitzGerald wrote in her nuanced account of Reagan's impossible missile defense dream.

Much of the enthusiasm for ambitious missile defense schemes originated in one way or another with DARPA, which had pursued the Seesaw particle beam in the 1960s and also sponsored a highly classified laser study called Eighth Card, so named as a trump card that could win a seven-card stud. The Eighth Card study, held in 1968, was sparked by excitement over an air force gas-dynamic laser; the study was intended to merely look at battlefield applications of future technology, but it captured the imagination of the hydrogen bomb inventor Edward Teller. By the time Reagan announced his missile defense plan in 1983, Teller was touting a far more ambitious project based on the Livermore lab's theoretical work: an X-ray laser powered by a thermonuclear explosion. The improbable scheme would have involved launching into space multiple X-ray laser weapons, which would be used to shoot down ICBMs mid-flight.

Over at the Pentagon, DARPA's director, Cooper, and other senior officials—including Secretary of Defense Weinberger—sat slack-jawed as they tried to digest the president's address. The president had just made one of the most significant military technology decisions of the past few decades without consulting the key people in the Pentagon responsible for that technology. "Everybody including DeLauer and myself were completely blindsided by it," Cooper recalled.

Weinberger had opposed rushing forward with a missile defense system. "Although we appreciate your optimism that technicians will find the way and quickly, we are unwilling to commit this nation to a course which calls for growing into a capability which does not currently exist," he had written to the retired general Daniel Graham, the founder of the pro-missile-shield group High Frontier, just months prior to Reagan's announcement. Now Weinberger was expected to build just such a capability, so he turned to DARPA's director. "I spent the following ten days after that—at least several hours a day—with Cap Weinberger, telling him what the President meant," Cooper said.

The irony was that the president's scientific adviser had been heading a study, funded in part by DARPA, about the feasibility of missile defense. The study, which was pessimistic about the chances of developing anything effective anytime in the foreseeable future, was nearing completion just as Reagan made his announcement. "When the

President made his announcement, the study disappeared in a flash of smoke, and suddenly there was wild enthusiasm in the Office of the Science Advisor to the President about a ballistic missile defense," Cooper said, "and things moved forward very rapidly after that."

Cooper opposed Star Wars, and he was even more opposed to the idea of turning DARPA into a Star Wars agency. Over the next few months, Pentagon officials, including Cooper, debated the best way to put together the president's vision and what DARPA's role in that vision would be. Robert Kahn, a DARPA computer scientist, recalled a meeting with Cooper and other officials where DeLauer, a devoted public servant, started tearing up in frustration over the missile defense plans. "He was not a happy camper about this at all," Kahn said. "There was not much he could do about it."

With DARPA's fate in flux, Cooper decided to hold an off-site meeting for office directors in Berkeley Springs, West Virginia, about eighty miles from the Pentagon. He would put the missile defense debate to a vote, allowing the agency's top officials to decide whether or not DARPA should work on a seemingly impossible scheme. The officials who worked on science were against it, worried it would eat up their budget, while those involved in more advanced weapons technology were in favor of it, seeing an opportunity for even more funding. Officials in DARPA's Directed Energy Office, which was home to the high-energy lasers being swept up into the president's program, were ambivalent, because their programs were likely to benefit regardless of whether they stayed in DARPA or left.

In the end, Cooper decided to hand off DARPA's missile defense programs. The Pentagon set up the Strategic Defense Initiative Organization in March 1984, sweeping up most of the Pentagon's missile defense research, including DARPA's laser programs. It was a smart decision; the laser programs, which were ballooning in costs, "were eating DARPA alive," Cooper said later. He told the exiled missile defense scientists they could come back to DARPA if they wanted, and "some of them did," he recalled, "after the craziness . . . got rolling."

Yet the loss of missile defense was not a bad thing for DARPA, which still benefited from the surge in defense spending. The Pentagon's budget more than doubled, and Cooper over the next four years presided over one of the largest expansions in DARPA's history, which fueled the agency's image as a technology factory. Soon, it was sponsoring dozens of aircraft and weapons projects, many of them highly classified. Not since DARPA was founded in 1958 did the agency have such a clear mission: create weapons that could outmatch the Soviet Union. Gone were concerns about insurgencies, advising foreign allies, or studying human behavior. DARPA would be building a weapons arsenal of the future.

Even if inadvertently, Reagan's enthusiasm for technology and defense spending had revived DARPA. The Tactical Technology Office, the division created from the remnants of AGILE and Vietnam, was the largest beneficiary of that increase, becoming the agency's new center of gravity. "We were spending money like it was going out of style," Cooper later said. "I mean it was fantastic." Ironically, the technology that would have the greatest impact on war was a DARPA project so small that many directors could not even remember its name. Its roots, like those of so many DARPA projects, stretched back to Vietnam.

In the early 1980s, Allen Atkins, now a DARPA official, was in Israel to give a talk at a conference and to brief the Israelis on some of the agency's unclassified projects, including its unmanned aerial vehicles. As Atkins recalled it, he was sitting in the hotel's lobby bar when a "little portly guy" came up to him and stuck out his hand. "I'm Abe Karem," the man announced. And without much more of an introduction, Karem plopped down next to him and began to describe his concept for an unmanned aircraft that could stay aloft for days at a time.

In the United States, drones on the battlefield had not progressed much since the Vietnam War, when DARPA sent the QH-50 to hunt Vietcong. DARPA had continued to sponsor work on small-scale drone projects, but without much interest from the military. The air force was run by pilots, who did not want to be put out of a job, and the army and the navy, though slightly more interested in drones, could not quite imagine how they should be used. For example, DARPA had handed the army a tactical drone called Prairie, which was powered by a lawn mower engine. The army eventually called its version the MQM-105 Aquila, which was supposed to be launched by a catapult and recovered in a net. Rather than keeping it simple, the army came up with more things the drone should be able to do. Cost estimates for Aquila grew to $2 billion before the army canceled it. The drone was never used in combat.

Compared with the United States, the Israelis had embraced unmanned aircraft, particularly during the 1973 Yom Kippur War, when the Israel Defense Forces used drones to draw out anti-aircraft artillery over the Golan Heights and to perform reconnaissance. Though the Israelis might have employed drones more readily than their American counterparts, the Israeli military did not have the money to fund ambitious new aviation projects, at least compared with the Pentagon. Frustrated, Karem, an Iraqi-born Jew who had worked as an aircraft designer in Israel, moved to the United States. He set up shop in California, eventually working from a garage attached to his home. There, Karem started work on unmanned aerial vehicles that could stay aloft for long periods of time, and that was the idea that Karem presented to Atkins that night in the hotel bar. "Karem wasn't really pushing any particular application," Atkins said, he was just interested in building drones.

Karem's proposal intrigued Atkins. Up to that point, one of the biggest problems with military drones was simply that they crashed too often. The QH-50, the unmanned helicopter that DARPA had sent to Vietnam, had a reputation for falling out of the sky; the Aquila did as well. Karem was pitching reliability, so Atkins decided it was worth at least having someone at DARPA check out the Israeli inventor's idea. Atkins assigned Robert Williams, a program manager who had a reputation as a brilliant aerodynamicist and a keen eye for spotting talent. "You are going to have to talk to him to find out if he has anything," Atkins told Williams.

There was another reason why Atkins thought DARPA might be interested, though he could not tell Karem, because it was classified. In 1980, DARPA had launched the top secret program Teal Rain, a code name for a series of unmanned aerial vehicles that would replace spy planes like the U-2 and the SR-71. Some of the projects under Teal Rain were classified, while others were conducted in the open. Even thirty years later, DARPA officials decline to speak about many aspects of Teal Rain, citing its secrecy.

As it turns out, Karem and Williams, both known for single-minded pursuits, meshed well. Karem was a rogue aircraft designer whose vision left little room for tact. "Gentlemen, everything I see in this room is nonsensical," he said at a meeting with a large defense firm. Williams was an engineer turned government bureaucrat with money to spend and whose love of visionaries could sometimes cloud his better judgment. As an initial step, Williams funded flight tests of the Albatross, a two-hundred-pound drone that Karem built in his garage. It flew an astounding fifty-six hours. When that design proved successful, DARPA then paid Karem to build Amber, an unmanned aircraft funded under the Teal Rain program. While Amber, which eventually flew 650 hours without a single crash, was ostensibly run in partnership with the navy, the real interest, according to Atkins, was from the CIA. "The navy was in it primarily for cover," he said.

By 1990, DARPA finished its work on Amber. As a research agency, it could only develop prototypes; it was up to the military to buy production aircraft. Lacking new orders, Karem was pushed into bankruptcy, forced to sell what was left of his company to General Atomics Aeronautical Systems. Then the CIA bought the Gnat, a derivative of the Amber that Karem built to sell abroad, and used it for surveillance during the war in Bosnia. Less than ten years later, the agency bought another derivative of Karem's work from General Atomics Aeronautical Systems, called the Predator. By that time Karem was not even involved in the company, though his work had led to a weapon that was about to change how the United States conducted war. The Predator was sent to Afghanistan in the days after 9/11, armed with Hellfire missiles and used to kill "high value targets," much as DARPA had wanted to do with the QH-50 back in the Vietnam War. This time, the plan worked, and the Predator ushered in an era of remote-control killing. For better or worse, the Predator, as Richard Whittle wrote in his history of the unmanned aircraft, "changed the world." So, too, had DARPA.

In the 1980s, DARPA was expanding rapidly into secret aviation projects. The success of Have Blue, the first stealth prototype, prompted DARPA to start another radar-evading aircraft program, called Tacit

Blue, a prototype plane based on Northrop Grumman's design, which had originally lost out to Lockheed's Skunk Works. Tacit Blue was a bizarre-looking aircraft that DARPA pursued, in part, to ensure that at least two companies would have the ability to build stealth aircraft. "Looking at it from the side, it looks just like a whale, with the fins," said Atkins, who was recruited to DARPA to manage the program because of his experience on Have Blue.

Tacit Blue's odd shape earned it the nickname the Whale, and many of those who worked on it wore gold tie tacks that had tiny whales on them—a way of acknowledging that they belonged to an elite club. Some of the tie tacks even included tiny diamonds to represent the aircraft's side-looking radar. As a spy plane, Tacit Blue was being used to test whether a narrow-band radar would be invisible outside its transmission range. "Most people didn't know about the radar," Atkins said. "They thought it was just building an airplane." Like Have Blue before it, Tacit Blue was kept completely in the "black" and flown at Area 51.

As DARPA's black aircraft programs boomed in the 1980s, the agency would often announce a research program in aeronautics as a cover for building a secret military prototype aircraft. This enabled DARPA to award contracts and buy needed equipment without raising suspicion. Atkins described one example of what he called a "White World" program, meaning unclassified, that was run jointly with NASA. Yet the technology was also being looked at for secret military applications. "We built some full-scale models to see how we would militarize it," said Atkins. The "black" project was a stealth rotorcraft.

The cover project was called the Rotor Systems Research Aircraft/X-Wing, or RSRA, a joint DARPA-NASA program that funded Sikorsky to design a hybrid of a helicopter and a fixed-wing aircraft. RSRA was a real program, but it was also a cover for one of DARPA's more significant attempts to develop stealth helicopters. The "black" program involved taking "the rotor head off of the RSRA and putting it onto a stealth vehicle," Atkins said. A helicopter's rotor blades typically produce a Doppler shift. Those shifts are difficult to mask from radar, but it could be done, as DARPA learned with the X-Wing.

One after another, "X-planes," or prototype aircraft, seemed to spill out of Atkins's office. One, called a CycloCrane, looked a bit like La Minerve, the nineteenth-century fantasy blimp dreamed up by the

Belgian physicist and magician Étienne-Gaspard Robert. The Cyclo-Crane was a hybrid lighter-than-air vehicle that incorporated elements of helicopter control. It looked as if someone had taken propellers off a beanie hat and placed them strategically around the body of the airship. Highly maneuverable and with the ability to haul large amounts of cargo, the CycloCrane was proposed by DARPA to the navy as a possible way of off-loading ships in areas that did not have developed ports. The navy was not interested in the strange-looking airship. "If you could ever get beyond the giggle factor, this would be a great aircraft," Atkins recalled one admiral telling him.

Another "doomed by looks" project was the X-29 forward-swept-wing aircraft; the wings look as if they were put on backward. The aircraft, which in theory would be highly maneuverable, never went beyond DARPA's investment, in part because the air force had no interest in it. "No, that's too ugly an aircraft," one four-star general told Atkins.

For Atkins, those were exciting days at DARPA as money flowed into the aviation programs, initially run out of the Tactical Technology Office. The programs grew so quickly that the office was "swallowing the agency's budget," said James Tegnelia, the deputy director of DARPA at the time. So, DARPA created a separate Aerospace Technology Office, headed by Atkins. Splitting aviation off into a new division did not necessarily solve the problem, because the new office eventually swelled to $1.5 billion, or more than half of DARPA's budget. "I had one program that lost"—Atkins then corrected himself—"spent $600 million, and it failed, but we knew where it failed at," he said. That program, like many others, remains a secret. Those aircraft might not have been successes, because they never led to something used by the military, but they were the types of high-risk concepts that were encouraged at the time. "You don't go after a minimum change or an incremental change," Atkins said. "You go after something that's going to force people to think outside of the box."

Secrecy and unbridled ambition were the hallmarks of aviation programs in the 1980s. DARPA had long managed classified programs, but secrecy seemed to be enveloping much of the agency as the black aerospace research burgeoned. As the head of the Aerospace Technology Office, Atkins enjoyed the cloak-and-dagger nature of those programs, even when they sometimes brought DARPA into conflict with the military services. DARPA was encouraged to charge forward, even if military officials were opposed.

One time, Atkins needed to get James Ambrose, the undersecretary of the army, to sign a classified agreement with DARPA on a major black aircraft program. Ambrose, a master of bureaucratic wrangling, was trying to avoid signing the agreement, even though the army had agreed to participate. Unperturbed, Atkins got Ambrose's travel schedule from his aides and hatched a plan to track him down at LaGuardia Airport. The only hitch was that the agreement was for a top secret code word program, so the documents had to travel with two individuals with appropriate security clearances. Atkins ended up taking his wife, Natalie, a DARPA secretary who was cleared onto the program. At LaGuardia, the husband-and-wife tag team ambushed Ambrose as he got off his flight.

Cornered, Ambrose agreed to sit down at a dimly lit airport restaurant. "It was Ambrose facing me, and [my wife] and I were sitting so we could look out to see what was going on, and we had his aides fix up a perimeter around it," Atkins said. "He's sitting there with a candle, holding it over the document, trying to read it. Finally he turns and signs it, and I get back up, put it in an envelope, seal it again."

Atkins recalled another run-in with Ambrose at a meeting of the Defense Resources Board. Ambrose was known as a fierce protector of the army budget, and DARPA programs were a direct threat to the army's own weapons development. Cooper, the DARPA director, believed Ambrose was blocking a top secret DARPA program, which the army was supposed to be supporting (Atkins declined to name the program).

Secretary of Defense Caspar Weinberger had a habit in meetings of closing his eyes, leading some to believe he was asleep or not listening. Yet at the end of protracted discussions, and even heated arguments, he would open his eyes and make a simple pronouncement. As Cooper argued his point at the meeting, he stood up, all six and a half feet of him, towering over everyone in the room, including the diminutive Weinberger, who sat back silently.

"There are over half a dozen ways you can undermine a program, and this wall-eyed son of a bitch has done all of them!" Cooper shouted, bending over Ambrose and pointing a finger in the army official's face.

"Bob, why don't you tell us what you really think?" Weinberger said.

Weinberger then turned to Ambrose and asked, "Is any of that true, Jim?"

Ambrose started to defend himself, explaining what he had done to protect the army's budget, which only bolstered Cooper's claims.

"Well, Jim, we're going to do the program, and the Army's going to do it with DARPA," Weinberger said.

To this day, Atkins will not say precisely which project was the subject of the contentious meeting, but three decades later an army stealth helicopter flew navy SEALs into Pakistan on a mission to kill Osama bin Laden. By that point, the descendant of another DARPA aircraft from Atkins's office—Karem's drone—had been hunting and killing suspected terrorists for nearly a decade.

As the 1980s progressed, Reagan's techno-optimism permeated the Pentagon and DARPA. Projects that might once have been discarded as pipe dreams suddenly seemed plausible, even to Cooper, who had been a missile defense skeptic. In April 1983, just a few weeks after Reagan's Star Wars announcement, a member of the duPont family came to talk to Cooper about a plan for a hypersonic space plane. It was fortuitous timing. A space-based missile shield would require putting satellites, weapons, and other technologies up in orbit quickly and cheaply, something that an aerospace plane could do, at least in theory. Rather than the months it takes to plan a rocket launch, the aerospace plane would just zip up to orbit and then return to earth, landing on a runway. "This is something DARPA ought to do," Cooper enthused.

Tony duPont was not involved in the family's eponymous business, E. I. du Pont de Nemours and Company, better known simply as DuPont, but he shared an entrepreneurial spirit. He was a former Pan Am pilot and then later an aerospace engineer who had worked for more than a decade at Douglas Aircraft. There, he specialized in missile and space systems, focusing on how to get those vehicles to reenter earth's atmosphere without burning up. In the 1970s, he struck out on his own, founding duPont Aerospace. He had moderate success working with NASA on a hypersonic engine but harbored grand ambitions of building an entirely new type of aircraft.

The route duPont took to DARPA went through Tony Tether, the head of the Strategic Technology Office and a future director of the agency. Tether had been attending government meetings on trans-atmospheric vehicles—vehicles that combine attributes of aircraft and spacecraft—when he met duPont. Some in the aerospace community regarded duPont as a snake oil salesman who oversold his ideas, but the soft-spoken, earnest-sounding engineer had a way of winning over even ardent skeptics. Tether, a science fiction fan, took an immediate liking to duPont and his ambitious ideas and sent him to Cooper, who in turn assigned the space plane to Robert Williams, the same DARPA program manager who had championed Abe Karem, the maverick drone designer. Cooper described Williams as an "imaginative fellow" and thus the "right guy," even if hypersonics was not his area of expertise. Cooper would later regret that judgment.

Space planes were an ambition that dated back to the very origins of DARPA, but there was reason to be skeptical of duPont's claim that he could design such a vehicle, which would need to travel at hypersonic speeds to reach orbit. Hypersonic vehicles—missiles or aircraft that can travel many times the speed of sound—have long been a dream of aerospace engineers. A hypersonic aircraft would make travel from the United States to Europe or Asia more like a brief train ride than a daylong journey. A hypersonic missile could strike an enemy halfway across the planet in just over an hour. And a hypersonic space plane, like what duPont was proposing, could lift things—people, satellites, or weapons—into orbit quickly and cheaply.

One of the keys to building the space plane was a supersonic combustion ramjet, known as a scramjet. A conventional rocket carries its own oxidizer, which for space vehicles, like the U.S. Space Shuttle, means taking off with a massive tank filled with liquid oxygen and liquid hydrogen. A scramjet, on the other hand, takes in air from the atmosphere. The difficulty is that a scramjet only works at high speeds, around Mach 6. Even then, keeping the engines going is "not unlike keeping a candle lit in a hurricane," as one writer described it.

But duPont believed he had a design that could do it. His idea was to build a space plane powered almost entirely by a hybrid scramjet surrounded by ejectors channeling rocket exhaust. Rather than a big external booster, the duPont engine would incorporate a small rocket to power the aircraft until it reached high enough speeds that the scramjet could kick in, boosting the plane into orbit. It was a novel concept but also incredibly complex.

One of the reasons that a space plane had not been pursued for several decades—along with the high price and complexity—was that it was not necessarily clear that such an exotic technology was needed. With Star Wars ascendant, DARPA now had a motivation to support duPont's idea. The Pentagon was geared toward space weapons for the foreseeable future, and duPont's aerospace plane could help put them in orbit. The Defense Department also had another highly classified mission for the space plane.

"Can you do this mission?" Williams asked duPont.

"I'll study it," duPont told Williams.

"It was exciting; someone was finally interested," duPont recounted. He stayed up all night, running the numbers, based on the hypersonic work he had done for NASA, and extrapolating out from the models he had built to determine whether ramjets could run to Mach 25, taking the hypersonic plane to orbit. A few days later, he called Williams back. "You can do it," he told him.

At that point, duPont was awarded a study contract for just $30,000 to draw up a theoretical design of an aerospace plane that could take off from a runway and then accelerate up to Mach 25. The study was to design the smallest plane possible that could reach polar orbit and then return to earth. The magic number was a fifty-thousand-pound plane with a twenty-five-hundred-pound payload, and that was the design that duPont submitted. DuPont delivered the results of his study at 6:00 p.m. on September 30, 1983, just as the fiscal year ended.

Cooper approved $5.5 million to expand on duPont's computer modeling and come up with an actual aircraft design. That was the beginning of Copper Canyon, a highly classified program to build a space plane. Soon, Williams and Cooper were making the rounds in Washington, briefing senior White House and Pentagon officials on DARPA's plan.

To this day, duPont will not divulge the secret mission for the theoretical space plane, but in an interview in 2013 he revealed details that provide strong evidence that Copper Canyon was intended to succeed the SR-71 Blackbird, a spy plane that could travel at more than three times the speed of sound. He confirmed that the secret mission required a polar orbit, which is the orbit typically used for spy satellites, because it allows them to image the entire earth. The secret mission would require two pilots "to be sure someone doesn't have a heart attack or indigestion," duPont quipped.

DARPA began promoting the project to lawmakers. In a congressional hearing, Cooper, the agency director, described it as a potential "globe-girdling reconnaissance system, a kind of super SR-71." Cooper said it could even be used as a "long-range air defense interceptor" against an incoming Soviet bomber. An SR-71 could travel at around Mach 3, an impressive speed, but that was nothing compared with a hypersonic space plane that could reach anywhere on the earth within an hour, de-orbit for ten minutes to spy, then re-orbit and return to the United States. The technology was nearly in hand, Cooper insisted, and an aircraft could be ready within a decade. "Over the past year we have convinced ourselves . . . that it is possible to operate at altitudes all the way up to 250,000 to 300,000 feet and at speeds of up to Mach 25, the speed required to exit Earth's gravitational field," Cooper told lawmakers.

The phrase "convinced ourselves" was an unintentionally apt description. At that point, the scramjet engines that would propel the plane into orbit had never actually been tested in flight. Yet officials gripped by the technological optimism inspired by Reagan were enthused. George Keyworth, Reagan's science adviser, recalled a 1984 briefing on Copper Canyon to the White House Science Council, where normally prolonged debate was replaced by declarations of immediate support, such as "Let's do it."

The following year Williams briefed the space plane project to Weinberger, who, after listening without uttering a word, replied at the end with a one-word pronouncement. "Interesting," he said. With Reagan's defense-spending spree, interesting was apparently enough. In 1985, the Pentagon chief approved Copper Canyon as a major project, which would soon be designated the X-30 National Aerospace

Plane. It became one of DARPA's best-known—and most disastrous—projects of the decade.

In early February 1986, Ray Colladay went to visit Pat Buchanan, the White House communications director, to talk about the National Aerospace Plane. Colladay at the time was the associate administrator of NASA, which was working with DARPA on the project.

Buchanan wanted to speak with Colladay about Ronald Reagan's upcoming State of the Union address. Buchanan showed Colladay a section of the president's speech, which included a key reference to the National Aerospace Plane. Colladay took one look and was horrified. DARPA's original concept for Copper Canyon had been for two pilots, and even that was considered ambitious. Reagan's draft speech referred to plans for a hypersonic passenger plane, something that was not anywhere close to what DARPA and NASA were working on, or even physically possible. "You can't say that," Colladay said. "That's just nonsense."

"Well, we're going to," Buchanan told Colladay. "We've got to relate this program to the American people and in a way that they can understand it."

On February 4, Ronald Reagan delivered his State of the Union, beginning with a tribute to the victims of the recent *Challenger* disaster; the space shuttle had blown up seventy-two seconds after takeoff, killing all seven crew members. The president assured the nation that the tragedy would not prevent the United States from moving forward in space. He then made the astonishing announcement that the government was "going forward with research on a new Orient Express that could, by the end of the next decade, take off from Dulles Airport, accelerate up to 25 times the speed of sound, attaining low Earth orbit or flying to Tokyo within 2 hours."

Copper Canyon's growth from a small space plane to an Orient Express was an appropriate reflection of Cold War excess under Reagan, whose vision of technology, whether space weapons or space planes, was never constrained by the laws of physics. The "Orient Express" announcement was greeted inside DARPA with pure dread. Most agencies would be happy to have their program singled out by the president in a State of the Union speech, but DARPA for years had benefited from operating under the radar, allowing its high-risk technology programs to succeed—or fail—without the risk of public humiliation and high-profile congressional inquiries. Now Reagan had put its small experimental space plane high on the national agenda.

DARPA's man in charge of the National Aerospace Plane was Robert Williams, who had also led the agency's successful drone program. Williams felt that for the space plane to succeed, it needed to be a big program, with large defense companies and NASA participation. "It's better to have them pissing outside the tent, than inside," Williams told duPont, explaining that to have big companies and labs involved would help the program. Soon, DARPA was contracting with five major defense and aerospace companies—McDonnell Douglas, Rockwell International, General Dynamics, Rocketdyne, and Pratt & Whitney—to develop the vehicle and its engines. "That's where Bob and I actually had a falling-out," Allen Atkins, the head of the Aerospace Technology Office, recalled. "I said, 'Don't get involved with the labs. Don't get involved with any other centers at NASA.'"

The National Aerospace Plane became everything that the stealth aircraft was not: big, bloated, and involving multiple government agencies and several large companies. With the stealth aircraft, George Heilmeier fought to have the air force contribute funding to its prototype but insisted that DARPA maintain managerial control. Williams did almost the exact opposite, believing that having multiple agencies and companies involved would help maintain a strong lobby to protect the program from budget cuts. He was right at first, as evidenced by the president's glowing endorsement.

Yet as the number of companies and agencies involved in the National Aerospace Plane grew, so too did the size of the aircraft. What started as Copper Canyon, an idea for a 50,000-pound design, soon ballooned to a monstrous 250,000-pound vehicle, a size the defense companies insisted was needed to achieve orbit without multiple rocket stages. In the meantime, the costs swelled to $17 billion to build two prototypes. Tony duPont, the man behind Copper Canyon, blamed the major defense contractors. "If we stuck with [the original design], we'd be flying into orbit for $10 a pound," he said.

Perhaps, but plenty of aerospace engineers were dubious of duPont's models. To save weight, his original design did not factor in landing gear. It packed just enough fuel to get to orbit, which meant the space plane would not be able to maneuver when it reentered the atmosphere. In fact, it could not maneuver period, because it had no rockets for maneuvering, let alone fuel to power them. As other designers addressed those shortcomings, the size and weight of the aerospace plane grew. So did the price tag.

In the fall of 1987, Williams broke all protocol by writing directly to the White House chief of staff, Howard Baker, to protest budget cuts to the National Aerospace Plane. As usually happens when a midlevel government official breaks protocol, the letter ricocheted back to the Pentagon's senior leadership, working its way down the chain until it landed on the DARPA director's desk. Furious, Robert Duncan, the director, immediately removed Williams as the head of the program. Watching from outside government, Bob Cooper, who had approved the space plane, was horrified. "When I saw that happen, it was like having one of your children start taking drugs," Cooper recalled.

In February 1988, DARPA handed over the reins of the National Aerospace Plane to the air force. The program would continue on for five more years, into the next administration, in part thanks to enthusiastic support from Vice President Dan Quayle, before being canceled. Almost $2 billion was spent trying to develop a prototype, making it one of DARPA's costliest failures. Star Wars, the missile defense system that the space plane was initially funded to help, ended even worse, though DARPA was not directly involved. The Strategic Defense Initiative Organization pursued a variety of far-fetched schemes that ranged from bouncing lasers off space-based mirrors to sending kamikaze mini-satellites into orbit (reminiscent of DARPA's "lunatic" BAMBI program). It would spend $30 billion of taxpayer money without ever deploying anything close to a shield that would render nuclear weapons obsolete.

By the mid-1980s, the Soviet Union was faltering amid its attempts to keep pace with American military spending, just as Caspar Weinberger had predicted. Consumer goods, always in short supply, were even harder to come by as the central government directed resources to its military. In the meantime, American Cold War defense spending peaked at more than $300 billion, and DARPA rode the military windfall, taking small prototype projects pursued during Vietnam, like drones, and transforming them into large-scale weapons projects. And yet it was not always the most ambitious or expensive projects that succeeded. Many of the aviation projects pursued during the Reagan years, like the forward-swept-wing aircraft and the oddly shaped X-wing aircraft (and its "black world" counterpart, the stealth rotorcraft), never even took flight, doomed by aerodynamics that confounded state-of-the-art technology. The long-endurance drone developed by the Israeli aircraft designer Abe Karem was a success, on the other hand, but that only became apparent years later, after DARPA was out of the picture.

In just over a decade, DARPA had completed yet another transformation. In the early 1970s, DARPA had reinvented its jungle warfare work as technologies to fight the Soviets, and now those technologies were at the center of the Pentagon's plans to develop a secret weapons arsenal. The agency benefited from generous funding and political support, and DARPA entered the latter stage of the Cold War seemingly stronger than it had ever been. There was just one problem: DARPA had helped build a phantasmagoria of weapons to fight an enemy that was about to collapse.

CHAPTER 16

Synthetic War

In the mid-1980s, the Warsaw Pact had two and a half tanks for every one NATO had, a grim statistic that haunted military war planners. Analysts debated just how important the Soviet advantage was; after all, the Soviets tended to favor quantity over quality, and the United States had focused on deploying advanced technology, including DARPA-driven innovations like stealth aircraft and precision weapons. Even then, the Soviet numerical advantage was hard to ignore, and despite claims that superior technology could trump numbers, the United States for more than a decade failed to win a NATO tank competition, called the Canadian Army Trophy, held in Germany. The implications were stark: If the United States could not even win in a mock battle against its own allies, what hope did it have in a real war against the Soviet Union?

In 1987, an American army armor officer at Fort Knox arranged to have new DARPA-built simulators sent to Germany, where the U.S. Army was training for the annual tank competition. The participants were not allowed to practice on the range ahead of time, but the contest did allow training devices. DARPA sent four new simulators it had developed, called SIMNET, over to Germany, and a complete graphic visual model of the range and targets.

The simulator's graphics were not particularly amazing; they were not much better than 1980s-era arcade games. Early on, DARPA had decided that in military simulation, fidelity—meaning essentially how realistic something looked—did not necessarily matter. Soldiers played video games; they could suspend disbelief. Instead, a simulator could have "selective fidelity," focusing on those elements that were critical for training. The key to these simulators was that they were networked together, allowing soldiers to practice against each other, the way people would play against unseen opponents over the Internet years later in the world of online gaming.

The networked simulators were the brainchild of Jack Thorpe, an air force officer with a PhD in industrial psychology. Thorpe had long been thinking about a way to get the air force to rely more on simulation. During the Cold War, the air force practiced for war, but rarely did it do so in the way that major battles would actually be fought, with hundreds of airplanes, which would need to coordinate and synchronize in ways that could not be planned ahead of time. Simulation had obvious benefits: it could save money by allowing pilots to train without paying for expensive flight hours; it would also allow pilots to practice tactics that might be too risky to carry out in a live exercise. What simulation had not done, however, was find ways to mimic the initial days of war, with large flight operations and a constantly changing battle plan.

Back in 1978, Thorpe had circulated a white paper to colleagues, speculating on what simulation might look like some twenty years in the future. In it, he predicted, "Significant breakthroughs in numerical processing will provide the resource of computational plenty. Cheap, powerful computers will proliferate training systems and their associated inter-connecting networks." The air force already had flight simulators, but Thorpe's idea was to link those simulators on a network so that pilots could practice being in combat together. The reaction, he recalled, was positive but not concrete. "Well, that's kind of a neat idea," he was told. "But how would you actually do that? How would you build a network of simulators to allow that to happen?"

Just a few years later, in 1981, Craig Fields, a longtime scientist at DARPA, recruited Thorpe to join the agency to work on simulation. The ARPANET, at that point, was in full swing, networking together computers across the country, allowing people to interact virtually. Fields, who was heavily involved in the agency's computer science work, realized that the same technology could be used to link Thorpe's simulators together. Simulators are fundamentally computers, and creating large networks of them was a computer-networking problem. In a single afternoon in 1983, Thorpe and Fields sketched out an idea for hooking together the simulators at different locations, creating a virtual world of combat. In the end, however, it was the army, not the air force, that signed onto DARPA's idea. Rather than aircraft, the first networked simulators would involve tanks. That year, DARPA, together with the army, launched SIMNET, or Simulation Networking, a $300 million research project that used packet switching and computer networking—DARPA innovations—to link tank simulators in a virtual environment.

The true revolution of SIMNET was not in creating an exact replica of the battlefield but in allowing people to interact in a virtual world. Prior to SIMNET, simulators had been like a one-player arcade game: you could practice warfare, but you were really just playing against a computer, with all of the limitations that entailed in the 1980s. Now tank operators could train on a simulated battlefield populated by other tanks, controlled by real soldiers.

Years before online gaming entered the commercial market, SIMNET allowed army tank operators to "play" in a virtual environment, and, in 1987, for the first time in the history of the Canadian Army Trophy competition, American forces won. That same year, the first networked tank simulators were fielded to the U.S. Army, and by the fall of 1989 half a dozen SIMNET sites had been established at bases around the country. SIMNET was fielded and being used to train tank operators just in time to see the collapse of the Warsaw Pact and the probable end of a potential tank warfare scenario in Europe. Yet SIMNET's technical success paved the way for a new direction for DARPA, which now sought to use computers to create synthetic versions of real war.

In the spring of 1989, a series of revolutions swept across the countries behind the Iron Curtain, ending nearly five decades of one-party communist rule. The Berlin Wall that divided East and West Germany fell, and the Fulda Gap went from being the hypothetical battlefield of World War III to just another stretch of bucolic German lowlands.

It was a watershed for the Pentagon, an institution that for decades had an almost maniacal focus on confronting Warsaw Pact forces on the European battlefield. No one in 1989 quite knew what was going to happen with the Soviet Union, but clearly its economic backbone was crumbling, and the Politburo was more preoccupied with reining in its increasingly raucous member states than trying to outpace the United States in technological weapons developments. The Soviet Union was still two years from its final collapse, but as a competitor to the United States in high-tech weaponry it was already dead.

For DARPA, an agency that had been created to help match Soviet technology, the shifting strategic landscape sparked internal changes as well. In July 1989, Craig Fields, who had been at the agency since 1974 and helped create SIMNET, was tapped to be the head of DARPA. The new director had a way of evoking strong reactions—both positive and negative. "Brilliant" was the word most often used to describe Fields, who would often dazzle military and intelligence officials with his command of science and encyclopedic knowledge of DARPA's programs. "Abrasive" was typically the second most common word used to describe him. Fields did not suffer fools gladly, and he saw fools all around him: in the Pentagon, on Capitol Hill, and in the White House. Fields had been at DARPA for fifteen years before becoming director, but he had resided in the "science" offices, which did not interact as much with Pentagon leadership. He was happy to distance DARPA from the Pentagon. "Moving out of the Pentagon strikes me as a great success, being far away from filling out all those forms," he recalled, in a later interview, of DARPA's post-Vietnam-era exile. "What a wonderful success!"

The future, in Fields's view, was not in warships or military planes but in electronics and computers. While DARPA in the 1960s and 1970s had laid the foundation for personal computing and the modern Internet, by the early 1980s the agency's computer science work had atrophied. The ARPANET had been transferred to the Defense Communications Agency, and DARPA's subsequent directors pushed the computer science office to work on immediate military technologies. Fields, who had been a close colleague of J. C. R. Licklider's, helped pioneer the agency's renewed push into computer science in the 1980s, including a billion-dollar initiative to create artificial intelligence. To justify the massive spending increase, which lacked an immediate Cold War justification, DARPA pointed to Japan. In 1981, Japan had announced plans for the Fifth Generation Computer project, aimed at creating artificial intelligence, and the country's economy, guided by the powerful Ministry of International Trade and Industry, was the new bogeyman. "We trundled out the Japanese as the archenemies and said we have to leap beyond the Japanese, you know, and whatever," recalled Bob Cooper, who was then the director.

"Whatever" was exactly right. The DARPA director privately acknowledged it was a ruse and he had selected the Japanese for convenience. DARPA formulated a ten-year, $1 billion effort to create artificial intelligence called the Strategic Computing Initiative; it would be DARPA's biggest and most ambitious investment in computing since the ARPANET. The plan would appeal to the "Atari Democrats," or the young tech-savvy Democrats who believed computing would save the American economy, and it would also pass muster with Republicans, who fretted over perceived threats to American hegemony. The director even went to Japan and traveled around the country, essentially gathering ammunition. "I came back and in private conversations with Congress and senators, I used it," Cooper bragged. "I mean unabashedly."

Despite the influx of money, the ambitious road to artificial intelligence would soon be lined with failed technology projects. DARPA funded Thinking Machines Corporation, a company building supercomputers based on massively parallel processing, only to see the company go bankrupt when its government contracts dried up. The Pilot's Associate, a talking, thinking computer program that would assist an aircraft pilot—as R2-D2 had served Luke Skywalker in the X-Wing fighter—ended in failure. A "smart truck" meant to navigate autonomously had problems differentiating between rocks and shadows. The billion-dollar artificial intelligence initiative failed to achieve any of its original objectives. By 1989, *The New York Times* was already writing the program's postmortem. Showing just how far afield the work had gone from the original vision of artificial intelligence, the article declared that DARPA was "giving up its work on its autonomous land vehicle," described as one of the "most publicized goals" of the Strategic Computing Initiative. Cooper, who had started and championed the program, was crushed. "It's over in my mind," he said.

DARPA had given up on creating artificial intelligence, but Japan was still a convenient bogeyman. With the collapse of the Warsaw Pact, Tokyo by 1989 had nearly eclipsed Moscow as the focus of Washington's policy wonks. Only rather than the nuclear balance, the pundits were warning of the United States' ballooning trade deficit with Japan, which had reached nearly $50 billion in 1989. Fears of an ascendant Japan (and a descendant United States) helped boost the Yale professor Paul Kennedy's book *The Rise and Fall of Great Powers* onto the best-seller list, based on the prediction that the United States was heading toward a downward spiral spearheaded by deficit spending and economic stagnation. "There is a basic conflict between Japanese and American interests notwithstanding that the two countries need each other as friends—and it would be better to face it directly than to pretend that it doesn't exist," the journalist James Fallows argued in *The Atlantic Monthly*. "That conflict arises from Japan's inability or unwillingness to restrain the one-sided and destructive expansion of its economic power."

The 1980s deregulation under President Ronald Reagan sparked a national debate over the role of government in managing industries. A group of vocal Democrats came out in favor of helping key industries. By the late 1980s, Congress and the White House, now under the administration of President George H. W. Bush, were sparring over the loaded term of "industrial policy," or the role of government in boosting the private sector through targeted investments. Senator Al Gore, a Democrat and a technology enthusiast, was a forceful advocate for having the government invest in key areas, like supercomputers, while Republicans blasted attempts to "pick winners and losers." The free market, the Republicans argued, would ultimately prevail.

DARPA, with Democratic congressional support, had quietly become a part of this debate. The agency was funding Sematech, a consortium of chip manufacturers formed in the 1980s to help advance research on semiconductors. A *New York Times* article from March 1989 compared DARPA to Japan's Ministry of International Trade and Industry, even though the latter had nothing to do with the military. "At a time when more industries are seeking Government help to hold their own against Asian and European competitors, DARPA is stepping into the void, becoming the closest thing this nation has to Japan's Ministry of International Trade and Industry, the agency that organizes the industrial programs that are credited with making Japan so competitive," the article said. DARPA was "being propelled, only partly by its own choosing, into the role of venture capitalist for America's high-technology industry."

Yet Fields, as the new DARPA director in 1989, embraced this venture capitalist role. He focused on the high-definition television market, arguing it would allow American industry to maintain its lead in superconductors. The new director wanted to place DARPA at the vanguard of industrial policy, pursuing dual-use technology to give the United States an edge in the global economy, even if it was a vision that ran counter to the White House. Congress, at DARPA's request, approved plans to spend $20 million to help the domestic high-definition television industry. Consumer electronics, Fields insisted, were critical for maintaining the American lead in semiconductors, which were also a key component in military systems. Fields was passionate about investing in consumer electronics but also naive, recalled Ray Colladay, a later DARPA director. Fields "had never run a private sector company or business, or product line, or anything that would give him experience that would help him appreciate how difficult it is to take a piece of technology and incorporate it into a product that people would want to buy."

When George H. W. Bush became president in 1989, he fashioned himself as a Republican idealist who opposed any government interference in the marketplace. Oblivious to the political mood in Washington, Fields became a vocal proponent of industrial policy. For those within DARPA who knew Fields, his determination to carve out a position in direct opposition to the presiding administration was not surprising. His undoing was a technology known as gallium arsenide, a potential replacement for silicon chips. Gallium arsenide chips were costly to make and the manufacturing base was still nascent, but compared with silicon, these new chips would be faster and more efficient and have properties, such as hardening against radiation, that made them particularly attractive to the military. DARPA had funded gallium arsenide in the past, but Fields wanted to use the technology as a test case for his ideas about supporting industry: DARPA, he decided, should invest in a gallium arsenide firm.

Around the same time that Fields was getting involved in industrial policy, Richard Dunn, DARPA's chief legal counsel, was exploring ways to enable the agency to cut through red tape. Dunn was in touch with a group of retired senior military officials who had been pushing the Pentagon to reform its contracting practices, and Dunn saw it as an opportunity to push through some of his own ideas. A few of those former officials, including the retired four-star general Bernard Schriever, regarded as the father of the American missile program, went to see Senator Sam Nunn, a powerful member of the Senate's military oversight committee. Soon, Congress gave DARPA legal authority to enter into something called "other transactions," which was, in the simplest sense, a way for the Pentagon to fund research companies, skipping the volumes of government regulations that accompany typical military contracts.

Fields saw the new legal authority as an opportunity for DARPA to act like a venture capital firm, and he turned his attention to Gazelle Microcircuits, a company working on gallium arsenide. Dunn argued it would be best to test a few small agreements—perhaps a few hundred thousand dollars—because DARPA had never used this new authority. But Fields wanted to make Gazelle the test case and took the idea to Charles Herzfeld, who by then was the director of defense research and engineering and DARPA's overseer. Herzfeld supported it enthusiastically.

Gazelle Microcircuits had already raised $10 million, and Fields wanted DARPA to be treated like an investor. Fields ordered Dunn not to write a statement of work. "If it was a venture-capital-supported firm, then he wanted this to be like a venture capital investment," Dunn said. Arati Prabhakar, a DARPA scientist in charge of the contract, sat in on company board meetings. "She operated in a completely different manner than a normal program manager. She was on the inside of that company," Dunn recalled. He was concerned about the vagueness, and after he consulted Prabhakar, they decided it would be best to specify the money was for research and development, distancing DARPA from the notion of investment. It was too late. On April 9, 1990, DARPA issued a press release announcing the Gazelle agreement, which gave the company $4 million over twelve months. "The agreement between DARPA and Gazelle was the first of its kind, based on DARPA's recently granted authority to support advanced research and development by innovative means other than traditional contracts and grants," the press release stated.

Gazelle would design a one-gigabit-per-second and faster gallium arsenide chip, and DARPA, for its investment, would get access to Gazelle's research and patents. Prabhakar hailed the award as a new way of doing business for DARPA. "Gazelle is typical of the companies we couldn't work fruitfully with in the past," she said. When *The New York Times* picked up the story, however, the deal was cast as an investment, as Fields had been advocating. "The Defense Department, through its advanced research arm, has for the first time made what is essentially a venture capital investment in a young Silicon Valley company," the paper reported. Suddenly calls started coming in from the White House, and Fields was in a panic. "Bring me the agreement quick," he ordered Dunn. Fields paged through the agreement and saw the statement of work that Dunn and Prabhakar had added. "Oh, thank goodness," Fields said.

Fields hoped that the statement of work—the very language he had originally not wanted included—would help distance him from the idea that DARPA was taking an equity stake in the company. At that point, it was too late. According to an unpublished DARPA history written by Dunn, Secretary of Defense Richard Cheney "was somewhat displeased by Dr. Fields' high-profile public support" for dual-use technologies, like gallium arsenide. More displeased, however, were the White House and the president.

Dennis McBride, a DARPA program manager, was waiting to see Fields when White House lawyers started to file into the director's office. "Dennis, you're postponed because these White House lawyers need to see Craig," the secretary told him. As McBride prepared to leave, he could hear Fields being fired. "Craig, you should not have had this press conference," one of the lawyers said. "President Bush loves you but you just spit right in his face." Pentagon officials initially claimed that Fields had been "offered an opportunity" for another job in the Pentagon. It was a face-saving measure; a few weeks after being removed from his DARPA position, Fields quietly left the Pentagon for good. He would never comment on the episode, and even when asked during an interview commissioned by DARPA nearly twenty years later, he tersely responded, "Well, I think you should check the public record."

As Fields was getting fired, Herzfeld was scuba diving in the Caribbean with his wife. Herzfeld had recently returned to the Pentagon, full of ambition. Once one of the most powerful positions in the Pentagon, the director of defense research and engineering was now just a high-level bureaucrat with a nice office. In the fall of 1986, the president had signed the Goldwater-Nichols Department of Defense Reorganization Act, the largest restructuring of the military since the National Security Act of 1947. The legislation was the culmination of several years of debate over the failure of the military services to work together cohesively. Among other changes, a newly created undersecretary of defense for acquisition would become the "weapons czar," while the director of defense research and engineering was downgraded to a second-tier job. Donald Hicks, who held the position at the time, resigned in protest. The change was not cosmetic; it meant DARPA reported to a lower level of the Pentagon.

Herzfeld was supposed to be DARPA's direct link to the Defense Department's leadership, but he found himself frozen out of major decisions and with no access to the defense secretary. "Did you hear Craig got fired?" Herzfeld's military assistant asked him when he returned from vacation, walking into his E Ring Pentagon office, several miles away from DARPA. Herzfeld had not, because no one had told him. It was a telling statement on the role of DARPA—and military science and technology generally—at the sunset of the Cold War. The agency once again had no agreed-upon mission, no political support, and, for the moment, no director. Even as its Cold War innovations were making their way to the battlefield, DARPA was edging away from war, or perhaps being edged out of it.

On August 2, 1990, some eighty-eight thousand Iraqi troops invaded oil-rich Kuwait. Less than six months later, after sanctions and international condemnation failed to persuade the Iraqi strongman, Saddam Hussein, to back down, the United States led an offensive that began in the predawn hours of January 17, 1991. Eight AH-64 army Apache helicopters slipped across the border from Saudi Arabia and into Iraq on a carefully rehearsed mission to destroy key Iraqi radar sites and open up a safe air corridor for air force aircraft. "There is no way we could build a corridor to get through safely with these jets without these surface-to-air missiles taking us down," explained Dennis McBride, who was in charge of DARPA's simulation work at the time. "We had to take out the surface-to-air sites."

In the days leading up to those attacks, DARPA-built simulation systems at Fort Rucker, in Alabama, were linked up to counterparts at Central Command, reviewing the tactics for taking out Iraq's air defense systems. It was, as McBride explained, a "nap of the earth" mission, meaning the helicopters had to fly very low and close to the ground to avoid detection by radar. Simulation was the only way to see where the mission might fail. At U.S. Central Command headquarters, General Norman Schwarzkopf, who was directing the war, reviewed the simulations. "We were instantiating it in the simulation, saying, 'This is a good idea but this one is not,' and 'Here's why,' and he would readjust," McBride recalled. "He personally planned that very first mission with us, with the simulation capability."

Shortly after 2:00 a.m. on January 17, the Apaches approached the Iraqi air defense sites—flying an approach first mapped out in DARPA's simulators and rehearsed in the Saudi desert—and destroyed the Iraqi air defense sites. A few hours later, an air force F-117 Nighthawk—a stealth aircraft derived from DARPA's Have Blue prototype—flew safely through that air corridor. The F-117's first bomb destroyed an Iraqi air force site, and the second leveled a telecommunications hub in the center of Baghdad.

Less than six weeks later, the first Joint STARS aircraft, a DARPA-sponsored airborne tracking radar, spotted a massive convoy of Iraqi vehicles fleeing Kuwait and passed the data directly to strike aircraft. The resulting destruction of some two thousand Iraqi vehicles earned the escape route the name "highway of death," and *Air Force Magazine* praised DARPA's airborne radar as "one of the more unlikely heroes of Operation Desert Storm." The Gulf War ended up as the proving ground for DARPA technologies that were developed to fight the massive military forces of the Warsaw Pact. Its simulators were helping plan the war. The F-117 Nighthawk, the air force's stealth attack jet and successor to DARPA's Have Blue, was conducting air strikes. Joint STARS, which emerged from DARPA's Assault Breaker program, was still a prototype but already changing how the military fought. Yet in the Pentagon, officials were not interested in any ambitious new DARPA projects.

"Look, we don't need now any new magic inventions," George Lee Butler, a senior official in the Joint Chiefs of Staff, told Victor Reis, who took over as DARPA director after Fields was fired. "What we need is something to basically drive costs down. We know our budgets are going to decrease." The Gulf War had just ended, and Reis admitted this view was a bit of a "culture shock" for an agency that had staked its reputation on developing revolutionary technologies, like stealth aircraft and advanced radar, but it was also the new reality. The Gulf War might have demonstrated DARPA's innovations of the past two decades, and bolstered its reputation, but it also came at a time when the Pentagon was mainly looking to save money.

Within days of the Gulf War's ending, Reis got a call from Jack Thorpe, the DARPA program manager who had pioneered the agency's simulation work. By 1990, the SIMNET program was done, and though Thorpe was long regarded as one of DARPA's most imaginative program managers, he had largely been sidelined. Even for an organization known for cutting through red tape, Thorpe had a reputation for bending the rules. When DARPA was told its simulator building at Fort Knox, Kentucky, was possibly in violation of the law because all new military construction required prior congressional authorization, he had a trailer hitch installed and claimed it was just a temporary structure. By 1991, Thorpe, who had managed to stay on as a DARPA employee for a decade, was working from a small office in Europe. "He had sort of been put there to get out of the way," Reis recalled.

Now Thorpe had an idea that seemed to match the "cost cutting" mission for DARPA. He said there had been a big tank battle during the Gulf War, just a few days prior. Thorpe wanted to re-create the battle in a virtual world of computers, something that had never been done before and could potentially save money on training. The key, however, was that he wanted to send scientists to a battleground in Iraq strewn with still smoldering Iraqi tanks. "I think we can do something with that," he told Reis, "but I would like to be able to go over."

The idea struck Reis as prescient; an inherent part of simulation is cost savings, because it is cheaper to practice in a video-game-like environment than to expend money on real fuel and practice ranges. "Sure," the director told him. And with that, DARPA set off on one of its most ambitious post–Cold War projects: to re-create real war in a virtual world, based on data gathered from the smoking ruins of a battlefield.

Normally, after major battles, army historians would be sent to interview participants and chronicle events as part of a written account of what took place. Thorpe wanted DARPA to send simulation experts to the battlefield, walk among the burned-out tanks, interview American soldiers who had fought there, and then plug that data into the virtual reality world of SIMNET. The entire battle could be re-created and played in a simulator, and more important, those simulators could be connected on a network, sending data packets back and forth, so that people could replay the battle as participants. "It would be like a living history," Thorpe suggested.

Thorpe proposed the idea to General Gordon Sullivan, the army's chief of staff. Soon, General Sullivan contacted Lieutenant General Frederick Franks, the Seventh Corps commander, who gave DARPA the go-ahead and selected 73 Easting, the largest tank battle since World War II. The battle had taken place on February 26, 1991, when the Second Armored Cavalry Regiment met up in the middle of a sandstorm with fleeing elements of Iraq's elite Republican Guard Corps. Over several hours, American forces destroyed dozens of tanks, armored personnel carriers, and trucks. The name 73 Easting was taken from the grid location in the desert where the Iraqi forces were decimated.

Two days later President Bush declared a cease-fire and the Gulf War ended. Within a week of the battle, a DARPA-sponsored team of researchers arrived in the Gulf and traveled to where the battle took place. The researchers interviewed the American soldiers who fought in the battle, asking them to recount what happened minute by minute. "There were still the tread marks in the sand. They were able to see all the blown-up Iraqi vehicles, which were still there," Thorpe said. "The army engineers had been in to annotate precisely where every one of those blown-up tanks was and the conditions of its demise. Was a turret blown off? And then what direction was it lying?"

When the team returned to the United States, its members worked in a room with Post-it notes stuck on different parts of a board. Like detectives tracing backward from a crime scene to re-create what happened, the scientists reconstructed the entire battle. The work took a year, but the results were unprecedented: a computerized, interactive reconstruction of a real-world battle, with a feature called "the magic carpet," which allowed users to zoom in and around to any place on the battlefield, at any moment of the battle. "You would be able to go in and visit, look at, and replay the simulation and see where everybody was, what they were shooting at, and even get inside their vehicles or ride a shell as it races downrange at a mile a second," Thorpe said.

By all accounts, the digital re-creation of 73 Easting was a technical success. Thorpe even commissioned a Hollywood-produced video showing the simulation in action, featuring interviews with some of the soldiers who fought in the actual battle; they praised the simulator for its realism. Reis, the director, showed the video to Cheney, the secretary of defense, and to General Colin Powell, chairman of the Joint Chiefs of Staff. He also took it to Capitol Hill and played it for lawmakers. Everyone loved it, and Cheney most of all. "Gee, if we had this earlier, I could have shown this to Hussein, and he might have realized in how bad shape he was and just given up," Cheney said, perhaps foreshadowing misplaced optimism he would have more than a decade later about Iraq.

Thorpe's SIMNET, which had been developed in the 1980s at DARPA, was already a widely regarded success. It had changed how the army used simulators and, for the civilian world, helped enable the technology that gave rise to online gaming. As a follow-up to SIMNET, the re-creation of 73 Easting was a hit, demonstrating what simulation could do when combined with real-world data. In technology circles, Thorpe became something of a folk hero, leading *Wired* magazine to declare that it was not the science fiction writer William

Gibson who invented cyberspace but Jack Thorpe. Yet it is less clear that SIMNET, and by extension 73 Easting, really had much practical effect on the military. The reality is that SIMNET was fielded too late to help the tank operators who fought at 73 Easting, though it was used in later years. "SIMNET was irrelevant to us," said Douglas Macgregor, who led a key squadron in the Battle of 73 Easting. "We never used it."

The question was, what could be done with 73 Easting, the follow-on to SIMNET? The answer, by the 1990s, turned out to be not much. The idea would have been to use the technology that emerged from the 73 Easting re-creation in other simulations, but DARPA's connections to the Pentagon's leadership were breaking down, and cutting-edge science and technology were no longer at the forefront of the Defense Department's agenda. Even the retired army general Paul Gorman, the intellectual godfather of simulated training, questioned whether the virtual re-creation of a battle contributed much to the military. "I don't know how to answer the question about 73 Easting," Gorman replied, when asked years later about its impact. "Elegant effort and delighted that it was done, but who is using it?"

For all the accolades heaped on 73 Easting, nothing concrete came out of it other than a re-creation of a tank battle that would likely not be fought again for at least another generation. Gorman had hoped it, and similar simulation efforts, would achieve much more, by changing how the military prepared for future wars. Using simulations to practice for the last conflict did not do anyone any good. "We were addressing in effect, force on force, something like symmetric opponents," Gorman said. For fighting insurgencies "you would need a superior formulation, and damn it, they didn't provide for that."

In its effort to develop simulation technology, DARPA found itself hitting a wall. When DARPA did try to move simulation beyond a Cold War battlefield, its efforts had limited success, because technology alone could not change policy. One never-publicized example involved the drug war. The White House's Office of National Drug Control Policy in the mid-1990s funded DARPA's simulation experts to create a model of drug trafficking to see if there might be ways of cutting off the drug cartels in South America. "The big issue was and still is the movement of cocaine from Central and South America into the United States," explained Dennis McBride, who was in charge of the effort. He named the project after Iolaus, who in Greek mythology had helped Heracles battle Hydra. The name ended up more appropriate than he had imagined.

"We built this incredibly complex end-to-end model from seed planting down in South America through the changing to a product at a wholesale level, the transportation across myriad modes of transportation, ultimately into warehouses in the United States of America," McBride said. Yet the more DARPA modeled the problem, the worse it looked. If one cartel was defeated, it ended up just strengthening another cartel. Like the Greek Hydra, if you cut off one head, two more rose in its place. DARPA came up with answers, but the answers did not fit what the White House wanted. If the Drug Enforcement Administration put more aircraft in the air, it did not help, because the cartels still had more planes. No matter which way DARPA modeled the drug war, it could not come up with a scenario that cut off the supply. "We built this very big model. We played with it every way we could. We said, 'Let's do this,' and, 'Let's do that.' At the end, this huge model would say here's the result and it was not good news."

The simulation showed the limits of technology to solve what was essentially a policy problem: simulation was not going to teach anyone how to win the drug war, it could only demonstrate that it was unwinnable, and that was not a message the government wanted to hear. The reaction was denial: law enforcement would just have to try harder. "I don't know if we're a hell of a lot better off that we now kind of understand the problem because we have the simulation," McBride reflected. "It's like massive wounds all over the body; blood is pouring out from everywhere. We can understand that, but there is nothing we can do about it."

The counter-drug simulation failed because technology hit up against the limits of policy. Other DARPA simulation efforts failed at a more basic level. One 1990s-era project, called War Breaker, was supposed to come up with a silver-bullet solution to the main threat American forces faced in the Gulf War: mobile Scud launchers. The Soviet-made tactical ballistic launchers had proved difficult to find and destroy. Despite Iraq's rapid defeat, its Scuds had proved deadly. One attack, on February 25, 1991, killed twenty-eight American soldiers in Saudi Arabia. DARPA contracted to have a simulation facility built outside Washington, D.C., even employing Herman Zimmerman, the set designer for *Star Trek: The Next Generation,* to build the laboratory, modeled after the bridge of the *Enterprise.* The simulations made for convincing theater when senior officials were brought in to view demonstrations, but the reality was far less impressive. "You could figure everything out beforehand if you wanted to show off a missile, an airplane, or whatever you wanted. You could put it in a situation where it would just look wonderful," said Ron Murphy, a DARPA program manager who worked on War Breaker.

The reality of mobile targets was too complex for manned simulation, according to Murphy, and for every simulation that turned out well, there were many more that did not. Despite the success of SIMNET, which focused on tank training, DARPA was simply never able to create a realistic simulation that would combine air and ground operations. War Breaker, once slated to get more than half a billion dollars in funding, and DARPA's most expensive post–Gulf War effort, was quietly ended. Almost all mention of it was expunged from later agency documents.

DARPA had spent the 1990s creating synthetic versions of war involving planes, tanks, and missiles. The agency had expanded beyond that on occasion, for example to the drug war, but even then it was focused on modeling concrete objects: drugs and money. L. Neale Cosby, a former army officer who was involved with SIMNET and 73 Easting, acknowledged that one thing DARPA's simulations could never do well was to model human beings. "It's easy to simulate this room," Cosby said. "To put all these people, doing the right things, all thinking independently, in the right place so you actually virtually copy this room is tough. That's the tough simulation part that we still don't do that well."

In the decade that followed the collapse of the Soviet Union, it became increasingly clear that people—not tanks or missiles—were precisely what needed to be modeled. In 1993, terrorists detonated a truck bomb in the North Tower of the World Trade Center, killing six people. Then, in 1998, al-Qaeda carried out near simultaneous attacks against two American embassies in Africa. Two years later, its operatives attacked the USS *Cole* in the port of Aden, in Yemen.

In 2000, the same year as the *Cole* bombing, Tom Armour, a former CIA officer who had recently joined DARPA, announced a new type of modeling and simulation work that the agency was beginning to pursue. The United States was facing a new threat, from terrorist groups, or what the Pentagon was calling an "asymmetric threat." This "asymmetric threat is physically small—perhaps even just a single person," Armour, the DARPA official, said. "To predict the potential range of actions, the analyst will need to model the group's beliefs and behavior patterns." The DARPA director that year authorized the start of a new program, called Total Information Awareness. It sought nothing less than to predict human behavior, in particular the behavior of terrorists.

Just a few months prior to Armour's talk, three al-Qaeda operatives based in Hamburg arrived in the United States to begin training as pilots. They were part of a larger terrorist cell sent by al-Qaeda to attack America. The very behavior pattern that Armour spoke about was emerging. DARPA had switched gears, but like the rest of the national security community it had done so just a little bit too late. By 2001, DARPA's simulations were still preparing the military for tank battles and air-to-air combat, while in the United States nineteen men were training to take over commercial airplanes armed with nothing more deadly than box cutters.

CHAPTER 17

Vanilla World

In July 2001, Scooter Libby, Vice President Richard Cheney's chief of staff, called up DARPA's new director, Tony Tether. The vice president wanted to visit DARPA to be briefed on its programs. Cheney knew Tether from his days serving as the secretary of defense, but even so, his call was a surprise. The White House had not taken an active interest in DARPA for years.

In the summer of 2001, the agency founded more than four decades earlier was adrift. Without a Cold War enemy to fight, DARPA during the 1990s was funding politically expedient projects, and over the previous decade that had led to a series of expensive flops. DarkStar, a stealthy drone sponsored by DARPA at the request of senior Pentagon leaders, suffered a software glitch on its second flight and crashed. With the navy, DARPA embarked on an ambitious ship project—a floating missile platform—only to have the work canceled following the suicide of an admiral who supported the program. Most disastrously, DARPA signed up to work on an ill-conceived army project called Future Combat Systems, which was supposed to link missiles, drones, and ground vehicles on a single network (the overly complex project was also eventually canceled). DARPA lurched from one project to the next, without any real strategic direction or plan. "DARPA had become a backwater type of organization in the latter part of the nineties," admitted Tether, who took over in June 2001. The vice president's sudden interest was a chance to change this.

Over the next three weeks, Tether and a handful of office directors worked days, nights, and weekends to prepare to brief the vice president. Cheney "was a person who liked pictures, he is a very visual person," Tether recounted. "So, having a slide with a whole bunch of words on it was just, 'oh, wow,' you don't ever do that to him. You know, you give him a cartoon." Tether recounted that he wanted something that would blow the vice president away, and so the cartoon he chose for Cheney was Superman.

The headlining act for the vice president would be super soldiers, the brainchild of Michael Goldblatt, a research manager who came to DARPA from McDonald's, the fast-food company. At McDonald's, Goldblatt had run the corporation's venture capital efforts, a sort of DARPA-like arm of the nation's biggest fast-food chain. There, he worked on projects like self-sterilizing food wrappers, which he had been trying to sell to the military. No one in the Pentagon would take his calls. Finally, he was referred to DARPA, where the then director, Larry Lynn, agreed to speak to him, thinking Goldblatt worked for McDonnell Aircraft Corporation rather than McDonald's. Within a few years, Goldblatt found himself working directly for DARPA, initially overseeing the agency's program for defense against biological weapons.

At DARPA, however, Goldblatt turned his attention to something far more ambitious than food wrappers: he wanted to work on enhancing human beings. Goldblatt was inspired by science fiction like *Firefox*, the 1982 movie starring Clint Eastwood, which featured weapons controlled by the human mind. Under Goldblatt, DARPA funded a group of researchers at Duke University to implant microelectrodes in monkeys' brains. The electrodes would read their brain signals, which could then be used to manipulate a real object, like a robotic arm. Goldblatt's other ongoing research programs sounded equally fantastic: humans who could survive severe blood loss in suspended animation, soldiers who could go for days without food or sleep, and warriors enhanced with superhuman strength and mental abilities. Superhuman soldiers with mind-controlled weapons sounded like the perfect cartoon for Cheney.

In late July, the vice president showed up at DARPA headquarters, accompanied by Secretary of Defense Donald Rumsfeld and Edward

"Pete" Aldridge, the Pentagon's chief weapons buyer. Starting with Goldblatt and his enhanced humans who could survive critical wounds and bitter cold, or go for days without sleep and food, Tether and his office directors briefed the three men for six hours. They focused on Cheney, and he loved it. "It was fantastic," Tether said.

When the three men left, Tether knew the briefings had been a hit. He got word later that both the vice president and the defense secretary were enthused. It was one thing to have the support of senior Pentagon and White House officials and another to figure out what to do with that support. The answer came just weeks later, and it turned out it had nothing to do with super soldiers.

If DARPA had been adrift over much of the past decade, so too was the rest of the national security establishment. The 1990s "peace dividend" had featured declining defense budgets, including for DARPA, and without the threat of an armed confrontation with the Soviet Union, no single threat generated sustained attention. Even as threat reports on al-Qaeda surged over the spring and summer of 2001, defense and intelligence officials were moving in slow motion. Around the time that DARPA was briefing Cheney and Rumsfeld on super soldiers, intelligence on al-Qaeda and its operations seemed to be exploding. Al-Qaeda was planning something "spectacular." An attack was "imminent." The reports on al-Qaeda activities had "reached a crescendo," Richard Clarke, the White House counterterrorism czar, warned in June 2001. In August, the FBI began investigating Zacarias Moussaoui, a French national with a fervent belief in jihad, $32,000 in a bank account, and an unexplained interest in learning to fly Boeing airliners.

That last clue bounced around the law enforcement and intelligence bureaucracy in August 2001, but officials with only pieces of the puzzle did not understand its meaning. An FBI supervisor in the Minneapolis field office, which had started the inquiry, got into an argument with headquarters about the importance of tracking Moussaoui. Pursuing the investigation was critical, the Minneapolis-based agent argued, "to keep someone from taking a plane and crashing into the World Trade Center."

FBI Washington headquarters rebuffed the request. When the CIA's director, George Tenet, was briefed on August 23 on Moussaoui, he said it was an FBI matter. Intelligence analysts and law enforcement officials were trying to piece the clues together, but none of them had enough of the picture to anticipate what was about to take place. Even if some CIA analysts understood this growing threat, the political leadership in the summer of 2001 did not see al-Qaeda as a critical issue. The bright lights that should have pointed straight to a terrorist plot to hijack and crash planes were only clear in hindsight.

At 8:46 a.m. on Tuesday, September 11, as many government workers in Washington were still fighting their way through rush hour traffic and others were just getting settled at their desks, American Airlines Flight 11 flew into the North Tower of the World Trade Center. Seventeen minutes later, United Airlines Flight 175 struck the South Tower.

Less than one hour later, radar at Dulles International Airport, and then Ronald Reagan Washington National Airport, picked up the track of an aircraft heading in the direction of the White House. The bureaucracies that had argued for months about what to do with a flood of confusing clues now had only minutes to respond. As the Secret Service prepared to evacuate the White House, the plane changed course abruptly. The hijacked aircraft was now flying toward the Pentagon, where workers had started receiving word about the attacks in New York from listening to the morning news but had no clue an airplane was closing in on them at more than five hundred miles an hour.

As the hijacked aircraft began its chaotic descent over Fort Myer Drive, Tony Tether was sitting in a top-floor conference room in DARPA's Northern Virginia headquarters, thumbing through messages on his phone. Tether had seen one message saying that a small plane had hit the World Trade Center in New York. "Interesting," he thought to himself, echoing a reaction that many people had, when getting that first, erroneous report. A bit later, he got another message, saying the other tower had been hit.

"Hey, looks like another plane has crashed into the towers," he said to the others in the conference room. "Does anybody know anything about it?"

"Nah," was the response. "It's nothing."

American Airlines Flight 77 struck the west side of the Pentagon at 9:37 a.m., a moment when many of those inside the building were just starting to watch television coverage of the Twin Towers burning. The crash killed all 64 people on board and 125 military and civilian workers in the building.

Moments later, Tether's secretary opened the conference room door and motioned wordlessly for the director to come out. As they walked around the corner together, the secretary pointed out the window, where billowing smoke was rising up from the Pentagon. Tether and other DARPA officials switched on a television in time to watch live coverage of the World Trade Center's South Tower in New York collapse—a surreal image of concrete morphing into dust. Just minutes later, United Airlines Flight 93 crashed in Pennsylvania after passengers on board realized their plane was on a suicide mission and rushed the cockpit. At 10:28 a.m., the North Tower in New York collapsed.

With concerns over more airplanes in the air over the capital, Tether sent the staff home and closed the agency. In a move that reflected how the new DARPA director would run the agency for the next seven years, Tether stayed in the building, routing all incoming calls to his office phone. Watching the day's news unfold, Tether was convinced that the problem had been not a lack of data but a failure to centralize and analyze the data.

DARPA had already begun exploring that very question, though on a small scale. Just two miles from the burning Pentagon, a small DARPA-financed laboratory on Washington Boulevard, across from the army's Fort Myer, had been quietly rehearsing terrorist scenarios for senior defense and intelligence officials, trying to convince them that terrorist attacks could be detected well before they are perpetrated. The key was to sift through vast amounts of data—both public information and intelligence records—to identify patterns of activities that might indicate that terrorists were preparing an attack. The intelligence data could be from intercepted phone calls, e-mails, or Internet traffic. The public data might include credit card transactions, doctor visits, and car rental records. The laboratory was designed to demonstrate what could be done if all of that data could be linked together and treated as a single database.

The "laboratory" was all smoke and mirrors, at least in the fall of 2001. A Hollywood set designer had been hired to create the futuristic-looking command-and-control center with large, sleek display screens and flashing lights. The humming computers were not churning any real data. There was research going on at companies and universities, but the laboratory was just a showcase to convince intelligence officials that data, or more important, spotting patterns in data, could help predict the next terrorist attack. Over the past few years, many of the officials who would rise to top levels in the intelligence community had passed through the lab, including Keith Alexander, the future head of the National Security Agency, and James Clapper, who would go on to become the director of national intelligence.

More notable than the Hollywood-inspired set was the wizard behind the curtain: John Poindexter, a retired admiral who had once been Ronald Reagan's national security adviser. A PhD physicist and technology enthusiast, Poindexter burnished his image in public memory as the relaxed, pipe-smoking witness during the 1987 Iran-contra hearings, which investigated the Reagan administration's sale of arms to Iran. Poindexter helped orchestrate the convoluted deal, which was also used to finance the contras in Nicaragua, in violation of law. He then systematically shredded evidence of the scheme once it came to light.

Within months of 9/11, Tether hired Poindexter to run an entirely new office in DARPA, called the Information Awareness Office, with plans to spend more than $200 million in its first two years, and a flagship project called Total Information Awareness. The reason Tether thought he could do something as audacious as hire a figure at the heart of one of the greatest modern national political scandals, let alone put him in charge of a high-profile counterterrorism program, had to do with Cheney's visit in July. "We really were armor-proofed," Tether said.

With Cheney and Rumsfeld's backing, Tether unwittingly embroiled the agency in what would become its most politically controversial work since the Vietnam War, putting his job and the agency in jeopardy. DARPA's forays into intelligence and data mining would shape the post-9/11 debate on surveillance and privacy, and the ensuing controversy would also shape DARPA for much of the next decade.

John Poindexter's circuitous route into DARPA was in some respects completely logical. His career, like DARPA, was a by-product of Sputnik. Poindexter graduated at the top of his class from the U.S. Naval Academy the same year DARPA was formed. He then pursued a doctorate in physics at the California Institute of Technology as part of a special program established by Admiral Arleigh Burke, then the chief of naval operations, who believed that Sputnik was a sign that the United States needed more scientific expertise among its officers.

After getting his PhD, Poindexter rose quickly through the ranks of the navy, earning a reputation as an early adopter of technology. At every stop along his career, he set about trying to bring—and sometimes drag—the government and the military into the information age. His reputation for being computer savvy helped propel him to the White House in the early 1980s, where he was assigned to modernize the Crisis Management Center in the Old Executive Office Building. Equipped with fiber-optic cable, the revamped center introduced videoconferencing to the White House. Poindexter also brought PROFS Notes, an early version of e-mail, to the National Security Council staff.

In 1985, Poindexter was promoted to vice admiral and appointed to be Ronald Reagan's national security adviser, which put him squarely in the middle of the Iran-contra investigation. He resigned November 25, 1986, when the Iranian arms sales went public. His deputy, the Marine Corps lieutenant colonel Oliver North, was fired. Poindexter was later indicted and found guilty of five counts of lying to, misleading, and obstructing Congress, but the conviction was tossed out when an appeals court ruled that the case relied on testimony to Congress for which Poindexter was granted immunity. Poindexter did not go to prison, but his reputation was badly tarnished. With his experience in computers, however, he found work in the private sector, fading into the background of Washington's technocracy.

Out of the public limelight, Poindexter pursued what had always been his passion, combining technology with intelligence to anticipate crises. It was an interest dating back to the bombing of the marine barracks in Lebanon in 1983, which killed 241 military personnel, an event he believed could have been prevented if there had been a way of sifting through all the data. In 1995, an opportunity came to try to prove this belief. Poindexter was introduced to Brian Sharkey, a program manager at DARPA who was interested in analyzing data to help predict political crises. Soon, Poindexter was working under contract to DARPA.

Together, Sharkey and Poindexter in 1996 launched a DARPA-sponsored data-analysis program called Collaborative Crisis Understanding and Management, later changed to Genoa (because both were former naval officers, they liked the idea of naming the program after a sail). "The idea with the technology was to take a much more systematic approach to the problem of anticipating crises and then managing them when they happen," Poindexter said. Over the next six years, Poindexter worked quietly on Genoa, which received more than $50 million from DARPA.

One of the things the Genoa project did was to establish what Poindexter described as a "laboratory" across from Fort Myer, "where we would run exercises and demonstrations for people in the national security community, primarily the intelligence community, but also DOD." There, they tried to demonstrate how computer algorithms could look for patterns in data that might signal a future terrorist attack. One scenario involved the Japanese terrorist cult Aum Shinrikyo, which in 1995 released sarin gas in the Tokyo subway system. Genoa used that scenario because a great deal of data was available about the attack. "Admittedly, we were looking in hindsight," Poindexter said.

By 2001, Poindexter felt that Genoa was proceeding along well, but it clearly was not on any fast track to adoption by the intelligence community. Some senior intelligence officials who attended demonstrations at the "laboratory" appeared to embrace what Poindexter was trying to show, but others not as much. Poindexter described one demonstration, held for the chairman of the National Intelligence Council, who nodded off partway through the hour-long visit. "When it was over, this chairman said, 'Well, John, this is very interesting, but we don't really have time to do all those things. The only thing I'm interested in: The day after [an attack], who knew what when?'"

For Poindexter, the response was not entirely unexpected. "I felt sorry for him, but I recognized there was a real cultural problem in that the intelligence community, especially the ones at CIA, just weren't taking advantage of the power that information technology could provide them in searching through data," he said. The September 11 terrorist attacks appeared to prove what Poindexter had been saying for years. The government had lots of data, it just did not have the means to make sense of it. On September 12, Poindexter went to see Brian Sharkey, who had moved from DARPA over to SAIC, a defense contractor. They agreed to approach Tony Tether together with a proposal for a major expansion of Genoa.

Just days after the 9/11 attacks, Poindexter was sitting in front of DARPA's new director with a briefing presentation titled "A Manhattan Project for Combating Terrorism." Poindexter pitched his vision of a massive technology program to combat terrorism on the scale of the World War II race to build the atomic bomb. Poindexter's idea was to create a massive data-mining system capable of aggregating databases across government and the private sector and then pulling out warnings of the next September 11 attack.

Poindexter proposed another Manhattan Project, made up of top researchers from government, academia, and industry. Poindexter even half joked that he would put everyone "in a compound with barbed wire around it" so the people inside could not leave until they had solved the terrorism problem. One of the slides Poindexter presented to Tether was particularly striking: it laid out a $100 million unclassified "white" program, called Total Information Awareness, and then a parallel secret "black" program that would have five times that budget. Operating in strict secrecy, this highly classified black program would be called Manhattan Project Terrorism.

The idea resonated with Tether, but a Manhattan Project was unrealistic; after all, it took several years to establish the World War II–era Manhattan Project even after Albert Einstein wrote to President Franklin D. Roosevelt, telling him that a nuclear bomb might be a possibility. But DARPA could quickly allocate tens or even hundreds of millions of dollars to a project, something no other agency could do. Tether suggested simply taking the Genoa project and putting it on a fast track with more money and resources. The catch was that Tether wanted either Sharkey or Poindexter to come into DARPA to run the program. Sharkey was earning good money as a defense industry executive and was not interested in taking a pay cut to enter government, so that left Poindexter, who was reluctant but willing. "I knew I was going to be controversial because of my experience in the White

House," Poindexter recalled. "In the end, I agreed to come back in for at most a couple years to get things started."

In retrospect, having someone whose name was synonymous with political scandal run a critical counterterrorism program should have rung alarm bells. But by 2001, Poindexter, at least in Tether's view, was just another former government official working in the Beltway contracting business. Moreover, no one had taken any notice that for the past six years Poindexter had been quietly working under contract to DARPA on a project designed to predict terrorist attacks. Now Poindexter was offering a way to place DARPA directly into a defining issue of the next decade—what President George W. Bush would eventually label the war on terror.

In January 2002, John Poindexter became a government employee for the first time since he retired from the navy in 1986 amid a pending indictment. Poindexter would head up an entirely new division at DARPA dedicated to counterterrorism, called the Information Awareness Office. The biggest program in that new office was Total Information Awareness, an umbrella name for a series of research projects, including Genoa, involved in sifting through data to identify potential signs of a terrorist attack.

Tether did not think a nation at war would agonize over the details of who was running a project. Besides, Poindexter "never really was convicted of anything," Tether later told a reporter, a reflection of just how deeply DARPA's new director was about to misjudge the situation. At first, Tether's belief about his new employee was confirmed; a brief item appeared in *The New York Times* the next month, noting Poindexter's new position. But nothing followed. An office director job at DARPA might be a nice position for a scientist, but in the Washington fishbowl it was hardly something to elicit much interest.

There were signs, however, that Tether, even more than Poindexter, was misjudging the danger of taking a modest research effort and turning it into a high-profile counterterrorism project. Poindexter knew from his prior experience that privacy would be a concern. Total Information Awareness was about finding ways to sift through large amounts of data, combining intelligence databases and public information. Tether encouraged Poindexter to go to credit card companies and collect commercial data, but Poindexter said he balked at the idea of using real-world data for a research program, knowing that could immediately raise public objection. Even though the idea was eventually to create a centralized database of real information, Poindexter decided, for the time being, to use made-up data.

Poindexter was guilty of his own misjudgments. One was the symbol he helped create for his new office. Prominent in the design was the all-seeing Eye of Providence, the familiar pyramid that appears on the $1 bill. The Information Awareness Office's seal featured the symbol and added a beam shooting out of the pyramid's eye focused on the globe; Poindexter had the Latin phrase *scientia est potentia,* or "knowledge is power," added to the seal, as a final flourish. The pyramid, though a familiar image, is also a symbol often linked to conspiracy theories. Yet no one at DARPA seemed to think that was a problem.

Shortly after Poindexter was hired, another familiar DARPA figure entered the picture. Stephen Lukasik, the former director who had pulled the agency out of the morass of Vietnam, had been working as an "idea man" for SAIC, which was involved in simulation and modeling work. Lukasik had always been interested in generating scenarios, just as he had back when he was director, looking at ways the Soviets might attack NATO. Now, in the wake of September 11, he was thinking about terrorist scenarios.

"I know six good ways to smuggle a nuclear weapon into the United States," Lukasik told Tether during a private meeting in the director's office. The timing was fortuitous, because Poindexter had just kicked off work on Total Information Awareness. Tether immediately took Lukasik down to Poindexter's office, and soon Lukasik was on contract as part of a "red team" of terrorists trying to perpetrate an attack on the United States. The red team's attack scenarios were incorporated into a "simulated world," which contained real addresses but used fictitious people living at those addresses. It was a simplified facsimile of the United States, a place populated by millions of law-abiding simulacra and a small group of former officials posing as terrorists. Poindexter called it Vanilla World.

If Tony Tether wanted to understand why DARPA's support for a centralized database might touch off a furor over privacy, he need not have looked far. In June 1975, more than twenty-five years before the Total Information Awareness imbroglio, a series of sensational news reports warned of a new Orwellian-sounding computer technology that would be used to create dossiers on individual Americans. "What this technology means to you is this: The Federal Government now has the means to put together a computer file on you, or almost any American, within a matter of minutes," Ford Rowan, a correspondent for *NBC Nightly News*, reported. "The key breakthrough in the new computer technology was made by a little known unit of the Defense Department—the Advanced Research Projects Agency, ARPA."

The technology involved was the ARPANET. Various news reports claimed that the government was using the ARPANET to create centralized dossiers through a secret network that linked the White House, the CIA, the Defense Department, the FBI, and the Treasury Department. None of that was true. Although government agencies were beginning to use elements of computer-networking technology, the ARPANET in 1975 was primarily linking academic institutions. But the reports came on the heels of several years of post–Vietnam War debate about surveillance abuses of the intelligence agencies, and the spread of computers and national data banks, which led to the Privacy Act of 1974. Almost three decades later, these same concerns would be played out following the 9/11 attacks, and it was a DARPA-sponsored computer scientist who first identified them.

On October 12, 2001, just one month after the attacks, the members of DARPA's Information Science and Technology study group, or ISAT, met to conduct their annual brainstorming session. The atmosphere at the meeting had darkened considerably from previous years. The group's chairman came into the room and cued up a slide. On the overhead screen flashed Osama bin Laden's 1998 fatwa against Jews and crusaders. "I don't want to affect anyone's thinking, but here's what we're up against this year," he told the assembled scientists.

ISAT was established in the 1980s, during the days of the Strategic Computing Initiative, as an advisory group for DARPA focused specifically on computer science. Unlike the JASONs, the elite group of scientists who had helped create the McNamara Line in Vietnam, ISAT did not operate as an independent group; it advised only DARPA. One of the ISAT members, Eric Horvitz, had already been thinking about how to use computers to help sift through large amounts of data to predict future events. But that year, Horvitz, a prominent artificial intelligence expert working at Microsoft, saw an opportunity to apply this work to the nexus of data mining and privacy. His vision was for something called selective revelation.

Under Horvitz's scheme, the government would collect data—be it intelligence, law enforcement, or commercial information—and put it into a centralized database. Humans would not have any direct access to it. Instead, computer algorithms would sift through the personal data, looking for patterns that might indicate a possible terrorist attack. When such a pattern was spotted, the government, in Horvitz's thinking, would obtain a search warrant that would allow for "selective revelation" of personal data. "The idea basically was, how could you minimize the revelation of personal data while supporting analysis that was deemed important to find needles in haystacks?" he said.

The data-mining system operated like a locked black box. Inside, "you can basically have standing queries of interest with automated computer agents walking over rafts of data," Horvitz said. "You collect data but no one is allowed to look at this data except computer programs, and whenever you find some troubling or concerning things, then the system alerts an operator and it says, 'I've found something which could be a problem.'"

Inevitably, a human would have to review the results, but the system would monitor and log every time someone peeked inside. Horvitz described his concept as a system that "watches the watchers," meaning that those with access to the database would themselves be monitored through random audits. It would function, he believed, like a "hall of mirrors," where anyone is potentially watching everyone else. Horvitz's idea for a "hall of mirrors" soon made its way to the desk of John Poindexter, who had just moved into his office at DARPA. Poindexter wanted DARPA to sponsor privacy research as part of Total Information Awareness, to see what safeguards could be incorporated into a data-mining system, and Horvitz's proposal was to study precisely that.

Horvitz's concept was also very close to how Poindexter saw the database searches working for Total Information Awareness. In Poindexter's vision, the computer algorithms would be focused not necessarily on a specific terrorist event—those were too rare to be anticipated—but on patterns of activities that might indicate that terrorists were preparing an attack. Rather than going to the Foreign Intelligence Surveillance Court, or some other judicial authority to seek data on a specific person, law enforcement or an intelligence agency could seek authorization to search for a specific pattern. As Poindexter described it: "If your search for that pattern comes back with 100,000 hits, it's not discriminating enough. You refine the pattern; you get that approved [by the judicial authority]. Maybe with that version you get ten hits. At that point, you go back to the court, through an automated system, and say, 'Okay, we got, ten hits. Now we want permission to find out the details of those ten hits.'"

Because Poindexter believed that having a privacy mechanism would be important for automated searches, he offered to fund the ISAT proposal. He even attended the summer study meeting in 2002 held at the offices of the Institute for Defense Analyses in Alexandria, Virginia. ISAT invited two privacy advocates to observe the meeting, including Marc Rotenberg, the head of the Electronic Privacy Information Center. Poindexter and Rotenberg had clashed previously in the 1980s, when Poindexter was in the White House and Rotenberg worked as a congressional staffer. Rotenberg, then a counsel to the Democratic Vermont senator Patrick Leahy, had opposed a White House National Security Decision Directive, which had offered NSA assistance to private companies on cyber security. Opponents on the Hill, including Rotenberg, saw it as an NSA intrusion. "It was a Big Brother is going to watch you, and all this crap," Poindexter said.

At the ISAT meeting in 2002, Rotenberg sat quietly, and when Poindexter approached him during the break, the conversation was civil. Poindexter thanked him for coming and said DARPA genuinely hoped to gather ideas on balancing security and privacy. According to Poindexter, Rotenberg said he understood, adding that he believed there needed to be more oversight. Poindexter took that as an encouraging sign. He was wrong, however.

Rotenberg saw the meeting from a completely different perspective. In his view, Poindexter and others involved in the project had no understanding of what privacy meant. Officials like Poindexter believed that internal audit mechanisms—the members of the surveillance state monitoring its own—would protect privacy. Rotenberg argued that the Privacy Act of 1974 was rooted in the public's right to have control over the data collected about them and not just how that data is used. His concern about the DARPA program only deepened when he participated in another workshop related to Total Information Awareness. This workshop, held at Stanford, looked at a proposed DARPA program to create "eDNA," which would make it possible to track every keystroke on the Internet back to a specific user. "Perfect surveillance" is how Rotenberg described it. "And completely crazy."

Despite Rotenberg's privacy concerns, DARPA's plans were moving forward. DARPA earlier that year issued an open call for proposals in areas of interest to the Information Awareness Office, including what Poindexter called the "privacy protection appliance." And with the floodgates open after 9/11 for counterterrorism spending, Genoa's budget alone was set to more than double, from $70 million in 2002 to about $150 million for 2003. Over the first half of 2002, Poindexter worked to put together the initial Total Information Awareness prototype, reaching out to other defense and intelligence agencies, which were invited to establish "nodes" on the data network. The central node would be controlled by DARPA, but the distributed nodes would allow the different agencies to access the network and test out the program's tools. The NSA, not surprisingly, had the most nodes of any agency.

By the late summer of 2002, the pieces were falling into place, although the picture from the outside probably looked different from how those inside saw it. The Information Awareness Office had as its head the bête noire of Iran-contra, an office seal that featured an icon of Illuminati-inspired conspiracy theories, and an ambitious vision for a centralized database. The new office was ready for its public debut, and for that it headed to Disneyland.

DARPATech, the semiannual DARPA conference, had once been a staid technical meeting held in cities like Denver and Dallas, but Tony Tether wanted to make a public splash, so in 2002 he moved the venue to Disneyland in California. "We have had a lot of fun creating this symposium," Tether said in the keynote speech. "Part of the fun was each office finding a Disneyland-like theme for what it does." (The theme for the Information Awareness Office featured the *Star Wars* androids, C-3PO and R2-D2.)

Just as Tether had not seen anything wrong with proposing that Poindexter go to credit card companies to collect financial data, he did not see a problem with allowing Poindexter to roll out the Information Awareness Office in a fantasy world populated by life-size versions of Goofy and Mickey Mouse. And so, in August 2002, in front of an audience of DARPA-supported researchers and a handful of journalists, Poindexter introduced the Information Awareness Office:

> One of the significant new data sources that needs to be mined to discover and track terrorists is the transaction space. If terrorist organizations are going to plan and execute attacks against the United States, their people must engage in transactions and they will leave signatures in this information space. This is a list of transaction categories, and it is meant to be inclusive. Currently, terrorists are able to move freely throughout the world, to hide when necessary, to find sponsorship and support, and to operate in small, independent cells, and to strike infrequently, exploiting weapons of mass effects and media response to influence governments. We are painfully aware of some of the tactics that they employ. This low-intensity/low-density form of warfare has an information signature. We must be able to pick this signal out of the noise. Certain agencies and apologists talk about connecting the dots, but one of the problems is to know which dots to connect. The relevant information extracted from this data must be made available in large-scale repositories with enhanced semantic content for easy analysis to accomplish this task. The transactional data will supplement our more conventional intelligence collection.

Poindexter's speech acknowledged potential concerns about privacy but promised to address them. "There are ways in which technology can help preserve rights and protect people's privacy while helping to make us all safer," he assured the audience.

DARPATech came and went, generating only modest press coverage, mostly focused on Tether's announcement that DARPA would sponsor a "robot race" in the California desert, a demonstration of self-driving cars. Nobody seemed to take note of Poindexter's speech, which did not necessarily introduce anything new. More than two years earlier, Brian Sharkey had talked about "Total Information Awareness" at a prior DARPATech meeting, though it was at the time a concept more than a program. As for Poindexter, his presence at a small defense conference hardly seemed newsworthy, particularly for the mostly science- and tech-focused members of the press who attended.

Yet Rotenberg, the privacy advocate, had been discussing his concerns with the *New York Times* technology reporter John Markoff. In November, Markoff's article described Total Information Awareness as "a vast electronic dragnet, searching for personal information as part of the hunt for terrorists around the globe—including the United States." The article quoted Poindexter's Disneyland speech, as well as Rotenberg. The DARPA program was described as building a "system of national surveillance of the American public."

Even that article did not garner much attention. The next week, the *New York Times* columnist William Safire declared war on Total Information Awareness, calling it an affront to the American way of life. "Every purchase you make with a credit card, every magazine subscription you buy and medical prescription you fill, every Web site you visit and e-mail you send or receive, every academic grade you receive, every bank deposit you make, every trip you book and every event you attend—all these transactions and communications will go into what the Defense Department describes as 'a virtual, centralized grand database,'" he wrote. "To this computerized dossier on your private life from commercial sources, add every piece of information that government has about you—passport application, driver's license and bridge toll records, judicial and divorce records, complaints from nosy neighbors to the F.B.I., your lifetime paper trail plus the latest hidden camera surveillance—and you have the supersnoop's dream: a 'Total Information Awareness' about every U.S. citizen."

Safire's column was a conflation of fact and fantasy: Total Information Awareness was a research project using make-believe data, and there were plenty of legal reasons why some of that data might be excluded even assuming the technology was eventually adopted. On the other hand, Safire's description was a reasonable portrayal of the scope of Poindexter's ambition. The column touched off the classic firestorm amplified by the Washington echo chamber: an avalanche of articles appeared, many quoting the Safire piece. The agency's spokesperson told Tether just to ignore it and the furor would eventually die down. DARPA officials steadfastly refused to comment on Poindexter's work or the programs. As more articles tumbled forth, Tether was shocked, a reflection of how oblivious he was to the potential public perception of DARPA's work. Total Information Awareness was, in his view, just a research program. Yet the articles were treating the project as if it were an operational system collecting everybody's medical records. "I am reading these things and I know what is going on, but I start to think, holy cow, maybe I don't know what is going on! But we didn't respond," he later told a group of reporters. "Where we screwed up is that we didn't respond. We literally did not take an active stance of coming out and saying, 'Hey, bullshit,' you know, and went around and made sure that everybody understood what we are doing until it was almost too late."

It was not just the press, however. Congress began to demand briefings, and that created a whole new problem. Poindexter, who had been convicted of lying to Congress, was not a good choice to send to the Hill to answer questions coming from incensed lawmakers. Poindexter suggested that his deputy, Bob Popp, do the briefings, but Tether insisted on speaking to lawmakers himself. It was a disaster. "Tony didn't understand the programs well enough to explain them in detail," Poindexter lamented. "It wound up looking like we were much more secretive than we wanted to be."

As the critics circled Poindexter, the controversy surrounding Total Information Awareness grew, yet Tether refused to budge an inch. After all, DARPA had top cover from the defense secretary and the vice president. For a while, the Pentagon's leadership backed DARPA, too. But as the articles proliferated, Congress demanded that the Pentagon provide a full report on the Information Awareness Office and all its projects. When congressional staffers combed through the report, one project in particular caught their attention: a small research study for something called the FutureMAP, which looked at the potential of using the "wisdom of crowds," represented by free-market investors, to predict future political events. FutureMAP actually had its origins with a researcher from the National Science Foundation, who was interested in the predictive capabilities of open markets. DARPA had awarded contracts for very preliminary work, which would fund researchers to test whether having people wager real money on future political events would make for accurate predictions.

Net Exchange, one of the companies involved, created a website that featured "colorful examples," like "the assassination of Arafat, and a missile attack from North Korea." It was an early effort at crowd-sourcing predictions, something that would be commonplace ten years later. For Congress, which was eager to come up with any excuse to get rid of Poindexter and his programs, FutureMAP was enough. "The idea of a federal betting parlor on atrocities and terrorism is ridiculous and it's grotesque," Senator Ron Wyden, a Democrat from Oregon, said.

The FutureMAP episode was almost a verbatim repeat of the opposition to AGILE toward the end of the Vietnam War, when members of Congress trawled DARPA reports in the hopes of finding something embarrassing. "The congressmen and senators that had been critics of the program all along became incensed that—as they described it—that I wanted to establish a betting parlor," Poindexter said. "At that point, I told Tony it was time for me to leave, and I did."

Poindexter's resignation letter to Tether in August 2003 ran five single-spaced pages, expounding on DARPA's origins, its purpose, and its achievements. He reviewed the history of the Information Awareness Office, providing a detailed and unapologetic defense of its activities, and blasted the "charged political environment of Washington, where glib phrases, 'sound bites,' and symbols" are used instead of debate. He expressed hope that Congress might salvage some of his office's work.

The following month, Congress voted to end DARPA's Total Information Awareness program and shut down the Information Awareness Office. Total Information Awareness was, at least officially, dead. Except it was not: it was just about to get transformed into something much different and arguably much worse.

In recounting his final days at DARPA, Poindexter stopped to puff on his trademark pipe, just as he had in the Iran-contra hearings three decades prior. "Congress claimed they had closed my office and shut down the TIA program, but in reality what happened was they moved all of the components of TIA out of DARPA and into the intelligence community, and that winds up in the classified part of the DOD budget," Poindexter said.

Was it possible, then, that Total Information Awareness was really just a cover for a massive black program, as Poindexter had first proposed to Tether in 2002? Poindexter said that in the end they had decided not to pursue the black program, because they wanted to include university researchers who did not have security clearances and it "would take too long to get all of the approvals required for such a classified program." However, moving the programs over to the intelligence community achieved much of the same effect: Total Information Awareness never died; it just went black, just as Poindexter had proposed at the start. The Advanced Research and Development Activity, part of the NSA, took over almost all the programs that had been in the Information Awareness Office, except the privacy protection research. "The critics we had got the worst of all worlds," Poindexter pointed out.

All too often, John Poindexter was, and often still is, portrayed as the paragon of the guileful government operative—the "ring-knocking master of deceit," as *The New York Times*'s Safire called him. The real Poindexter—unapologetic, but gracious, even with skeptics—was never quite the bogeyman he was made out to be by the critics. His vision, however reprehensible to privacy advocates, was born of a deep, if misguided, concern about national security.

Had Total Information Awareness proceeded at DARPA, the plan was to expand Vanilla World, the simulated world where officials practiced spotting terrorists. Poindexter envisioned Cherry Vanilla World and then French Vanilla World, each with added layers of complexity and realism. But DARPA's research never got beyond a simplified make-believe world. The fundamental scientific question—whether computer algorithms can be used to spot patterns of terrorist activity from a combination of private and public data—has never been proven, at least publicly.

Total Information Awareness, Poindexter insisted ten years later, was a success that fundamentally changed the focus and direction of how the intelligence community dealt with data. "Although TIA was cut short in 2003 by the Congress, I think basically we accomplished really what I had set out to accomplish, which was to propagate ideas of technology, develop some early versions that could later be improved, and look at a new process for intelligence analysis," Poindexter said.

Poindexter's belief in the need to balance privacy with security was sincere but also handicapped by a fundamental misconception of what privacy means to a significant portion of the American public. The concerns about government data collection, computers, and privacy dated back as far as the 1970s, and those concerns had only grown as the scale of data available had increased. "Some government officials attempt to finesse the privacy issue by insisting that individual records and data will not normally be shared or subject to examination by a human observer so, they argue, there is no real infringement on privacy," wrote Steven Aftergood, a privacy advocate at the Federation of American Scientists, shortly after the debacle. "But that doesn't get to the heart of the issue. Personal privacy is compromised whenever one is subject to unwanted surveillance, even by a machine."

In the final accounting, the legacy of Total Information Awareness ran much deeper than any single DARPA program. The controversy not only ended the related privacy research; it also pushed data mining deep into the classified world of the intelligence community, laying the intellectual groundwork for the massive analysis and collection system that would be revealed ten years later by an NSA contractor named Edward Snowden.

For DARPA, the repercussions were also long lasting. Michael Goldblatt, the former McDonald's scientist turned DARPA manager, found that his super-soldier research, once the darling of Richard Cheney, was also under fire by congressional staffers eager to find more dirt on DARPA. His work to develop vaccines for pain, and soldiers who could survive blood loss, was interpreted as a sinister plot hatched by mad scientists. "They thought we were making people robotic. They totally misunderstood," he recalled. "I took a lot of heat on that." But as with Total Information Awareness, the work did not actually end. "The beauty of it is we changed the name, and the program went on," Goldblatt said, laughing.

Goldblatt might have found the episode funny with hindsight, but at the time he was deeply disappointed. His efforts to build the soldier of the future—embraced by the vice president two years earlier—were now a liability for DARPA. He resigned in 2003, the same year that Poindexter left over Total Information Awareness. Goldblatt even told Tether that the controversy over robo-soldiers could be blamed on him. "We got rid of that crazy Goldblatt," he instructed Tether to tell Congress. "He was out of control, a cowboy, looking for ways to kill us all, pump us full of drugs."

The DARPA director soon faced his own crisis. Shortly after the controversy, Newt Gingrich, the former Speaker of the House, paid him a visit with a direct message from Rumsfeld, who rarely communicated directly with DARPA's director: "I came over to tell you that the Chief says you're getting close [to getting fired]." Tether kept his job, but DARPA lost what he called the agency's "greatest strategic thrust." It also meant that at least for the foreseeable future, DARPA would not be involved in high-profile research related to the war on terror. Futuristic aircraft were fine, but applying the agency's expertise in computer science to counterterrorism research was not.

For those who believed that DARPA should focus only on futuristic developments, ten or twenty years out, losing a critical counterterrorism mission was no tragedy. But for an agency that pioneered protection for the presidential vehicle in a matter of months, built a revolutionary nuclear test detection system in less than a few years, and led a worldwide counterinsurgency program, it marked a dramatic curtailment of the agency's mission. The DARPA that emerged from Total Information Awareness would still be revered for its technological accomplishments but largely left out of frontline work in national security. Fearful of criticism, but eager for attention, DARPA turned instead to fantasy.

CHAPTER 18

Fantasy World

I believe strongly that the best DARPA program managers must have inside them the desire to be a science fiction writer," wrote Tony Tether, when asked about his philosophy for running DARPA. "Writers such as H. G. Wells, who for example wrote in his 1914 novel *The World Set Free* about nuclear power and talked about the atomic bomb and gave it the name used today, would have been a great DARPA [program manager]."

Tether's passion for science fiction had been why he promoted Tony duPont's fantastical space plane back in the 1980s. And it was also the reason why he selected Michael Goldblatt's super soldiers and mind-controlled weapons to be the opening act for Richard Cheney's visit back in 2001. That sort of techno-thriller material was exactly what Tether liked. "Imagine 25 years from now, where old guys like me put on a pair of glasses or a helmet and open our eyes," Tether said in a DARPATech speech, referring to the work in Goldblatt's office. "Somewhere there will be a robot that will open its eyes, and we will be able to see what the robot sees. We will be able to remotely look down on a cave and think to ourselves, 'Let's go down there and kick some butt.'"

More than anything, Tether loved Disneyland. The man who would become the agency's longest-serving director believed the home of Mickey Mouse and Walt Disney's Enchanted Tiki Room represented everything he envisioned for DARPA. The amusement park was where

DARPA had debuted Total Information Awareness in 2002, and it continued as DARPA's meeting spot. Over his nearly eight years as agency director, Tether would hold all four DARPATech conventions at Disneyland, featuring speeches punctuated by *Star Wars* theme music and schwag that included DARPA-embellished playing cards, custom DARPA-logo golf balls, and T-shirts decorated with armed drones. "Welcome to our world!" Tony Tether said, beaming to the audience in the opening ceremony in March 2004. "A world where science fiction morphs into reality."

The second DARPATech in March 2004 also marked the one-year anniversary of the U.S.-led invasion of Iraq. What was expected to be a year of rapid reconstruction had instead turned into a bloody conflict, pitting American forces against not the conventional army they were prepared to take on but nearly invisible insurgents. March ended up being a particularly bad month: militants released a video showing the beheading of Nicholas Berg, an American hostage; four employees of the American security firm Blackwater Worldwide were killed in Fallujah, their charred corpses dragged through the streets; and a bomb killed dozens at a hotel in central Baghdad. Fifty-two American soldiers died over the course of that month in Iraq, many while driving in vehicles hit by improvised explosive devices, the homemade bombs being used with increasing frequency.

At Disneyland, however, where program managers were handing out DARPA-embossed M&M's, the mood was upbeat. Tether believed his mandate was to make the agency "like it used to be," a place where its employees "were always getting the director in trouble." That was exactly what had happened with John Poindexter and Total Information Awareness. Tether survived the congressional and public backlash, even though it meant ceding DARPA's role in counterterrorism research to the intelligence community.

While his hopes of being involved in top-level national security issues were dashed, Tether's other vision—equal parts fantasy and spy lore—flourished, creating an entirely new perception of the agency. He funded ideas like Blackswift, an unmanned hypersonic fighter plane that could travel six times the speed of sound—a successor to the failed National Aerospace Plane. He also embraced novelties like "polymer ice," a synthetic substance that could be thrown from the back of Humvees to make an enemy slip off the road. Some of his ideas, however, were more scientifically dubious. He supported work, for example, on a controversial "hafnium bomb," which used a radioactive material that would potentially be tens of thousands of times more powerful per gram than conventional explosives, if scientists could figure out a way to trigger it. None could.

Tether knew he needed something big to recover from the Orwellian data-mining controversy, and even in the midst of the scandal he had come up with a new science-fiction-sounding idea: a robotic race. The competition would pit robotic cars on a 150-mile course of rough desert that stretched from Barstow, California, to Primm, Nevada, just forty miles short of Las Vegas. The Grand Challenge, as the race was called, would give $1 million to the winning robot creator. As Tether prepared to travel to California to kick off the competition, he had in the back of his mind the realization that the robotic car race, if it went well, might save DARPA from the circling critics. If it did not go well, DARPA's future was at stake.

When Tony Tether flew into Los Angeles on a Saturday to attend an "industry day" for potential participants in the Grand Challenge in February 2003, he was in a panic. The race was a gamble, and DARPA, at huge expense, had rented out the entire Petersen Automotive Museum on Wilshire Boulevard, which housed luxury and vintage cars. "Gee, probably all the people are going to be our own. If there are five other people there, I'm going to have to go out on the streets of L.A. and get the homeless people to come in to fill this place up to make it look like we have people there," he recalled.

Tether belonged to a generation of engineers who grew up watching Gene Roddenberry's *Star Trek* but ended up working on Ronald Reagan's Star Wars. The future DARPA director graduated with a PhD in electrical engineering from Stanford University in 1969, the year of the Apollo 11 moon mission, and went to work for the growing Cold War military-industrial complex, which was absorbing engineers as quickly as universities could produce them. There was still no Silicon

Valley to compete for top graduates; the money, jobs, and excitement were in defense and aerospace. Tether worked for a series of defense companies until he was recruited into DARPA in the 1980s during Ronald Reagan's military buildup, running the Strategic Technology Office, which was heavily involved in missile defense.

A relentless worker who treated Saturday as just another day, Tether wanted to become director of DARPA in the 1990s, only to lose out to a political appointee. When he got another chance, in 2001, he jumped at the opportunity. With his oversize engineer's glasses and pomaded hair, Tether seamlessly blended into the Washington Beltway's techno-industrial complex. But he also had a true, almost childlike love for futuristic technology, which was reflected in his first stint at DARPA, when he pushed people and concepts like Tony duPont and the space plane. His favorite phrase when describing something exciting or surprising was "holy cow!"

Before he became the head of DARPA, or even an engineer, Tether worked full-time for a period as a Fuller Brush salesman, going door-to-door selling personal care products. "I always say that [was] the best education I ever had," Tether later told an interviewer hired by DARPA. "When you're a Fuller Brush man and you knock on the door, you only have a second or two to really assess who's answering that door and how are you going to get in that house." Selling Fuller Brushes was about reading your audience, or as Tether put it, "telling the right story."

Tether desperately hoped the Grand Challenge was the right story, but he worried that even a robotic car race, with no obvious connection to weapons or the military, might provoke a backlash, or at least put people off from participating. Despite attending Stanford for graduate school, Tether viewed the West Coast as a strange land, filled with liberals who simply did not comprehend how the world had changed after 9/11. "People, especially people from the Western part of the United States, really, I think, believe that the attack on New York was like a movie, 'Godzilla Invades,' you know, 'New York City,' or something like that, and didn't really appreciate that we really were at war," he later reflected.

At least for the car race, however, his concerns were overblown. When Tether showed up at the car museum, there was a line twisting around the block. Some eight thousand people attended. "Holy cow!" Tether thought to himself. "This can really turn out to be quite something."

DARPA's Grand Challenge was the brainchild not of a DARPA scientist but of the agency's onetime chief legal counsel, Richard Dunn, who had been inventing creative mechanisms to evade bureaucracy. Whether it was finding ways to hire employees on special contracts or circumventing normal government procedures to work with small companies, he had become something of a one-man "fix it" shop for DARPA's red-tape cutting. The Grand Challenge, or at least the framework of it, was another Dunn innovation, modeled after the Orteig Prize, offered by Raymond Orteig, a New York hotelier who put up $25,000 for the first solo flight across the Atlantic by plane.

Shortly before Dunn retired from DARPA in 2000, he persuaded Congress to give DARPA authority for "incentive prizes," although there was no specification of the type of contest. Robotics was one of the early suggestions, Dunn recalled, although the idea had been robots that could scale buildings. When Tether took over DARPA, the prize authority was sitting around unused, and he had a different idea: a robot car race. "Well, everybody in this country owns a car. Everybody can buy these computers," he reasoned. "The sensors you can buy. The actuators are even available through the handicapped market. So, this is something that has a low hurdle for the average guy."

Originally, Tether had wanted the race to be in Anaheim, part of the Los Angeles metropolitan area. He envisioned a 250-mile car race from Los Angeles all the way to Las Vegas along public highways. As with the crossing of the Atlantic by plane in the 1930s, the individual technologies needed to build self-driving cars were all available, in theory, but pulling them together seemed like a nearly impossible feat, or at least something that had simply never been done before. It would be a mental advance, as much as a technical one, and Tether hoped the contest would spark nationwide interest in robotics. In the end, however, an air force colonel hired to manage the logistics told him there was no way to shut down a highway so close to Los Angeles, even at night. Instead, the colonel suggested Barstow, a dying town in the California desert dominated by desert shrubs and methamphetamine labs. Shutting down a road in Barstow would not be too difficult.

On March 13, 2004, fifteen competitors pulled up to a starting line in Barstow: At stake was the $1 million jackpot for the vehicle that could achieve the best time along a treacherous, 150-mile course through the Mojave Desert. National press from around the country had descended on Barstow to document the historic event. The anticipation lasted less than eight miles. One by one, the competitors dropped out, stuck on a rock, caught in an embankment, stumped by software, or, in one case, dramatically barreling through a fence. The race favorite, a Humvee, made it the farthest, logging 7.32 miles before stopping, its "belly straddling the outer edge of a drop-off, front wheels spinning freely, on fire," as the magazine *Popular Science* reported, headlining news of the race as "Debacle in the Desert."

The Grand Challenge was off to an inauspicious start. No one claimed the $1 million prize. Then again, it had taken eight years from the time the Orteig Prize was offered until Charles Lindbergh won it. The saving grace was that unlike the Orteig Prize, which claimed the lives of several contenders, no one died in the desert race. Tether played down the failure, promising to hold another Grand Challenge and blaming the press for hyping the first race. "I know you guys built it up so much [and then] it only went seven miles, you were embarrassed about it," he said. "But don't think we were."

While Tether was pushing the frontier of robotics with the Grand Challenge, DARPA-supported scientists were seeing the flip side of science fiction fandom. Tether was a disciple of George Heilmeier's and his "catechism"—the seven questions the then director used to evaluate whether to start a program—to the point that chocolate bars with wrappers imprinted with the catechism were distributed at DARPATech. Like Heilmeier, Tether wanted breakthroughs, and he wanted those breakthroughs on a schedule, which he called go/no-go points. Programs that could not reach specific goals on schedule, such as six months or a year, were swiftly ended. University researchers accustomed to long-term funding took a hit. *The New York Times* reported in 2005 that DARPA had slashed its computer science funding for academics; by 2004, it had dropped to $123 million, down from $214 million in 2001. Tether defended the cuts, saying that he had not seen any fresh ideas from computer science departments. "The message of the complaints seems to be that the computer science community did good work in the past and, therefore, is entitled to be funded at the levels to which it has become accustomed," Tether shot back, when faced with criticism.

This was not Tether's first clash with the scientific community. In 2002, he suddenly ended DARPA's four-decade-long relationship with the JASONs, the independent scientific advisory group, by taking away their funding. Tether never publicly commented on what precisely led to his decision to sever ties, but according to several accounts the conflict was over the group's membership, which some Pentagon officials perceived as being weighted toward older physicists. Secretary of Defense Donald Rumsfeld wanted to add specific members to the JASONs, "a couple of young Silicon Valley 30-year-old types," he later told *Fortune* magazine in an interview. The JASONs, who had long prided themselves on selecting their own members, refused, and Tether canceled their contract. After a congressional outcry, the Pentagon's director of defense research and engineering funded the contract; the JASONs lived, but the break with DARPA was permanent.

The JASON controversy, combined with the cuts to university computer science departments, was beginning to paint Tether as antiacademic. But Tether insisted DARPA was not cutting back on funding for universities; it was merely redirecting funds into interdisciplinary research that he believed would result in major breakthroughs. One of the examples he highlighted at a congressional hearing was computers that could read people's thoughts, or what DARPA ended up calling "augmented cognition."

The term "augmented cognition" came from Eric Horvitz, the Microsoft scientist who had theorized the "hall of mirrors" privacy device for Total Information Awareness. At a DARPA-sponsored meeting in 2000, Horvitz proposed a computer that could adapt directly to a person's mental state, an extension of J. C. R. Licklider's dream of man-computer symbiosis. Licklider had wanted computers to help with decision making, because computers could do calculations faster than the human brain. The focus for years had been on making computers more powerful, while the human brain remained the same.

Now the computers were much faster and smarter than they had been in Licklider's time, and the problem was the human brain could not keep up with them. Horvitz wanted to combine cognitive psychology with computer science to find ways to allow the human brain to work faster with computers.

Horvitz envisioned computers that would sense when a person was tired, overtaxed, or forgetful, reorienting its display, for example, or providing auditory cues to alert an inattentive user. To demonstrate this vision, he and his colleagues in the Information Science and Technology study group even arranged at one meeting for "mind reading" helmets, to demonstrate one aspect of augmented cognition: the use of sensors to detect brain signals. The idea was that helmets with electrodes could be placed on someone's head to detect neural signals—or point infrared sensors at the brain to look for changes in blood flow—and then use that information to adjust information provided by the computer.

Augmented cognition captured the attention of Dylan Schmorrow, a new DARPA program manager, who funded a formal study. Horvitz envisioned augmented cognition as a mix of basic science and engineering—a broad research program studying how people integrate information from across the senses. That research could improve how people interact with computers, such as by creating displays that adapt to the user by highlighting important information. Schmorrow liked the idea, but he liked the helmets even better. The Augmented Cognition program DARPA created ended up focusing on a single application—a device that would detect and respond to someone's cognitive state. "We actually thought it was interesting still, but we were surprised at just the narrow-focused course," Horvitz said. "Then again, the program resonated with DARPA's interest in hardware, devices, and putting a cap on someone's head."

Horvitz and Schmorrow did not realize initially that augmented cognition had a predecessor at DARPA known as biocybernetics, the research program George Lawrence had led in the 1970s. When word of a prior DARPA program eventually filtered down to Schmorrow, he called Emanuel Donchin, one of the original DARPA-funded biocybernetics researchers, asking him to come to DARPA to discuss his earlier work. Donchin was shocked when he showed up at DARPA's headquarters in Northern Virginia. Back in the 1970s, Donchin might drop in at DARPA to see Lawrence about brain-driven computers, but he would also poke his head in Licklider's office or chat with other program managers with similar interests. DARPA back then was an open office building, at least for its unclassified projects, and visiting one research manager was an invitation to chat and meet with other officials and swap ideas. "When I came to see Dylan [Schmorrow], there were security people in the lobby," Donchin said. "I couldn't speak to anybody about anything except Dylan Schmorrow. It was an amazing transformation." More shocking for Donchin was that DARPA officials had no idea that the agency had been involved in similar work in the past. "They had zero information about the biocybernetics program," he said. "DARPA had no capacity for institutional memory. It was very strange."

Biocybernetics had focused not on working devices but on investing in a new field of science. In the late 1960s and early 1970s, the technology for detecting neural signals was rudimentary; forty years after that initial DARPA program, the technology had evolved, but scientists still disagreed over the interpretation of those signals. For example, scientists had developed better ways to detect the P300, a brain signal that occurs about three hundred milliseconds after a stimulus, like a specific sight or sound. Yet DARPA's vision for augmented cognition assumed that such brain signals, which were still being studied in the lab, could be used to start immediately building the equivalent of mind-reading caps for the military. It was science fiction, quite literally.

To illustrate its vision, the agency enlisted the services of Alexander Singer, a Hollywood television director best known for the new Star Trek series, *Deep Space Nine,* to create a half-hour mini-film depicting augmented cognition. The video, inspired by the *Star Trek* holodeck, opened with lingering shots of groundbreaking scientists: Charles Darwin, father of evolution theory; B. F. Skinner, famous for operant conditioning; and Hans Berger, inventor of electroencephalography. It then flashed to DARPA's Dylan Schmorrow, credited as the father of augmented cognition. The science fiction story line featured a cybersecurity officer, Claudia, who must head off a cyber attack designed to destabilize Africa. Claudia is outfitted with a headpiece that monitors her cognitive state, and the computer parses out information to speed her decisions, with occasional interruptions by a Yoda-like cyber chief teleconferenced in from a fishing vacation. "We may be looking at an anomaly, big time," Claudia declares.

Alan Gevins, a neuroscientist and longtime researcher in brain-computer interface, was perplexed. Gevins, whose forty-year career had spanned the original DARPA biocybernetics program, was invited to take part in the Augmented Cognition program but was disillusioned with the emphasis on science fiction visions over experimentation. "I'm a data guy, not a philosopher," he said. DARPA was calling the researchers "performers," which was accurate, Gevins joked, because some of those under contract to DARPA were acting like performers at a circus. Gevins recalled watching one DARPA-funded researcher demonstrate the use of a brain signal to move a cursor on a computer screen. Picking up these sorts of signals required careful controls and knowledge of the equipment, but the researcher simply stomped his foot when he wanted the cursor to move, introducing a deliberate "artifact," or error. (Ironically, this was the same method that Uri Geller, who claimed to have psychic powers, was accused of using three decades earlier.) "It clearly was fake, and it wasn't subtle at all," Gevins said. "I pointed that out, but it didn't seem to make any difference. It was astounding actually." Aghast that DARPA was spending huge amounts of money on efforts that yielded dubious scientific results without using any sort of peer review, Gevins soon dropped out of the program.

The Augmented Cognition program was less interested in exploring a field of science than in producing hardware. The first phase of the program ended with what DARPA called the Augmented Cognition Technical Integration Experiment, which tested some twenty different gauges of cognitive state, ranging from electroencephalography to pupil tracking. Subjects were monitored as they played a video game called *Warship Commander Task*, which tested a person's ability to respond to threatening aircraft. In a sense, it was like playing an old Atari game, where the primary goal was to spot and destroy an enemy aircraft without shooting down friendly aircraft. As the game progressed, sensors would monitor the player's cognitive state to identify when the brain was overloaded and then parse out information in the most efficient way possible. In an overview paper describing the work, Schmorrow and two colleagues called the results "promising" and said they pointed to "great potential" of using such sensors for applications.

Not everyone was so optimistic. Reviewing the 2003 experiment, Mary Cummings, a human-computer interface expert, noted that even the published results indicated that none of the signs of "overload" that the researchers were testing were consistent across all three variations of the test (the researchers varied the number of aircraft, the level of difficulty, and authority). And the two measurable signals that worked consistently across two variations of the game—mouse clicks and pressure—were only indirectly related to someone's cognitive state. In a published critique of DARPA's claimed success, she noted the experiment's errors, data problems, and, most critically, the potential absurdity of developing military equipment that would require someone in combat to carry a thirty-five-pound device and wear an EEG cap with gel sensors attached to the scalp.

Cummings's criticisms were particularly stinging. She had been one of the navy's first female fighter pilots and later went on to get a PhD in systems engineering, then worked at MIT. As an experienced pilot and specialist in human-computer interaction, she knew more than most about creating military technology that could pass scientific muster while also being usable in a realistic military environment, and she was underwhelmed on both fronts by the DARPA Augmented Cognition program. Cummings laughed when asked if she saw scripted demonstrations, like foot stomping. "What didn't I see?" she replied.

She recalled being briefed by a company that had already spent several million dollars of DARPA funding on eye tracking—looking at gaze and blink rate—which can be used to gauge someone's attention to features on a computer screen or even determine whether someone is mentally overloaded. The company officials were claiming an order of magnitude improvement in reaction time using eye tracking, which sounded impressive. "When I asked them to show the experimental results—the results, which were the basis of millions of dollars of funding, I found out that it had only been tested on two people—the creators of the system."

The Augmented Cognition program wound down by 2007, although DARPA continued work on two related technologies, including brain-reading goggles that were supposed to help soldiers detect possible threats and a wearable head device that would allow intelligence analysts to sort through imagery quickly. Both programs were based on detecting the P300, the neural signal sparked when someone has unconsciously recognized an object. In the case of the goggles, the device would act as a sort of "sixth sense," alerting the wearer to a possible threat, like a sniper, or someone planting a bomb, before the conscious brain has recognized it. For the intelligence analyst, it would tap his or her unconscious thoughts to sort through thousands of images quickly.

Todd Hughes, a DARPA official who ran the program that created wearable brain-reading devices for imagery analysts, admitted it required some imagination to see the applications. The technology required electrodes attached to the scalp with gel—not the way most government employees would like to work. Hughes joked that his vision was a special team of analysts: "There would be a dozen guys with shaved heads; they would wear special armbands. When the plane crashes and they don't know where it is, they run into the lab, put on their headgear, and start searching imagery until they find it. Then they walk out of the room heroes."

DARPA eventually turned over the imagery analyst gear to the National Geospatial-Intelligence Agency, and the brain-reading goggles went to the army's Night Vision Laboratory. Technically, that meant both programs "transitioned," DARPA-speak for technologies that have successfully gone to the military, but it appears that neither was ever used outside of a lab.

As a research program, augmented cognition was a great idea, Cummings maintained. The problem with the program, she said, was that researchers were being asked to show concrete results in an area that was still basic science. "Where DARPA started to fall overboard is when they started to try and make it applied, ready for some sort of operational results," she said. The allure of science fiction, without the checks and balances of rigorous science, had led the promising field of augmented cognition down a rabbit hole. The question was whether the same would be true for robotic cars.

Back in the 1980s, DARPA as part of the Strategic Computing Initiative had funded an autonomous land vehicle, dubbed the "smart truck," which the historian Alex Roland described as a "large ungainly, box-shaped monster." Instead of a windshield, the front of the vehicle sported a "large Cyclopean eye" that housed the robot's sensors. It looked more 1950s camp science fiction than *Terminator*, but the exterior was not important. What mattered were the rows of computers stacked inside the fiberglass shell of the truck and the algorithms that were supposed to make sense of the outside world. Those algorithms did not work very well.

The truck was equipped with television cameras, whose pictures were analyzed by the onboard computers to create what is known as "computer vision," the term for how computers process and analyze images. The human brain does this well, letting someone know, for example, the difference between a tree and the shadow of a tree. The smart truck did this badly, so researchers found it was best to test it in the noonday sun, when there were no shadows. When Carnegie Mellon researchers took the truck out for a spin in Pittsburgh's Schenley Park, they had to use masking tape to denote borders, because the truck's computer vision would confuse inanimate objects, like a tree trunk, for the edge of the pavement. If robots were really going to take to the road, computer vision had to be improved well beyond the smart truck.

The year of the first Grand Challenge, Larry Jackel, a physicist by training, came to DARPA to take over the agency's robotics programs. One of the first things he did was buy himself a Roomba, the autonomous vacuum cleaner that has spawned thousands of YouTube videos, many involving the robot's interaction with people's pets. The Roomba was made by iRobot, a company that also produced military robots, including its flagship PackBot, which had been developed with DARPA funding in the 1990s. The PackBot showed up in Afghanistan in 2002 to help clear caves and was not particularly effective (the robots lost communications and got stuck). The robot soon found a higher calling in explosive ordnance disposal. Eventually, thousands of modified PackBots were sent to Iraq and Afghanistan to help defuse roadside bombs. But the PackBot's civilian cousin, the Roomba, frustrated Jackel: it got stuck on the modern shag area rug in his New Jersey home; it was flummoxed by computer cords; and it could not navigate inside the four legs of a chair, or if it did, it got stuck there, as if trapped in a virtual jail of invisible walls. Frustrated, he finally got rid of the Roomba and went back to a regular vacuum cleaner.

In popular culture, robotic vehicles—or even robotic soldiers—are often portrayed as right over the horizon. The threat of armed Terminators is debated as if the Pentagon were already building armies of them. Most of DARPA's programs were focusing on advancing different aspects of robotics, rather than building war robots. For example, DARPA sponsored Boston Dynamics to build LittleDog, a four-legged vehicle (which actually looked more like a bug than a canine) that was designed to travel on rough terrain. LittleDog was followed by BigDog, a larger version that could carry supplies for troops, like a robotic mule. While tech blogs and popular magazines often called the headless BigDog a "war robot," it was actually more appropriately a lab robot. BigDog was meant to demonstrate a specific ability, in this case how to move a legged robot over rough terrain. BigDog was not destined for the battlefield.

Even Congress was possessed by technological optimism in 2000, writing into law that by 2015 one-third of all military ground vehicles should be unmanned. It was an ambitious if misinformed goal. The enthusiasm stemmed from growing use of unmanned aerial vehicles. Drones in the first decade of the twenty-first century were rapidly replacing manned aircraft, so unmanned ground vehicles sounded like the next logical step. What was not immediately apparent to Congress was how different an unmanned aerial vehicle was from an unmanned ground vehicle. Drones, particularly at high altitudes, are mostly in danger of hitting other aircraft. On the ground, robots have to contend with every type and size of obstacle. Differentiating between a rock and its shadow, as the 1980s DARPA smart truck demonstrated, could be difficult for even the most advanced robots. The 2004 Grand Challenge demonstrated all of the limitations of technology in gory, tire-burning detail.

Separate from the Grand Challenge, however, were DARPA's regular robotics programs, now run by Jackel, who was fast learning the limitations of state-of-the-art autonomous vehicles. One DARPA-funded robot, nicknamed Spinner, was essentially a giant sport-utility vehicle that was designed for "extreme mobility," meaning it could traverse some of the most rugged terrain. "It was meant to be able to flip over and run upside down," Jackel said. "The cargo bay was on a pivot: if the vehicle went upside down the cargo bay could flip over." That sounded great until Spinner went out to the desert for testing and everyone realized that the more than ten-thousand-pound vehicle was so hard to tip over there was really no reason for all those complicated mechanisms to allow it to operate that way. "It just wasn't needed," Jackel said.

The bigger problem for robots was not agility but brains. Vision, or lack of it, is what had flummoxed DARPA's 1980s-era smart truck in Pittsburgh's Schenley Park. Twenty years later, DARPA was still trying to solve the fundamental problem of providing robots with the ability to process what they see and navigate around obstacles. Robotic vehicles over the years had added all kinds of sensors, such as lidar, which sends out a laser and then measures reflected light to sense objects. But with most ground robots, Jackel found, if they encountered an obstacle, they would back up, often hitting yet another obstacle, move forward into the original obstacle, essentially getting stuck in a loop, just like his Roomba trapped by chair legs.

Jackel had inherited a program called PerceptOR, which was supposed to improve robotic navigation, but it was not clear DARPA was making much progress. Then, one day, Jackel was walking his two dogs, American Eskimos, in the woods behind his house and watched as they bounded forward. The dog's stereovision was similar to a human's, limited to about forty or fifty feet. But he watched with fascination as his dogs would spot something of interest, like a possible animal, and then dart forward at full speed, navigating around trees with ease. "I thought, 'Gee, I don't know what these dogs are doing, but they're not running on lidar and they aren't running on stereo.' They're somehow interpreting the image. The dog doesn't go around and label that this is a tree and this is a bush."

That became Jackel's inspiration for a new program called Learning Applied to Ground Vehicles, or LAGR, focusing on machine learning. Rather than having to identify each specific object, the LAGR robots would learn by experience how to navigate the terrain, mapping out a path in the distance. The robots did this using stereo cameras that created three-dimensional models looking out to about nine meters ahead, where obstacles could be more easily identified, and then comparing that with the color and shading of more distant scenes, where objects are not as easily identified. In that way, the robots could identify a clear path. The program ended up enabling robots to extend their effective vision out to a hundred meters. "We never got to the point where they were as good as the dogs, but they were a whole lot better than when they started," said Jackel.

In 2005, Tony Tether returned to Barstow to kick off the second Grand Challenge competition. While spectators and press were still excited, Tether was even more nervous than he had been in 2003 when he showed up at the Petersen Automotive Museum. "I never said anything to anybody in DARPA, but I knew that we had to get somebody across that finish line, or, at least we had to get them really close," he later recounted.

Many of the same competitors from the first Grand Challenge lined up for the second race. Carnegie Mellon, a longtime leader in robotics and a race favorite, entered two vehicles. The course was even more challenging the second time around and included routes like Beer Bottle Pass, a treacherous strip sandwiched between a rock face and a sheer cliff. There was also a new entrant from a group at Stanford University led by the German-born computer scientist Sebastian Thrun. The Stanford team's vehicle, an unassuming blue Volkswagen Touareg named Stanley, was outfitted with lidar, cameras, GPS, and an inertial guidance system. Thrun, along with several other competitors, had participated in Jackel's computer vision program.

As in the first Grand Challenge, DARPA provided GPS waypoints just prior to the race to guide the cars along the course. GPS, however, was of little use to vehicles when they had to navigate rocks, shrubbery, and other desert obstacles or even manage sharp turns and an occasional cliff. While some vehicles focused on identifying each obstacle, Stanley's sensors scanned the road, not only concentrating on specific objects, but looking ahead and identifying the best course.

Like Jackel's bounding dogs, Stanley did not need to identify every single obstacle; it just needed to choose a good enough course to allow it to move along at a decent pace. That application of machine learning is what Thrun and his team had been practicing in the desert. "It was our secret weapon," he told a reporter from *The New Yorker*.

Even then, the second Grand Challenge was no high-speed robotic drag race. Stanley's average speed was about nineteen miles per hour, but that was enough to pull ahead of its main competitors, the two Carnegie Mellon vehicles. Stanley nabbed first place, while Carnegie Mellon took second and third prizes. A dark horse candidate headed by an IT manager for a Louisiana insurance company took fourth place. When Stanley crossed the finish line, Tether took a deep breath. "Holy cow, we did it," he said to himself.

The Stanford team took home the $2 million jackpot. In all, five vehicles crossed the finish line, compared with none in the first event. What exactly enabled the winning teams to pull ahead of the others is hard to pinpoint. All the teams learned from studying the experience of the first Grand Challenge, according to Jackel, and knew what to expect the second time around. But it is impossible to ignore that the winning teams, Stanford and Carnegie Mellon, had received significant DARPA support for their robotics programs over the years.

The Grand Challenge did not produce any new technology; its success was simply in demonstrating that self-driving cars would work. Though that demonstration was itself a critical achievement, Jackel was hesitant to give it an unwavering endorsement. There were benefits to incentive prizes, but they should not replace funded research, he argued. In the first two competitions, people had to fund themselves or find corporate sponsorship. Jackel was concerned about the long-term implications of such competitions for research and the survival of institutions that support research. "At some place, money had to flow into the system," he said.

Jackel knew how precarious the support for research institutions could be, even those that were nationally revered. Jackel was a refugee from Bell Labs, the storied research and development division of the Bell Telephone Company. Ma Bell, as the monopoly was affectionately called, operated its lab as a quasi-academic institution, allowing its scientists to work with a large degree of independence. The scientists were encouraged to work on problems facing the telecommunications industry, but their research was judged by its scientific merit, not by the dollar figure their innovations generated. "Basically, the U.S. population funded Bell Labs through their phone bills," Jackel said.

In its heyday, Bell Labs gave birth to the first transistor, enabling a revolution in electronic devices. The lab was also home to scientific giants, like the father of information theory, Claude Shannon, whose work contributed to digital computers. In a rough sense, Bell Labs was to Ma Bell what DARPA was to the Pentagon—a problem-solving organization afforded wide latitude to explore scientific and technological solutions. And that worked well as long as Bell had a monopoly on telecommunications, the way the Pentagon has a monopoly on running the military. When the telephone monopoly was broken up, the lab was downsized, and its autonomy all but eliminated.

The Grand Challenge was good publicity for robotics and for DARPA, but Jackel worried that it would overshadow the need to support long-term research. Without funding for early scientific exploration, challenges would not accomplish much. The contests cost much more than a $1 or $2 million prize; DARPA also had to pay for logistics, which was the most expensive part of the competition. And yet no money went to research. "It's not self-sustaining," Jackel said. "You can do it based on something that already exists, but if all we did was have challenges, then at some point we'd just stagnate."

In 2007, DARPA held yet a third and final Grand Challenge, called Urban Challenge, which took place on a former military base in Victorville, California. Instead of just going along a single road, teams had six hours to navigate a course in a city-like environment. It was not exactly *The Fast and the Furious;* the emphasis was on avoiding collisions while obeying traffic laws, so the average speed was around fourteen miles per hour. At one point, competing vehicles by MIT and Cornell University ended up in a bizarre slow-motion collision, with both cars creeping along at just five miles per hour.

Carnegie Mellon University took first place. By that point, the Grand Challenge was already a national sensation, featured on magazine covers and in television documentaries. It has also ended up, by tragic happenstance, being prescient. By 2007, roadside bombs were the leading cause of casualties for American and coalition troops fighting in Iraq and Afghanistan. "Imagine if we had convoys being driven by robots," Tether said.

The Grand Challenge was about the future of DARPA more than about robots. In 2003, in the midst of the Total Information Awareness imbroglio, the agency had been a hairbreadth away from congressional intervention that would have permanently ended its independence. "Total Information Awareness got to the point where, quite frankly, I almost lost the agency," Tether said. "The Grand Challenge really saved DARPA." The agency that just a few years earlier was accused by one California senator of paving the way for a "George Orwell America" was now the hero of politicians, techies, and science fiction enthusiasts. "The Grand Challenge was one of the greatest public relations efforts, I mean worldwide, and that instantly changed the whole image of DARPA back to where it was," Tether said.

The Grand Challenge did more than restore the agency's image. Tether would soon be presiding over the largest expansion of DARPA's budget since the agency's creation. When Tether took over in 2001, the DARPA budget had been stable at about $2 billion a year, but it started to climb dramatically along with the rest of the Pentagon budget, rising to $3 billion a year by 2005. Tether, the former Fuller Brush salesman, had read the mood correctly.

By the middle of the first decade of the twenty-first century, DARPA was facing a paradox: it was billing itself as a science fiction agency in the midst of a war with mounting casualties and military leaders demanding immediate solutions. The Grand Challenge might have saved DARPA, or at least the agency's image, but it had no immediate effect on the wars in Afghanistan and Iraq, nor was it intended to, because robotic vehicles that could go beyond a racecourse were still years in the future.

The Pentagon's immediate response to casualties, mostly caused by homemade bombs, was to establish an agency known as the Joint Improvised Explosive Device Defeat Organization. A few former officials wondered why the Pentagon did not turn to DARPA, whose resident technical expertise and ability to work quickly outside bureaucracy would have made it ideal as a place for the bomb-fighting mission. No one in the Pentagon appears to have even considered the option.

If during the Vietnam War DARPA had sent social scientists to the battlefield, this time around it employed them at home, designing computer programs to predict future conflicts. The army sent anthropologists to Iraq and Afghanistan but without DARPA's support or involvement. DARPA did contribute to the war, but in piecemeal fashion. It deployed a few technologies, like the Wasp, a handheld drone that troops could put in a backpack. But the most public face of DARPA's war effort was the Phraselator, a handheld translation device that was rushed to Afghanistan after the U.S. invasion in 2001. In hearings and interviews over the next several years, Tether touted the Phraselator as a prime example of DARPA's battlefield innovations. The technology press also lavished praise on DARPA's "universal translator," even though the device did not really translate; it essentially had preloaded phrases that could be activated by voice recognition of the English equivalent or manual selection. The Phraselator fast became one of DARPA's most public accomplishments in Afghanistan.

Some eight thousand miles away from Disneyland, Ken Zemach was on foot patrol with American troops in Afghanistan, when the soldiers he was with decided it was time to test the Phraselator with an Afghan villager they encountered. Zemach, a PhD engineer from MIT, was working for the Rapid Equipping Force, an army organization that was created in 2002 to rush technology to soldiers in Afghanistan, and then later in Iraq, without having to go through the typical military bureaucracy that could take years or decades. Without even realizing it, Zemach and his colleagues were doing a bit of what DARPA's AGILE program had done in the 1960s in Vietnam, which was field-testing off-the-shelf or rapidly developed technology in a war zone.

The Phraselator's road to war had started shortly after the September 11 attacks, when Tether called for possible DARPA technologies that might be deployed quickly to troops in the field. A DARPA program manager working on automated speech recognition had suggested a handheld translator, and the agency awarded a $1 million contract to a Maryland-based company called Voxtec to build what became known as the Phraselator. By 2002, the clunky-looking devices started showing up in Afghanistan, and at DARPATech two years later Tether praised the Phraselator as a prime example of what DARPA was doing to help troops.

DARPA was once again trying to send technology into war, only this time around it had no deployed personnel or any sort of larger strategy. And Zemach was quickly growing disillusioned with what would become the most public face of DARPA's wartime efforts to field technology in Afghanistan and Iraq. The Phraselator was held up in Washington as a grand success, but Zemach had a different assessment: "It sucked."

On patrol in the Afghan village, Zemach held up the Phraselator, which looked more like a *Star Trek* tricorder than a universal translator. The device spit out a few sentences in the local language. The Phraselator had just said it was going to ask some questions and instructed the man being addressed to raise one hand for yes and two hands for no. The first question was whether the man understood this. The Afghan smiled and raised one hand. The next question was whether there were any foreign fighters in the area. The man raised two hands. No. Were there any minefields in the area? The man raised two hands. No, again.

Then the group brought over a local interpreter. Suddenly the man's answers changed. There was a minefield in the area, he reported. It was not that he was lying, Zemach concluded after many similar experiences; it was just that Afghans did not feel comfortable giving information to an electronic device. This scenario repeated itself in village after village, Zemach recalled a decade later.

The device was designed to recognize a select set of English phrases and then translate them into different languages, like Pashto, Dari, and Arabic. Though the hope was to eventually build a two-way device that could translate the replies, the Phraselator was one-way, limiting it to simple commands and questions. Even with those limitations, the Phraselator was fielded to troops in Afghanistan, spitting out questions to confused Afghans, who often found themselves faced with a device speaking a dialect they did not understand. Even in the rarefied atmosphere of a government office building, military and law enforcement officials trying out the Phraselator were perplexed. One navy tester expressed frustration that the Phraselator, even after five tries, failed to translate a simple question like "Do you speak English?" instead rendering it into phrases like "Follow me," "Drop it," and "Can you walk?"

More important, the Phraselator lacked what troops really needed. Most of the preloaded phrases in the Phraselator were either yes or no questions, asking about the presence of foreign fighters, or direct orders, such as telling people to put up their hands. What troops usually needed were simple instructions to help defuse a potential confrontation when clearing villages. "You are effectively the invading army," Zemach said. "You are going into a man's home in front of his family with weapons and going through his stuff. It's emasculating."

What they needed were phrases explaining that the soldiers were Americans and needed to search the village and its homes. Of course, those phrases could be loaded onto a Phraselator, but there was really no need for a custom device that cost thousands of dollars. Zemach had an interpreter record the phrases they needed on a pocket computer, and then he used a web page interface to call up the phrases as needed. It cost nothing and required no additional technology. "You didn't need this power," he said of the Phraselator. "You needed something simple."

Yet for months, and even years, after the introduction of the Phraselator, the talking phrase book was hauled out at Capitol Hill hearings and in Pentagon meeting rooms, usually by people who did not speak the foreign languages it was programmed with and had never used it in an operational situation. In 2009, an army report collected surveys from soldiers in the field, which failed to garner a single positive comment about the device. "Took too long to translate the correct phrase." "Translation wrong more often than not." "It translated the wrong words." "Is not adequate for 'heat of the moment' situations."

Zemach said he encountered similar problems a few years later with another DARPA quick-reaction technology fielded in Iraq. The device, called Boomerang, was an acoustic sniper detection system. "DARPA swore up and down they never had a single false positive in the seven months of testing," he said. "By the time they got from Kuwait to Iraq, they had over five thousand false positives." DARPA suggested downloading updates on a weekly basis, without realizing that many of the soldiers using it had no easy access at the time to the Internet. DARPA eventually made the fixes to Boomerang, but the agency's lack of knowledge about war made the process tortuous. "You have no right deploying the stuff" is what Zemach said he wanted to tell DARPA. "You have no idea how war works."

In fact, DARPA's funding of natural language processing in the first decade of the twenty-first century did have one major success. DARPA funded wide-ranging artificial intelligence research under a program called Personalized Assistant That Learns, which sponsored work at SRI International. The military was not interested in the work, and the DARPA program was terminated, but SRI International spun off the technology as a company called Siri, which was eventually bought by Apple and incorporated into the iPhone. DARPA's work on natural language processing was not necessarily going to help soldiers talk to Afghans, but it could help Americans find the closest Starbucks.

Zemach did not object to DARPA's supporting research into something like the Phraselator; he just did not think it belonged on the battlefield. "When the war kicked off, DARPA got a lot of pressure to be relevant, which is a problem, because DARPA is not good at being relevant," Zemach concluded.

That view reflected DARPA's image in the early twenty-first century: a great science fiction agency, but not a place that the Pentagon turned to during wartime. It was a view at odds with the first two decades of DARPA's existence and a reflection of how much had changed in the ensuing years. When it came to modern threats central to American forces fighting in wars, like the proliferation of roadside bombs, DARPA's role was tangential or aimed at developments, like driverless cars, that would only help years in the future.

In April 2008, Vice President Richard Cheney, one of the chief architects of the Iraq War, feted DARPA at a black-tie dinner for sixteen hundred guests at the Washington Hilton to celebrate the agency's fiftieth anniversary. At that point, Tether was the longest-serving director, thanks to the political support he had from Cheney and Rumsfeld. Cheney, in fact, gave the keynote speech at the dinner, praising the agency's work in areas, such as drones, that had become a hallmark of the administration's ongoing war on terror. "One thing we didn't have a lot of in Desert Storm was the unmanned aerial vehicle," Cheney said. "But thanks to DARPA, that technology was advancing rapidly in the early '90s. And we've been able to use it all the time in both Afghanistan and Iraq—for reconnaissance, for remote sensing, and to strike the enemy."

Outside the Washington Beltway, what had cemented DARPA's reputation for innovation was not necessarily drones or stealth aircraft but the Internet. The agency's most important creation had ensured DARPA's place in history, even if it had emerged from a tiny effort four decades prior. Whether the DARPA of 2008 was capable of producing the types of innovation that had emerged in 1968—when Robert Taylor had published plans for the ARPANET—was not something that was widely debated. Introspection was not a characteristic of the modern DARPA. When Tony Tether wanted to kick off a celebration for the agency's fiftieth anniversary, he commissioned a series of articles describing DARPA's successes. The final product, published by a private company, was interspersed with paid advertisements from defense manufacturers. DARPA also hired a video production company to interview all of the living former directors for a brief promotional video as part of the anniversary celebration. The unedited interviews, which were only released after a Freedom of Information Act lawsuit, offered insights into how much the agency had changed over the past few decades.

When asked about DARPA's role in the war on terror, which was in full swing in 2008, Tether seemed slightly stumped. "While we're containing them, you do need an ability to go in and really start with the three- and four-year-olds to try to get into their minds to basically teach them, 'Hey, it isn't a bad thing. It isn't a bad thing to deal with non-Muslims,'" he said. "And in this country, basically programs like *Sesame Street* really went a long ways toward integrating this country, toward our kids growing up, saying, and 'Hey, blacks and whites and pinks and everybody—it's okay! It's okay to play with them!'"

In vastly simplified form, Tether had hit on the fundamental problem of the war on terror: there was no way to win the war with technology. DARPA's sensors and drones could find and kill terrorists, but it was not stemming the growth of support for extremism. And yet DARPA could not seem to find any other course to pursue other than creating weapons and technology. DARPA in the post-9/11 era was praised for its gizmos and gadgets and portrayed in the press as a supercool science-fiction-inspired agency that was building universal translators, driverless cars, and brain-controlled prosthetics. But DARPA was no longer, as it had once been, the go-to place for national-level problems. That DARPA's leaders had once advised presidents and defense chiefs, sent hundreds of employees to work in field offices in Thailand, Vietnam, Lebanon, and Iran, or helped armor the president's vehicle in the wake of an assassination, was all but forgotten.

DARPA had its high points: its work on the monkey with the mind-controlled cursor grew into a program to develop a prosthetic controlled by the human mind. While a true neuroprosthetic is still many years away, DARPA-funded researchers by 2008 were at least demonstrating a very rudimentary arm that could be controlled by a person twitching a chest muscle. At that point, DARPA had spent more than $150 million, and the product was nowhere near a Luke Skywalker–inspired prosthetic. Still, given the thousands of amputees coming back from the wars in Iraq and Afghanistan, the DARPA program made sense.

More immediately successful was the Grand Challenge. In 2012, Google debuted its driverless car, based on the work by Sebastian Thrun, the Stanford professor who led the winning team in the 2005 competition. The unmanned car competition had done exactly what DARPA had hoped it would do: take a bold technical goal, and prove it was possible. If the Orteig Prize ushered in the modern era of transatlantic aviation, then the Grand Challenge can rightfully take credit for the dawn of autonomous cars. It was Tether's greatest legacy, even if it did nothing for Iraq and Afghanistan.

The Disneyfication of DARPA was hard to reconcile with the realities of the modern battlefield. DARPA had reimagined itself as an agency of future war, producing technologies that might be used decades later. It was still successful, but it was no longer relevant to war, or at least not the type of war the United States was facing in Iraq and Afghanistan. In his final year of congressional testimony, Tether's only direct mention of the agency's current contributions to counter-insurgency was a reference to translation technology. The larger issues facing the United States were beyond the modern DARPA's mandate. "The counter to this problem is a long, long war," Tether said in the internally commissioned interview. "I'm not sure whose job this is, but, basically, what we're trying to do is develop technology to contain them, to contain this threat, which is worldwide."

He offered one additional thought: "We can't go and kill them all, you know."

CHAPTER 19

Return of Voldemort

The only tiki bar in eastern Afghanistan had an unusual payment program. A sign inside the establishment in Jalalabad read simply, "If you supply data, you will get beer." The idea was that anyone—or any foreigner, because Afghans were not allowed—could upload data on a one-terabyte hard drive kept at the bar, located in the Taj Mahal Guest House. In exchange, they would get free beer courtesy of the Synergy Strike Force, the informal name of the group of American civilians who ran the establishment.

Patrons could contribute any sort of data—maps, PowerPoint slides, videos, or photographs. They could also copy data from the drive. The "Beer for Data" program, as the exchange was called, was about merging data from humanitarian workers, private security contractors, the military, and anyone else willing to contribute. The Synergy Strike Force was not a military unit, a government division, or even a private company; it was just the self-chosen name of the odd assortment of Westerners who worked—or in some cases volunteered—on the development projects run out of the hotel where the tiki bar was located.

The Synergy Strike Force's Beer for Data exchange was a pure embodiment of the techno-utopian dream of free information and citizen empowerment that had emerged in recent years from the hacker community. Only no one would have guessed that this utopia was being created in the chaos of Afghanistan, let alone in Jalalabad, a city that had once been home to Osama bin Laden. Or even more unlikely, that the Synergy Strike Force would soon attract the attention of DARPA, which was reaching back in its history to resurrect elements of AGILE, its most ambitious, and controversial, wartime research effort.

DARPA's interest in the potential of open-source information came at a critical juncture in Afghanistan. In January 2009, Barack Obama was sworn in as the forty-fourth president of the United States, the war in Afghanistan was in its eighth year, and a resurgent Taliban, whose regime had rapidly collapsed after the American invasion in 2001, was challenging the central government's tenuous authority in the provinces. Counterinsurgency, the doctrine that had been promoted in Vietnam decades earlier, was back in vogue.

Nearly forty years after Vietnam, General David Petraeus and his followers resurrected elements of ARPA's "glorious failure" as modernized counterinsurgency, or COIN, a "hearts and minds" campaign that emphasized providing security for the local population. The writing that had the "greatest influence on Petraeus's thinking," according to the journalist Fred Kaplan, was a counterinsurgency book by David Galula, the French officer sponsored by DARPA in the early 1960s under Project AGILE. Petraeus lifted Galula out of decades of obscurity, dusting off his writing and incorporating elements of it in a new counterinsurgency manual. "The objective is the population" became an often-cited sentence from Galula's DARPA-funded study of Algeria. In Iraq, COIN was heralded as a success, elevating Petraeus and a new generation of "COINdinistas," as the newly minted experts were called, to near-rock-star status. "By 2009 [counterinsurgency] was being celebrated as the answer to America's mounting woes in Afghanistan as well," wrote Greg Jaffe, a *Washington Post* reporter.

DARPA in 2009 also returned to counterinsurgency. That year, the agency launched an ambitious initiative in data mining, the very work that had created one of the agency's most public debacles of the past decade. Only rather than rooting out terrorists in the United States, as Total Information Awareness had attempted, DARPA's new focus was on Afghanistan. DARPA eventually brought two data-mining programs to Afghanistan: a highly secretive data analysis program meant to predict insurgent attacks based on "big data" science used by companies like Amazon to predict customers' purchases; and a program based on the emerging science of social networks, attempting under the guise of humanitarian work to enlist an unwitting army of Afghan civilians to spy for the American military. And so DARPA's first deployment to a war zone since Vietnam began with a group of well-intentioned hacktivists trading beer for data at Afghanistan's only tiki bar.

In February 2009, the month after Obama took office, Tony Tether was ordered to resign to make way for a new DARPA director. Even though Tether had already been at DARPA longer than any other director, he recalled being devastated by the order to leave. Obama's pick to head DARPA, announced in July 2009, was a public watershed for the agency: Regina Dugan became the agency's first female director.

As a program manager at DARPA in the 1990s, Dugan had earned a reputation for boldness—and some alleged recklessness—visiting minefields and combat zones to test bomb detection technology. When she started making the rounds in the Washington Beltway as DARPA director, her choice of attire—short skirts, stiletto heels, and leather jackets—generated as much buzz as her credentials. Dugan's financial ties to a family-owned firm receiving DARPA contracts for bomb detectors proved fodder for critics, though she insisted Pentagon lawyers signed off on her recusal from dealing with the company. A self-styled technology enthusiast, Dugan liked talking about theories of innovation, such as Pasteur's quadrant, an approach that emphasizes finding an idealized type of research that combines scientific exploration with practical applications. Her lecture style was often better suited to the world of TED, the popular technology conference, than to old-school military briefing rooms. In a marked departure, Dugan canceled DARPATech, the Disney fest that had come to symbolize the Tether era. "There is a time and a place for daydreaming. But it is not at DARPA," she told Congress. "DARPA is not the place of dreamlike musings or fantasies, not a place for self-indulging in wishes and hopes."

One of Dugan's first moves was to hire Peter Lee, a prominent university computer scientist, to run a key DARPA office. Lee, the head of Carnegie Mellon's computer science department, was reluctant at first, even though he and Dugan were already friends. The Information Processing Techniques Office, which paved the way for the ARPANET, had long since moved away from supporting basic computer science research, focusing instead on more traditional weapons technology like ARGUS-IS, a 1.8-billion-pixel camera that could be flown on a drone and used to keep watch on entire cities. As a member of the Computing Community Consortium, Lee had co-authored a paper titled "Re-envisioning DARPA" that forwarded ideas for how to bring the agency back to its golden age roots of the ARPANET and computer networking. Dugan "started to lean on me" to take the job as head of DARPA's Information Processing Techniques Office, Lee recalled. He agreed, but a month before he was supposed to take over, Lee had dinner with Dugan to talk about DARPA. "You know, Peter. I don't think you should take over IPTO," she said, just as he was dropping her off after dinner. "You should just start a new office." Dugan did not say what the new office would do, other than it should be a "pure expression of what DARPA could be." Lee had no idea what that meant.

Then, the week before Lee was supposed to start at DARPA, Dugan called him on the phone and asked him to come to Washington, D.C., immediately. She wanted him to meet with Secretary of Defense Robert Gates, who was scheduled to visit DARPA's headquarters. As he drove from Pittsburgh toward Washington, Lee grew nervous. He was an ivory tower academic who had written a paper telling DARPA what he thought it should do, but now that he was actually about to become a senior DARPA official, he realized he had no plan of his own. His apprehension grew worse when he was pulled over for speeding. He had visions of multiple tickets and eventually having his license suspended. "Oh my God, I'll be driving back and forth to Pittsburgh and D.C., and I won't have my license," he thought.

The speeding incident, however, provided Lee with inspiration for the new office. He had recently learned about Trapster, a smartphone app that allows users to map and share information on speed traps using GPS. Trapster enabled a virtual army of tipsters to create a real-time map warning drivers of areas where the police may be lying in wait. For Lee, who was interested in social networking technologies, and in avoiding speed traps, Trapster finally provided him with an idea that he thought might interest the Pentagon chief. Instead of plotting speed traps, he imagined a Trapster-like application that could track potential bomb attacks in Afghanistan. Crowdsourced data was allowing millions of people to monitor events in real time. Already, communities of people were collaborating online to track nuclear proliferation, spotting potential test sites in North Korea. Humanitarians were using crowdsourcing to monitor elections and respond to natural disasters. If crowdsourcing could plot speed traps and spot election fraud, perhaps it could be used in war zones. When Lee presented the idea, the defense secretary seemed to like it. So did Dugan, who encouraged Lee to pursue it. That was the beginning of Lee's Transformation Convergence Technology Office, a name so incredibly awkward it seemed tailor made for hiding secret programs.

Dugan greeted Lee's idea with enthusiasm for a specific reason. Crowdsourcing was part of the booming field of "big data," an area that DARPA had avoided for the better part of a decade following the public brouhaha over John Poindexter's Total Information Awareness. Poindexter's work had become the agency's Voldemort, the "program that shall not be named," as Dugan put it. "It was an area that the agency wasn't doing a lot of work in. It didn't make any sense," she said. "This field was explosive."

Big data was indeed exploding, not just in the private sector, where it was being used to predict consumers' movie rentals and book purchases, but also with the military, which was parsing reams of data being sucked up by sensors monitoring Iraq and Afghanistan. One classified program, described opaquely by Bob Woodward in his book *The War Within,* was used to comb through data to "locate, target, and kill key individuals" in Iraq. Woodward touted the technology in interviews as "very top secret" and "one of the true breakthroughs" of the war in Iraq. He later revealed in a subsequent book more details, including the name, the Real Time Regional Gateway, an NSA computer program better known by its acronym, RTRG. It was designed to pull together many feeds of information—everything from intercepted phone calls to information on bomb attacks—and analyze that data to identify insurgent networks and predict attacks.

Once Woodward went public with RTRG, senior intelligence officials began to drop more details about this top secret program. According to the retired air force colonel Pedro "Pete" Rustan, who helped develop the technology, RTRG started as an intelligence program in Iraq that tracked insurgents by intercepting phone calls and triangulating the location of insurgents. The system worked by collecting and analyzing streams of data in real time and then pinpointing insurgents by location. "If you're smart enough to combine all that data in real time, you can determine where Dick is out there," said Rustan, then at the National Reconnaissance Office. "He's in block 23 down there, and he just said he's going to place a bomb."

Back at DARPA, Lee was not anywhere near hunting down insurgents; he was still working to get his ideas off the ground. When he arrived at the agency, he faced, as he put it, "a steady stream of not necessarily trustworthy defense contractors just one after another coming through my little office with ideas." It was, as he described it, a "turbulent time."

Lee had a fortunate alternative to the slick defense executives lining up outside his office. Dugan had assigned him a group of military officers working at DARPA on a short-term basis, a sort of professional internship called the Service Chiefs Fellows Program. Normally, the officers toured military laboratories and did not do much substantive work, but Dugan wanted them to actually create a project with Lee. Soon, Lee and the fellows brainstormed a contest based on the DARPA Grand Challenge, only instead of racing robotic cars, contestants would use social media in something resembling a national treasure hunt. The fellows proposed having teams compete to locate red weather balloons that DARPA would release across the United States. Lee was not sure about the idea: having people hunt for balloons sounded a little odd, even for DARPA, but Dugan encouraged him. "That idea might be stupid, but that's what you came up with yesterday, so you're going to execute," he recalled her telling him.

Like the Grand Challenge, the Network Challenge, as it was called, was a contest, but on a smaller scale. DARPA offered a $40,000 prize to the first team that could, on a specific day, identify the locations of the ten red weather balloons placed across the United States. The idea was that teams would use social media to help locate the balloons. The contest would test the teams' ability to leverage a network, figuring out how to motivate people to participate while weeding out possible fake sightings, and to do it quicker than other competitors. On December 5, 2009, the day of the challenge, Lee's biggest fear was that no team would identify all the balloons, undermining the point of the challenge. In the end, it took only nine hours for a team from MIT to win. They beat the competitors by using a sliding scale of financial incentives that rewarded not just those who spotted balloons but those who recruited others who successfully spotted balloons. Alex "Sandy" Pentland, an MIT computer science professor who headed the winning team, called the task "trivial."

Pentland had reason to be self-confident. He had already established a reputation as one of the nation's leading "big data" scientists. Long before Google Glass, Pentland had written about wearable sensors that would record everything the user sees, hears, and experiences. His specialty was sifting through data to predict patterns of human behavior, an area he called "social physics." Pentland's team had created a novel financial incentive system based on the assumption that people's actions are dictated not purely by profit but by intangible benefits that come from exchanges that strengthen someone's position in his or her social network. "If you look at the models for incentives, or for management through the army, through companies, through economics, they're all about individual incentives, and they ignore the social fabric. What I just said about red balloons was that it wasn't about economics; it was about the social fabric," Pentland said.

Pentland theorized that someone's position in the network—his or her social standing—was the primary motivator. In his calculation, people act to make their social fabric stronger, not necessarily just to earn a bit of money. "I give you a favor. Maybe in the future you'll give me a favor. That's what drove this thing," he said. "That's a very different way of thinking about things. Instead of paying attention to individuals, you pay attention to relationships."

Taking what was learned from the Network Challenge and moving it into a formal DARPA program was the next step, and Lee again had a bit of good fortune. Randy Garrett, an NSA official involved in big data, had recently moved over to DARPA. At the NSA, Garrett had been a key official for RTRG, the program that had helped track and kill insurgents in Iraq. Garrett had also been working on creating a data cloud that would allow analysts to search through real-time data as it was vacuumed up by intelligence agencies. This cloud would include "essentially every kind of data there is," Garrett said. There were some obvious parallels between the work of the NSA and the Network Challenge. Garrett's NSA work had focused on integrating large streams of data in real time to spot something of interest, such as insurgents. The Network Challenge did roughly the same thing using social media data and red balloons. The national security establishment has an enormous amount of real-time data at its fingertips, and the biggest source of that, of course, is the NSA, which intercepts millions of calls a day around the world, in addition to various forms of Internet traffic, from e-mails to Skype calls. Afghanistan, after ten years of war, was one of the NSA's top targets for cell phone interception. Now some of that data was about to be made available to DARPA.

"Someone made the observation and brought it to my attention that it might be possible for DARPA to get direct near-real-time access to several hundred data intelligence feeds from the theater in Afghanistan," Lee recalled. "I thought that was very interesting. Most of the data feeds were classified only at the secret level. Some were even unclassified. One immediate question was, what might be possible if we did large-scale data mining on all of those feeds?" Lee began to contact all the experts he knew in data mining, including Werner Vogels, the chief technology officer at Amazon, who "provided a lot of framing for how we would approach this problem because it's very similar to the kinds of data mining that Amazon does on its customers."

What Lee eventually formulated was a data-mining program based on the latest predictive analysis work being done in the commercial sector, but using military data from Afghanistan. "For example, we were trying to understand if the price of potatoes at local markets was correlated with subsequent Taliban activity, insurgent activity, in the same way that Amazon might want to know if certain kinds of click behaviors on Amazon.com would correlate to higher sales of clothing versus handbags versus computers," Lee said.

Big data was about to be enlisted in a program to predict whether a village in Afghanistan was being taken over by the Taliban or when insurgents might plan the next attack. More important, big data was going to take DARPA back to war.

In February 2010, just two months after Lee's red balloon contest, Dugan did something that no DARPA director had done since the Vietnam War: she traveled to a war zone to see what the agency might be able to contribute. General Michael Oates, the head of the Joint Improvised Explosive Device Defeat Organization, the Pentagon's bomb-fighting agency, invited Dugan on a three-day tour around Afghanistan. Military personnel expressed surprise to see her. "You're from DARPA," she recalled their general reaction. "We call you when we have three- to five-year problems."

When Dugan got back to Washington, D.C., she assembled the office directors and their deputies and gave them a month to come up with ideas for technologies DARPA could contribute immediately to the war in Afghanistan. Dugan already had her own ideas for projects in Afghanistan as well. At RedXDefense, the family-run company Dugan founded, she had developed a theory of bomb detection with a copyrighted slogan called the "Bookends." The "books" were the weapons that insurgents used, while the "ends" were the terrorist organizations that built and placed the bombs. Her theory was that defeating IEDs required identifying bomb makers and bomb-making facilities, rather than trying just to detect the bomb. (It was not a particularly unique theory: the Pentagon's bomb-fighting agency's own slogan was "Defeat the Network.") After her return from Afghanistan, Dugan developed a slide briefing summarizing her ideas:

> BIG BREAKTHROUGH . . . Bookends suggests that fighting in the books is wrong . . . [The] [o]nly thing that works there is humans/dogs . . . What if the key to boosting performance in the books is simply that we get more eyes on target?
>
> More noses?

Dugan described several proposed DARPA programs for Afghanistan in that same briefing. One, called More Noses, was a plan to send several hundred dogs outfitted with sensors and GPS trackers. Normally, when bomb dogs smell explosives, they are trained to sit down, alerting their handler to a possible threat. With Dugan's proposed program, hundreds of dogs would fan out over a specific area in Afghanistan and operate off leash, sniffing out possible bombs. When a dog sat down, that is, sensed a possible bomb, the sensors the dogs wore would send a signal back to the person monitoring the data remotely. More Noses equipped dogs with sensors; More Eyes, another new program, equipped people with sensors. The people, or Afghans to be specific, would be given smart phones, which they could use to send back information about possible threats. More Eyes, according to Dugan, would use the "newest social networking" techniques to create "a civilian populace reporting capability." More Eyes, together with More Noses, would create an "offense" system to track down IEDs.

By April, DARPA had identified about a dozen projects that could have an immediate effect on the war, and then Dugan narrowed those down to a final list. The technologies ranged from a blast gauge that would go in soldiers' helmets to detect exposure to possible blast waves from IEDs to an imaging sensor, called the High Altitude LIDAR Operations Experiment, that could be used to create three-dimensional maps of Afghanistan. Dugan's priority, however, was a new program based on Lee's big data work, called Nexus 7, which would help predict insurgency in Afghanistan. In August, Dugan met with the chairman of the Joint Chiefs of Staff and laid out DARPA's plan for Afghanistan. The data-mining project, her briefing noted, would "sequester [a] team of the Nation's leading researchers in large scale computation techniques and social science." She called Nexus 7 "the potentially big win."

Key members of the Nexus 7 team came from Sandy Pentland's Human Dynamics Lab at MIT, drawing on the same ideas that drove his team's win with the balloon contest and applying them to an entire society. Pentland described his contribution as informal, providing more of an intellectual framework than nuts-and-bolts work. "When Nexus 7 started up, some of my students went and joined it," he said. "My role was to cause people to realize that there was something different that could be done, something qualitatively different than anything they'd ever done before."

Forty years earlier, another MIT scientist, Ithiel Pool, had promised DARPA he could use science to help the Pentagon understand the dynamics of insurgency. Dugan might not have realized it, but DARPA had just created a new version of Simulmatics' 1960s-era "people machine." Pentland called his version "computational counterinsurgency."

In the summer of 2010, Nexus 7, a data-mining program named after a humanoid robot in the movie *Blade Runner,* was launched with a former NSA official as its head. Similar to his work at NSA, Garrett's goal with the DARPA program was to "actually build this big data aggregated environment, a cloud, and then see how you would use it." The program was a direct carryover of work started at the NSA, according to one scientist involved in creating Nexus 7.

In budget documents, Nexus 7 was obliquely described as a program that combined data analysis and forecasting with social network analysis. "For the military, social networks provide a promising model for understanding terrorist cells, insurgent groups, and other stateless actors whose connectedness is established not on the basis of shared geography but rather through the correlation of their participation in coordinated activities," the description stated. "Nexus 7 supports emerging military missions using both traditional and non-traditional data sources for those areas of the world and mission sets with limited conventional Intelligence, Surveillance and Reconnaissance."

DARPA brought in the researchers and contractors working on Nexus 7—about two dozen computer scientists, social scientists, economists, and counterinsurgency experts—and took over the tenth floor of the agency's headquarters for a brainstorming session. The meeting included big data gurus, like Sandy Pentland, to provide technical advice, and L. Neale Cosby, a retired army officer, who was there to help provide an operational perspective. The question, said Cosby, was, "how can we take all that data that comes in and streams, minute by minute, second by second, into [the NSA at] Fort Meade and other places and use that data to make sense in assessing the actual security of a village in a place like Afghanistan?"

The direct relationship between the NSA and DARPA was one of the hallmarks of Nexus 7, but it was also the most problematic, because working with data from the NSA required navigating a maze of legal and statutory requirements that often prevent sharing and aggregating data among government agencies. As for why DARPA wanted the NSA data, Cosby invoked the famed bank robber Willie Sutton: "Why rob a bank? Because that's where the money is." The NSA was the bank; it had all the data.

Dugan believed she could avoid the scandal of Total Information Awareness, which had dragged DARPA into a national privacy debate in 2003, by restricting the work to a war zone. Total Information Awareness wanted to find the terrorists that might be lurking in the United States, which was bound to concern privacy advocates, while Nexus 7 was focused on Afghanistan. More important, she also made the entire program secret. DARPA workers one office down had little idea what was going on when the group of young computer scientists set up shop in the agency's headquarters. "We had to have cover stories to tell people if various Beltway people came to visit me in my office and they were walking through this pandemonium," Lee said.

The Nexus 7 program was different in many respects from a normal DARPA project. Typically, DARPA contracts work to universities or businesses, but the core of Nexus 7 was Peter Lee, who created the program and ran it from his office. "Nexus 7 turned out to be a bunch of desks, laptops, and secure computers literally in the hallway outside of my office," Lee said. "It was just a zoo."

Not all the work was done at DARPA, however. The agency recruited David Kilcullen, an Australian counterinsurgency expert who had advised American government officials, including General Petraeus. Kilcullen by 2010 had moved to the private sector and headed a company called Caerus Associates, selling his services to government clients. Nexus 7 meshed well with Kilcullen's belief that metrics, ranging from the cost of transportation to the price of exotic vegetables, could be used to gauge a population's susceptibility to insurgency.

DARPA was creating an intelligence program far more ambitious than anything John Poindexter had attempted with Total Information Awareness ten years prior. In those days, the work was focused on predicting large-scale terrorist events—plots that might require complex, long-term planning. Nexus 7 was in the weeds, looking at patterns of daily life, to make specific predictions on the ground in Afghanistan. "We were really using the latest research in quasi-experimental design, in machine learning, and data mining literally on hundreds of intelligence feeds to make inferences about what would happen next," Lee said.

According to Dugan, Nexus 7 started making its "first discoveries"—or meaningful predictions—eighty-two days into operation. Over the weekend that those results came in, Dugan briefed the marine general James Cartwright, the vice chairman of the Joint Chiefs of Staff, to get an official green light for Nexus 7 and its personnel to deploy to Afghanistan. Cartwright, a technology enthusiast, embraced her ideas and her approach. His response, said Dugan, was "Go and go faster."

Before Nexus 7 made it to Afghanistan, its creator, Peter Lee, abruptly left DARPA after less than a year to become the head of research at Microsoft. On the day he left for Seattle in September 2010 to start his new job, the Nexus 7 team, some members as young as their mid-twenties, was departing for Afghanistan. "I should have been with them," he said regretfully.

DARPA would eventually deploy more than a hundred people across Afghanistan, working on Nexus 7 and other technology programs. "It was the first operational deployment from DARPA since the Vietnam War," Dugan later recounted. The program also became Dugan's top priority as she shuttled back and forth to Afghanistan with General Cartwright. In congressional testimony in 2011, Dugan did not use the Nexus 7 name, but simply described "a 90-day Skunk Works activity" that involved scientists and counterinsurgency experts working on "crowd sourcing and social-networking technologies."

Nexus 7 went from inception to execution in just a few months, and it was not without hiccups. General Stanley McChrystal, the head of the International Security Assistance Force in Afghanistan when Nexus 7 started, was interested in the data-driven work promoted by DARPA. But in 2010 he was forced to resign after a *Rolling Stone* magazine profile, which depicted him and his staff as mocking senior White House leaders. General Petraeus returned to Afghanistan to take over, but he was not enthusiastic about Nexus 7. A disastrous meeting between Petraeus and Dugan in Afghanistan almost brought it to a halt. DARPA's proposal for algorithms did not sit well with a general who believed he wrote the book, metaphorically and literally, on counterinsurgency.

At that point, however, Nexus 7 had support from General Cartwright, and soon the DARPA team, or what Dugan called the "DARPA army of technogeeks," started showing up in Afghanistan. They were young and had no military experience, and the culture shock soon became apparent. Military officials in Kabul were reluctant to share intelligence with computer scientists just out of graduate school, and the intelligence they did provide was not nice and neat, like consumer data. Once in Afghanistan, the analysts began to gather up as much intelligence as they could: phone records from the NSA, radar feeds from the military, and intelligence reports. But much of the data that came into Nexus 7 was qualitative, rather than quantitative, which was not easy to plug into a computer program. Even when the data was quantitative, like from radar, it rarely covered the exact same place over time.

By late 2010, DARPA was touting Nexus 7's successes within the Pentagon, but it was not clear what it had accomplished, if anything. As members of the team worked on a base crunching numbers from military and intelligence data feeds, another team of contractors, the Synergy Strike Force, was working in the provinces of Afghanistan, swapping beer for data and using crowdsourcing techniques honed in the red balloon hunt.

The Synergy Strike Force was always more a concept than a formal organization, an improbable mix of humanitarians, hacktivists, and technophiles who had set up shop in Jalalabad, Afghanistan, at the Taj Guesthouse, which had previously been occupied by Australian mercenaries. The eclectic group included techno-enthusiasts who wanted to bring the Silicon Valley ethos to Afghanistan. There were a few "burners," attendees of the annual Burning Man festival, but there were also scientists, security contractors, and a dedicated group from the MIT Fab Lab, short for "fabrication laboratory," which was building technology, like solar power and Wi-Fi networks, using do-it-yourself engineering.

For a while the Taj was something of an informal meeting place for Westerners in Afghanistan, or "the tiki bar at the edge of the universe," as Smári McCarthy, who was part of the Synergy Strike Force, explained in a video interview. McCarthy, a self-described information freedom activist, called the Taj "a little oasis on the edge of Jalalabad where you've got this strange mixture of military people, private security contractors, NGO people, and crazy people who are out there to try and build infrastructure in their time off. All sorts of people who under normal circumstances would never meet."

When the Synergy Strike Force took over the Taj, the tiki bar began attracting a mix of misfits, artists, and do-gooders, if for no other reason than that it was probably the only bar in all of Nangarhar province. The motley group, described as "super-powered geeks," set about building do-it-yourself Wi-Fi networks and other small-scale tech projects for Afghans. It was sometimes hard to see what united them, other than the belief that open-source technologies could, if not save the world, then at least substantially improve it.

In 2010, around the time that DARPA was thinking about data mining in Afghanistan, Todd Huffman, one of the leaders of the Synergy Strike Force, was introduced to DARPA officials at a chance meeting in Washington. Huffman had recently returned from Haiti, where he had been working with Ushahidi, the open-source mapping organization that helped locate victims of the 2010 earthquake. Huffman, a bearded devotee of Burning Man, whose hair on any given day might be died in shades of red and yellow, started talking about crowdsourcing in Haiti and similar work he had done in Afghanistan during the elections. It sounded similar to what DARPA wanted to do with the More Eyes project. Soon, Ryan Paterson, an official in charge of the agency's newly formed field unit in Afghanistan, showed up at the Taj to spend a month with the burners and anarchists; he even tended bar.

The Synergy Strike Force was perhaps best known for its Beer for Data program, but it had also done crowdsourcing work in Afghanistan to spot election fraud. Unlike the young computer scientists who sifted through Nexus 7 data from the confines of a military base, the Synergy Strike Force would be on the ground—outside the wire, as the saying went—working with Afghans to collect data. "We were referred to as those weird DARPA people," said a regional coordinator for the More Eyes program. "Weird for DARPA is a real accomplishment."

Soon, DARPA was sponsoring mini-versions of the Network Challenge in Afghanistan. Under More Eyes, members of the Synergy Strike Force in 2011 fanned out over Afghanistan, handing out cell phones to participants in contests to map out areas in the provinces of Nangarhar and Bamiyan. Afghan participants, often drawn from the humanitarian and development community, were provided with GPS-enabled phones and instructed to mark the location of buildings and streets. Like with the red balloon contest, the experiments often had an economic incentive: winning teams got to keep their cell phones. Participants were not told that More Eyes was intended to provide the military with intelligence, and DARPA never publicly announced the program.

While some of the experiments involved collecting information on politics or health care, the focus was on gathering data useful to military operations. "Generally speaking, U.S. forces have been very successful at intercepting cellular calls and incorporating them into our intelligence framework," an undated report by one Beltway defense contractor noted. "However, these operations are just the tip of the iceberg of what can be done through cooperative techniques such as crowdsourcing."

The crowdsourcing projects promoted by groups like Ushahidi in Haiti were dedicated to humanitarian operations, sometimes even in cooperation with the military. But More Eyes laid bare the overlap between crowdsourcing and intelligence collection. According to an unpublished white paper written by DARPA's Paterson, crowdsourcing would, for example, allow an Afghan citizen to report an attack on a convoy. That report might then cue a drone and eventually a military strike. Paterson also noted that More Eyes worked directly with the Defense Intelligence Agency on a project called "Afghanistan Atmospherics," which involved using "selected local persons to passively observe and report on things they see and hear in the course of everyday activities."

Paterson described More Eyes as a way "to catalyze the local population to generate 'white' data useful for assessing stability at multiple levels (regional, provincial, district and village)." The advantage of this white data, as opposed to the black world of intelligence, is that it is "generated spontaneously by the local population . . . untainted by influence of outsiders." In other words, More Eyes was recruiting unwitting spies.

DARPA was clearly concerned that recruiting local Afghans to provide intelligence could be viewed as citizen spying. More Eyes documents warned against using foreign phones "that stand out due to their appearance or advanced functionality and can be an indicator of collusion with foreigners and can invite threats from local insurgents." The DARPA white paper suggested that the phones be equipped with an application that can delete data, either by the user or remotely, presumably protecting the information from discovery by insurgents. The members of the Synergy Strike Force at the Taj, some of whom blogged regularly about their experiences, never publicly mentioned the Defense Department, perhaps because many of the team's hacktivists and technophiles found it difficult to reconcile their self-image as development workers with being military contractors.

The Synergy Strike Force was a bizarre cultural convergence. As hackers around the world were chafing at the American government's attempts to crack down on self-proclaimed information freedom organizations, like WikiLeaks, the Synergy Strike Force's information activists were working on a project to help the defense and intelligence communities collect data in Afghanistan. The group touted its work with headlines like "Afghanistan's DIY Internet Brings the Web to War-Torn Towns," and it used DARPA money to pay for solar panels for local universities, but More Eyes was really about intelligence collection. Though the Beer for Data effort was never formally part of the DARPA program, the Synergy Strike Force happily offered the one-terabyte hard drive to the Pentagon. Even the DIY Internet was an opportunity to mine data, providing a treasure trove of Internet traffic in Afghanistan that the NSA could only dream of collecting, according to a scientist leading the project. The program bought laptops, which could be accessed remotely, for provincial Afghan government officials, including the governor of Jalalabad. "Was the More Eyes program successful?" the scientist asked rhetorically. "Well, let's see, I just put a foreign electronic sensor into the governor's bedroom."

In the end, however, the program fell short of its hopes to demonstrate crowdsourcing in Afghanistan. According to the white paper by DARPA's Ryan Paterson, a series of experiments showed that More

Eyes overestimated the ability of Afghans to access the Internet and the reach of mobile phone services in Afghanistan. "The More Eyes Team quickly learned that only 4 percent of the population had access and skills necessary to access and exploit the Internet," he wrote. "Rural populations had even less." The DARPA contract, which ran out toward the end of 2011, was not renewed.

The Synergy Strike Force and its tiki bar "oasis" also soon came to an end. Violence in Jalalabad grew steadily worse over the course of 2010 and 2011. Afghans who had worked and socialized with the foreigners at the Taj received death threats, and the insurgency that Western visitors to the bar were trying to forestall enveloped the establishment. On August 11, 2012, two men on motorcycles intercepted a car driven by Mehrab Saraj, the manager of the Taj and a friend to many who had worked on More Eyes. Saraj, who had survived the Soviet invasion, Taliban rule, and then the American invasion, was shot in the chest and killed.

Almost fifty years after William Godel's ill-fated trip to Vietnam to collect data for the strategic hamlet program, hacktivists and humanitarians tried and failed to map out Afghanistan. Godel gave cash and gifts for data; the Synergy Strike Force offered free beer and mobile phones. Ultimately, however, what is so striking about DARPA's efforts in Afghanistan was how marginal they were compared with the Vietnam War. DARPA's deployment to Afghanistan echoed elements of Project AGILE but never approached its scope. What was forgotten, moreover, was that Godel's version of counterinsurgency was designed to help local military forces in order to avoid sending American troops. In Afghanistan, counterinsurgency—and DARPA's contribution—were viewed as tactics for aiding American troops already entangled in a foreign war.

For several years, Petraeus's reintroduction of counterinsurgency was hailed as a success, at least in Iraq. The accolades were ultimately short-lived. Similar to Vietnam, the local government's failure to govern effectively in Iraq ended up fueling the insurgency. In Afghanistan, where the central government was even weaker, counterinsurgency by 2013 was being widely derided as a failed strategy. In the end, Petraeus's approach suffered from the same fatal flaw as counterinsurgency in the latter days of Vietnam: the local government, not foreign forces, must ultimately provide security. Iraq and Afghanistan were counterinsurgency turned on its head.

DARPA never publicly discussed More Eyes, and although the Pentagon later touted Nexus 7 as a success, there is no evidence that it had any useful impact on operations. "There are no models and there are no algorithms," one anonymous official told *Wired,* griping about the program's deployment to Afghanistan. A more sanguine assessment was published in *The Wall Street Journal,* which quoted an unnamed former official claiming that Nexus 7's predictions about attacks in Afghanistan were accurate between 60 and 70 percent of the time. "It's the ultimate correlation tool," the official told the newspaper. "It is literally being able to predict the future." Neither statement added substance to the debate. One thing was clear, however: like the McNamara Line, Nexus 7 did not change the course of the war.

In March 2012, Dugan left to take a job with Google's Motorola unit, where she began working to create a DARPA-like entity within the company. She had been director for less than three years—a not atypical tenure, although brief compared with that of her predecessor, Tony Tether. A Pentagon inspector general investigation into her continuing financial ties to RedXDefense, the family-owned explosive detection company receiving DARPA contracts, put her departure under a cloud. The inspector general's report, released two years later, determined that Dugan had violated ethics by promoting the company's proprietary work, though it found no evidence she had tried to direct funding to the company.

By 2013, DARPA's work in Afghanistan had drawn to a close, though the Taj, the onetime staging base for More Eyes, lived on as a symbol of the agency's well-intentioned but ultimately failed efforts to harness science in the service of counterinsurgency. The Taj that year was nominally still open, serving stale cornflakes to the rare guest, who could lounge on a rusting lawn chair perched on cracked concrete overlooking the long-empty pool. At the edge of the pool, solar panels, which powered nothing at all, tilted up wistfully toward the sun. All that was left of the only tiki bar in Afghanistan was a collection of moldering business cards and a "Synergy Strike Force" combat patch stapled to a piece of wood, the last remnants of the Western patrons who had swapped their data for beer. Outside this onetime oasis of DARPA-funded techno-optimism, Afghans lived and fought much as they had for more than a thousand years. Inside, the bar stood empty, an enduring testament to science's ability to transform warfare but not to end it.

Epilogue

GLORIOUS FAILURE, INGLORIOUS SUCCESS

> The intelligence aspects of military operations have grown and grown and grown until it seems the entire operation must collapse of its own mass and sheer awkwardness. In this arena science and technology have seemingly run amok and it is not at all clear that the volume of information has not, by itself inundated the Cray computers, confused the analysts, sacrificed credibility with the consumers, and virtually destroyed the capacity of the system to understand itself.
>
> —WILLIAM H. GODEL

When I arrived at DARPA's new headquarters building in February 2013 to meet with Arati Prabhakar, the agency's director, I was struck by its location in the backwaters of Pentagon real estate. The building on North Randolph Street in Arlington, Virginia, is DARPA's fourth headquarters; each move has taken the agency farther away from the Pentagon's leadership physically and psychologically.

The agency's first location was inside the Pentagon's elite E Ring, steps away from the Office of the Secretary of Defense. The second was in the Architects Building in Rosslyn, Virginia, not far from the Pentagon. The third was in Virginia Square, several stops farther out on the Washington Metro. DARPA in 2012 moved into a custom-designed building near the Ballston station, a project several years in the making. The new building features enhanced security and a customized conference center on the ground level, and each DARPA office now has its own floor. The top floor houses the executive suites for DARPA's director and staff, who work behind frosted-glass doors. As with all grand projects, construction was plagued by rumors of snafus: bathroom doors that would not close, and windows that at first were not properly tinted to safeguard classified materials. Later, once they moved in, some employees complained that the building's layout led people to work behind closed doors, interacting less with each other, let alone outside researchers.

For an agency that claims paternity to some of the past few decades' most exotic technologies, its offices have in the past been unassuming, even ramshackle. For several decades, visiting DARPA—at least the "white world," or unclassified parts—was as simple as walking into the building and heading up to the office. That came to a halt in the 1980s, when a man walked in off the street and into one of the offices, dropped his pants, and mooned an unsuspecting DARPA official.

Those days are long gone, not just for DARPA, but for all government offices. In the era following 9/11, visiting such offices involves metal detectors, X-rays, and, in many cases, confiscation of all electronic devices. My visit to DARPA's headquarters began with a ritual familiar to those who do business inside the Washington Beltway with its layers of security, badges, and restrictions: I emptied my bag of electronics.

"We'll need to disable your camera," the guard said, examining my iPad. I watched, worried that the security professionals from the agency that created the foundations of the Internet and boasts a number of classified cyber programs would perform some procedure on my iPad that would be impossible to reverse. Instead, the guard took a piece of masking tape, placed it over the camera lens, and handed the iPad back to me.

"There's another camera in back," I volunteered.

"We'll disable that, too," he said, placing another piece of tape over the back camera.

In DARPA's top-floor director's suite, Arati Prabhakar invited me to sit down at a conference table. She had been director less than a year but was well liked and respected, particularly after what some felt was a tumultuous few years under Regina Dugan. If there was a criticism of Prabhakar, it was only that no one quite knew what she planned to do with the agency. So, I asked her what she thought was DARPA's current mission. "I think it is unchanged," she replied. "It has been and will be to prevent and create technological surprise."

She acknowledged, however, that DARPA's mission has become more difficult in recent years. In the mid-1980s, during her first stint at DARPA, the military was focused on the Soviet Union, and the presumption was that the Cold War would continue indefinitely. "Facing that kind of existential threat, that one monolithic threat sort of drowned out all of the rest of the complexity of national security," she said. "Of course DARPA, along with everyone else, was almost purely focused on thinking about that one threat. It is not that the world was not more complex, we just treated it as somewhat less complex." Her perspective is understandable. When Prabhakar worked at DARPA in the 1980s, the agency was already following a path that was narrowing by the year.

The irony of DARPA is that even as its mandate has shrunk, its reputation has ballooned. The agency that created the foundation of the Internet and stealth aircraft is hailed today as the "gem of the Pentagon," touted as a model for government innovation, and praised by Democrats and Republicans alike. Prabhakar also knew firsthand the dangers of trying to redirect DARPA. In the 1980s, she had unwittingly presided over DARPA's failed attempt to act like a venture capital firm, taking an equity stake in a gallium arsenide start-up. When asked about the episode, which had led to the only firing of a director in DARPA's history, Prabhakar quickly changed the subject.

The danger facing the agency today is irrelevance to national security. In 2003, when the American military in Iraq found that its greatest threat was not from tanks and missiles but from roadside bombs, the Pentagon did not, as it once did, turn to DARPA, an agency stacked with top-notch technical personnel and decades of experience in bomb detection. Instead, it created an entirely new organization that was largely bereft of the type of science and technology expertise long resident in DARPA. What followed is hardly surprising: billions of dollars were spent, and yet casualties from bombs continued to increase.

Today, the agency's past investments populate the battlefield: The Predator, the descendant of Amber, has enabled the United States to conduct push-button warfare from afar, killing enemies from the comfort of air-conditioned trailers in the United States. Stealth aircraft, another DARPA innovation, are used to slip across borders to conduct precision strikes and covert operations. Networked computers have shortened the "kill chain" to just seconds, and precision weapons allow the United States to conduct strikes anywhere, even in heavily populated urban areas.

The question, however, is whether those novelties have successfully achieved what DARPA at one time intended: to create technologies that ensure the United States would not have to go to war and, if it did go to war, that it would achieve a swift victory. Nothing illustrates that disparity better perhaps than the 1960s-era investment in counterinsurgency research, intended to prevent large-scale conventional engagements in so-called limited wars. By 2006, counterinsurgency theory was resurrected as a tool to help conventional forces wage war against insurgents, the exact thing the original DARPA program was meant to prevent. The allure of applying the wizardry of science and technology to warfare seems only to have made the temptation to engage in armed conflict more inviting and to have entangled the United States in a "forever war."

If one glances only at the headlines, any concerns about DARPA appear at face value unwarranted, and the agency would seem to be at its peak. Its projects, like the driverless car, are splashed across magazine covers and tech websites, which run breathless stories about the agency's plans for brain implants to cure mental illness. Lawmakers over the past decade have rarely done more in congressional hearings than laud the agency. Republicans and Democrats praise DARPA, calling it a model for government-sponsored innovation. Politicians, economists, and techies regularly exalt the "DARPA model," even though it is unclear what the model is.

Former directors do worry about DARPA and its bureaucratic growth. When DARPA was born in 1958, it had no building. Its senior officials were allotted a few offices in the Pentagon, and members of its technical staff were given windowless offices in the interior rings of the building. For the first few years, the staff directory fit on a standard index card. Today, the directory approaches the size of a small phone book. DARPA may tout a technical staff of only 140 scientists, but they are aided by an army of contract personnel, many who serve as the near equivalent of permanent employees, the very thing it is supposed to avoid. Even its new headquarters seems to run counter to the agency's once ad hoc, minimalist existence.

Victor Reis, a former DARPA director, cited one of Parkinson's laws, not the one on work expanding to fit available time, but the one on the correlation between the decline of organizations and the construction of a "perfect" headquarters. The Pentagon was not completed until World War II was almost over. St. Peter's Basilica in Vatican City took more than a hundred years to build and was finished well after the papacy's influence was waning. "By the time you get the real building done, you're finished as an organization," Reis said.

More than the building, Reis expressed concerns about DARPA's ceaseless touting of its accomplishments. "I'm just a little uneasy that they're starting to say how wonderful they are, you know what I mean?" Reis told me, when I met with him in his modest government office in downtown Washington, D.C.

Yet DARPA's reputation is now so entrenched that the government in recent years has made a series of flawed attempts to "replicate" the DARPA model in other government agencies. The Department of Homeland Security bungled its own version to the point that it exists in name only. The Department of Energy's ARPA, or ARPA-E, has a budget merely a fraction of what DARPA receives. The intelligence ARPA, called IARPA, has been constrained by bureaucracy. None of them have approached the scope or ambition of their namesake.

The attempts to replicate DARPA belie the temptation to draw some fantastical lessons about management science. Should organizations get rid of all their employees every three to five years, as DARPA does? Should science agencies do away with peer review, as DARPA often does, in order to pursue revolutionary ideas? In a culture that reduces analysis to bullet points, celebrates the intellectual reductionism of TED talks, and prays at the altar of PowerPoint, it is important to remember that not everything can be reduced to an organizational chart. DARPA should not be treated as a black box management tool that can be dropped on top of any organization to make it more innovative.

It is tempting to reduce DARPA to a caricature. The truth is that DARPA's legacy cannot be easily packaged as "innovation in a box." Its successes—and failures—have always been a function of its unique bureaucratic form, which arose from its historical role as a problem-solving agency for national security. Rearranging boxes on an org chart, or cubicles in an office, will not produce another ARPANET. With the exception of having technical staff managing research, and a director, the agency has never had a fixed organizational structure.

In fact, DARPA's style often runs counter to fuzzy management theories of collaboration. So-called kumbaya moments at DARPA are few and far between. With some notable exceptions, the program managers often know little of what their colleagues in other offices are doing. In one case, a program manager described to me being shocked to meet a stranger on a Pentagon shuttle bus, only to learn that both men had worked for several years in DARPA. In a cast of thousands, this would not be surprising, but in a relatively small agency it is more unusual. DARPA, as one former director called it, is "140 program managers all bound together by a common travel agent."

For every management lesson that DARPA might hold, there is a counter lesson: the agency pioneered a novel system for detecting nuclear tests that simultaneously modernized the field of seismology. It was a huge program, with high-level White House attention. Yet the ARPANET was started by a psychologist hired to run research programs that the agency's leadership did not particularly care about. He isolated himself, pursuing his own grand plan for an "intergalactic computer network" that led to the ARPANET, the foundations of the modern Internet. That both programs could exist in a single agency should give pause to anyone looking for easy answers about management and innovation.

Just a few miles away from DARPA, Stephen Lukasik's house in Northern Virginia looks out over Lake Barcroft, the same body of water where, some fifty years earlier, William Godel's young daughters had tried to navigate in the "Jesus shoes" intended for Southeast Asia's waterways. Tony Tether is one of Lukasik's Lake Barcroft neighbors. Though cordial, the two former directors are not quite friends.

On one of their encounters, Lukasik asked Tether, "What are you trying to do in DARPA?"

"We want to fix it so that if we can find them, we can kill them," Tether replied.

Tether was likely being facetious; even his science fiction version of DARPA was about much more than creating killing machines. Yet his answer struck at the heart of Lukasik's concern: there did not seem to be any thought given to the overarching problems that DARPA was supposed to solve; it was just generating technology.

Lukasik has spent the past four decades contemplating DARPA's legacy in national security. The walls of his basement are neatly lined with books that span topics ranging from Stalinism to cyber warfare. Notably absent are any books on management theory, a topic that Lukasik openly mocks, even though it is often what people want to discuss with a former DARPA director. Now in his eighties, Lukasik is sometimes perplexed that his grandchildren's friends are in awe when they find out he was once the head of DARPA, the agency whose inventions regularly appear in whiz-bang television shows featuring futuristic weapons. Lukasik is deeply proud of DARPA's legacy but openly disappointed that the agency he helped mold is regarded as a science-fiction-inspired gadget lab. His vision for DARPA is still as an agency that solves important national security problems. "That is my growing discomfort with DARPA," Lukasik told me. "It is not because DARPA is irrelevant as an institution, but is DARPA doing what it ought to be doing for the security of this country?"

While some officials argue that DARPA is as good now as it has ever been—and that could be true in terms of the quality of science and technology—there is no denying that the agency has largely been absent from the past ten years of national security debates, which have centered on terrorism and insurgency. In late 2014, Lukasik advised the Defense Department to have DARPA start a new long-range strategic planning study to look at future technology for national defense, as the agency had done during his tenure. The Pentagon agreed, even co-opting the old name the Long Range R&D Planning Program (it was shortened to "Long Range Research Development Plan"). Rather tellingly, this new study was launched without DARPA.

The current DARPA is so narrowly focused on technical problems it is hard to see how its mandate could have allowed it to come up with anything more creative in Afghanistan than computer algorithms. Nexus 7 was significant as an attempt to address current problems of warfare and insurgency by exploiting cutting-edge science and technology. And it was the first time since the Vietnam War that DARPA deployed personnel to a war zone. Yet its narrow scope also highlights how much had changed. In the Vietnam War, DARPA had sought to understand the fundamentals of society and the causes of insurgency; by 2011, DARPA in Afghanistan was seeking simply to predict the next IED attack. The wide-ranging exploration of human behavior that led to the hiring of J. C. R. Licklider seems unlikely today in an agency whose notion of social science is limited to computer programs that spit out predictions like a Zoltar Fortune Teller machine. "This may be more like an entropy process," Lukasik said. "Once you move in that direction, you move in the direction of more detail, and if that's the case, you run the risk of becoming irrelevant, because your measure of survival is political adroitness rather than technical excellence and solving important problems."

Evidence of this entropy process was highlighted in June 2013 when *The Guardian* and *The Washington Post* published reports about the NSA based on leaked documents provided by Edward Snowden. The documents revealed the depth and scope of the agency's post-9/11 mass surveillance. Lukasik blamed DARPA's loss of Total Information Awareness, John Poindexter's data-mining program, as a contributing factor to the fiasco. Research that could have and should have been conducted by DARPA in the open was instead "transformed into intrusive government policies," Lukasik wrote in his personal memoir.

One can only guess what William Godel, who launched DARPA's original counterinsurgency program, would have made of the current agency, whose press releases tout devices that can help soldiers scale glass skyscrapers, while American forces fight in a country dominated by mud houses. For Godel, technology was part of a larger strategy, not a narrow operational tactic. Project AGILE failed, but it was, as Charles Herzfeld proclaimed it, a "glorious failure." By comparison, Nexus 7 was a failure regardless, not even because the technology was faulty, but because the national security problem it was trying to address—insurgency—could not possibly have been solved by any algorithm, no matter how elegantly designed. Even if it worked, it would be, at best, an inglorious success.

More than fifteen years after the 9/11 attacks, and over two decades since the end of the Cold War, the dilemma for DARPA is finding a new mission worthy of its past accomplishments and cognizant of its darker failures. In 2014, the agency announced the creation of the Biological Technologies Office, with a heavy focus on neuroscience, building on work DARPA has sponsored since the 1960s. Its new research, part of a larger White House initiative in brain science, has received wide attention. Helping soldiers recover from the devastating effects of traumatic brain injury, one of the more noble goals of the new office, is worthy of DARPA's attention and one of the most exciting areas of research that it is currently pursuing.

When I interviewed Justin Sanchez, the acting deputy director for the office, in 2016, he was refreshingly aware of DARPA's earlier work in this field. The science and technology, he noted, had evolved in the forty years since the days of biocybernetics, as had the agency's focus. In past years, DARPA had openly said that its goal was to develop ways to have the brain directly control weapons. DARPA officials are careful now to couch the agency's work in terms of medical applications. The focus is "on restoring the injured war fighter," Sanchez told me. "That has been the recent motivation for trying to understand the kinds of brain function research that we're doing."

One of the agency's programs, called Restoring Active Memory, is developing neuroprosthetics, essentially neural implants, which can help repair injured brains. After just two years, DARPA had developed prototype medical devices, and the work has moved to studies involving human subjects. "We already have some initial tests to see how interfacing with a neuroprosthetic device will affect the ability to form and recall memories," Sanchez said. Similarly, another program, called SUBNETS, short for Systems-Based Neurotechnology for Emerging Therapies, is building implantable medical devices for people suffering from a variety of neuropsychiatric conditions, ranging from post-traumatic stress disorder to depression. There, again, Sanchez cited progress. "We're also already in human subjects and have some preliminary evidence that is showing that we can understand some neurosignatures related to anxiety and we can modulate the brain with respect to anxiety," he said.

Modulating the human brain with neural implants holds the potential to treat any number of illnesses and injuries, but it also rightfully raises questions with ethicists, who wonder about the potential pitfalls of having the Pentagon involved in an area that touches upon the core of what makes us human. When I asked Sanchez, for example, whether DARPA would ever consider funding classified work in neurotechnology, he provided a careful answer. "I think we can say none of this is classified at this point in time," he replied. "We always just keep our eyes open and never want to be caught off guard on that front. We're actively just looking at the space and seeing where the opportunities are. I think we'll have to make those determinations as we learn more about how neurotechnology plays out."

The potential for secret neuroscience work should give pause to anyone familiar with the Pentagon's history of human testing. Moreover, though Sanchez and other DARPA officials downplay the potential for the agency's neuroscience research to lead to weapons, it is impossible to ignore the reality that if the current work were successful, it would have applications in those areas. The world is still adapting to the drone revolution; is it really ready for brain-controlled aircraft? Such technology may be decades away, but there are other issues to be considered with this work. If DARPA succeeds, it could indeed revolutionize neuroscience. Yet if it fails, or is involved in scandal, such as a human subjects study gone wrong, the potential backlash against the agency could have repercussions as serious as those that resulted from Total Information Awareness.

The ultimate question with this research, as with so much of DARPA's work, is whether the agency will be allowed—or should be allowed—to pursue something with such truly high risk. Like many ambitious areas that DARPA has pursued in the past, from counterinsurgency to computer networking, its neuroscience work could transform the world by revolutionizing medicine, and it could lead to weapons that change the way we fight in future wars. Whether that world will be a better place is unclear.

Acknowledgments

If I could trace the idea of writing a DARPA history back to any particular moment in time, it would probably be to a conversation in Washington, D.C., in 2004 with my friend Robert Wall. We debated what a careful examination of DARPA might reveal about its legacy. Is it a genius factory? A Pentagon boondoggle? A refuge for crackpots? More than a decade later, I do not have an unequivocal answer, but many of the questions I raise in this book were guided by that initial conversation with Robert.

Credit for moving that idea from a coffee shop to a formal book proposal goes to my extraordinary agent, Michelle Tessler, who gently nudged me until I committed thoughts to paper. The person most responsible for taking that proposal and turning it into a cohesive book is Andrew Miller, my editor at Alfred A. Knopf, who guided me through several rounds of revisions to help craft a narrative for an agency that has lived its life episodically. I also thank Emma Dries, editorial assistant at Knopf, for providing valuable comments on the draft.

A number of friends and colleagues, in addition to Robert, read and critiqued versions of the manuscript, shared contacts, and pointed me in helpful directions. My friend Ann Finkbeiner, co-president for life of the Garwin Fan Club, provided critical comments on the rough draft. Steven Lee Myers offered support during some particularly tough times and provided thoughtful edits on the completed manuscript. I am also indebted to Richard Whittle, who has helped me in countless ways, and to Noah Shachtman, who will forever be my "work spouse."

The writing of the book spanned several nomadic years living in

Washington, D.C.; Kraków, Poland; New York City; and Cambridge, Massachusetts. In Washington, I benefited from conversations with Jonathan Moreno of the University of Pennsylvania. I also learned a great deal from informal exchanges with Mark Lewis, Richard Van Atta, and David Sparrow, all of the Institute for Defense Analyses, and Steven Aftergood of the Federation of American Scientists. Shane Harris, one of the nicest and smartest national security reporters in town, generously shared knowledge and contacts. After I moved from Washington, my friends Askold Krushelnycky and Irena Chalupa hosted me during several return visits while I was conducting archival research. I also thank John Schidlovsky of the International Reporting Project for his support, and I am particularly grateful to the fearless Washington attorney Jeffrey D. Light, who took my Freedom of Information Act case to court and won.

In Kraków, Marek Vetulani was an important sounding board for ideas, and Wojciech Kolarski and Agata Kolarska made me feel at home in the city. In New York, Nina Burleigh of *Newsweek* made sure I did not become a book-writing hermit. I am also grateful to the entire staff of *The Intercept,* particularly the editor in chief, Betsy Reed, for allowing me to hide out for nine months in Cambridge, Massachusetts, so that I could finish the manuscript. In Cambridge, I thank Subrata Ghoshroy of MIT and Kevin Kit Parker of Harvard University for their ideas and interest in my topic. Finally, I thank Loretta Oliver, who transcribed the majority of the interviews conducted for this book; she has been a virtual companion on a long journey.

My family has been a source of support, in ways big and small. I am grateful to my brother Marc Weinberger, and his wife, Kacey, for their encouragement in California while I conducted interviews there, and to their children, Eli and Talia, who like to read books; I hope they will want to read mine someday. And I thank my father, to whom this book is dedicated, for teaching me to love ideas.

I am also deeply indebted to Nathan Hodge, who provided advice and encouragement during the critical early stages of this book, as well as to his father, Brien Hodge, who shared his thoughts and recollections on being a military adviser in Vietnam. Nathan also sent me pictures and notes chronicling the sad fate of the Taj Guest House and its tiki bar, a poignant symbol of America's failure in Afghanistan.

This book was the first opportunity I had to work extensively with the incredible staff at the National Archives and Records Administration, including at the presidential libraries. Many staff there assisted me, but I am particularly grateful to David Fort, at the National Archives in College Park, for his help in processing my FOIA requests.

As I mention in the note on sources, Kathleen "Kay" Godel-Gengenbach was generous enough to share significant portions of her father's unpublished memoir as they pertained to DARPA. She also answered countless questions about her father's career and pointed me to archival sources I would never have found on my own. I admire Kay's devotion to the historical record and the protectiveness she has toward her father's legacy. I have tried to balance respecting the Godel family's desire for privacy with the historical obligation to shed light on his neglected contributions to DARPA's history.

I also thank Dick Davis of the Kyoto Symposium Organization and Jay Scovie of the Inamori Foundation, for the generous fellowship that allowed me to attend the Kyoto Prize Symposium in Japan, where I interviewed Ivan Sutherland. I also thank Robert Dujarric, director of the Institute of Contemporary Asian Studies, Temple University, in Tokyo, and Yuri Ota, my translator in Japan, whose idea it was to interview a Nagasaki survivor. And I am grateful to Richard Weiss, DARPA's director of strategic communications, who knows it is more important to have the story right than favorable.

Many former employees of DARPA agreed to speak to me, even knowing that this book was intended as a critical history of the agency and its legacy, and I am grateful for their time and candor. I regret that a number of those whom I interviewed, including Lee Huff, co-author of the 1975 history of DARPA, passed away before I could finish this book. I will always be grateful to the late Seymour Deitchman, who spent his final days making sure I had as much information as he could provide. Finally, I am incredibly thankful to Stephen Lukasik, who more than any other former DARPA official gave me countless hours of his time, with no expectation that the story I wrote would reflect his views or thoughts. It is no coincidence that he commissioned the 1975 history and contributed significantly to this current book.

I also benefited from the support of two great institutions: the Woodrow Wilson International Center for Scholars and the Radcliffe Institute for Advanced Study at Harvard University. The bulk of the research and interviews for this book was conducted in 2012–2013, while I was a fellow at the Wilson Center. I am grateful to the entire staff, and particularly Kent Hughes and Robert Litwak for believing in my project. Cole Thomas and Ryan Ricks, my interns there, contributed invaluably to the research, and my colleague from that year, Laura Gomez-Mera, has been a continuing source of moral support and friendship.

The Radcliffe Institute hosted me during the final stages of this book, in 2015–2016. The staff, including its director, Judith Vichniac, and the dean, Lizabeth Cohen, has created an unparalleled atmosphere for scholars, writers, and artists. At Radcliffe, I was fortunate to have unbelievably talented research partners, particularly Paul Banks, whose careful fact-checking saved me from many embarrassing errors (and he is in no way responsible for any that remain). I also thank my other partners, Caleb Lewis, Pat O'Hara, and Jordan Feri.

Special thanks go to my friends and colleagues at Radcliffe, and in particular the second-floor "sherry hour" crew, including Ayesha Chaudhry, Elliott Colla, Ann-Christine Duhaime, Wendy Gan, Sarah Howe, William Hurst, Raúl Jiménez, Philip Klein, Valérie Massadian, Scott Milner, Michael Pollan, and Licia Verde. They taught me about issues ranging from Arabic poetry to polymer physics. They encouraged me to pursue my obsessions. Most important, they made me laugh. For those nine months, I owe them perhaps the greatest debt.

Sharon Weinberger
Cambridge, Massachusetts
March 2016

Notes

PROLOGUE: GUNS AND MONEY

3 *In June 1961:* Godel/Wylie trial transcript, *United States of America v. William Hermann Godel, John Archibald Wylie, and James Robert Loftis,* R. Criminal No. 4171, National Archives, Atlanta.

3 *called Godel an "operator":* Bundy, interview with William Moss, John F. Kennedy Library.

4 *"He was one of the more glamorous":* Huff, interview with author.

6 *more than five hundred in Thailand alone:* This number is based on a directory of ARPA personnel in Thailand, dated 1968, provided to the author. The number is also supported by documents held in the National Archives, though the directory provides the most specific count.

6 *whose seminal work:* David Galula, *Pacification in Algeria, 1956–1958* (Santa Monica, Calif.: Rand, 1963). Rand released a new addition in 2006 to respond to a surge of interest in Galula's work on counterinsurgency. Both the original and the reprint acknowledge that the research for the book was supported by the Advanced Research Projects Agency.

7 *personally signed off:* J. Ruina, ARPA Order 471, April 15, 1963; J. C. R. Licklider, ARPA Program Plan 93, April 5, 1963. It is worth noting that these documents, posted online by the National Archives, College Park, are part of a record series not yet declassified.

10 *The final document was regarded:* Huff, interview with author.

10 *the new director was aghast:* George Heilmeier, interview with author.

CHAPTER 1: *SCIENTIA POTENTIA EST*

15 *Michiaki Ikeda was a chubby-faced:* Ikeda, interview with author.

15 *more than twenty kilotons:* The most reliable estimate appears to be twenty-one kilotons. John Malik, *The Yields of the Hiroshima and Nagasaki Nuclear Explosions* (Los Alamos, N.M.: Los Alamos National Laboratory, 1985).

15 *the majority of the people:* U.S. Strategic Bombing Survey, *The Effects of Atomic Bombs on Hiroshima and Nagasaki* (Washington, D.C.: Government Printing Office, 1946).

16 *seventy thousand people:* The estimates vary widely and a precise number would be impossible to know. This figure comes from the Department of Energy's *Manhattan Project: An Interactive History,* www.osti.gov.

16 *"You have known for several years":* Alvarez (Headquarters Atomic Bomb Command) to Ryokichi Sagane, Aug. 9, 1945.

17 *"We have spent two billion dollars":* Statement by President Harry Truman, White House, Aug. 6, 1945.

18 *"We have made war obsolete":* York, *Making Weapons, Talking Peace,* 25.

18 *The V-2 , a liquid-propelled rocket:* Chertok, *Rockets and People,* 2:242.

20 *"This is absolutely intolerable":* Quoted in G. A. Tokaty, "Soviet Rocket Technology," *Technology and Culture* 4, no. 4, *The History of Rocket Technology* (Autumn 1963): 523.

20 *The Soviets' hunt for technical expertise:* "While inspecting German factories and laboratories, don't get carried away with intellectual achievements, but first and foremost compile a list of the types and number of machine tools, industrial engineering equipment, and instruments," was one directive. Chertok, *Rockets and People,* 2:218.

20 *"The Americans looked for brains":* Paul H. Satterfield and David S. Akens, *Historical Monograph: Army Ordnance Satellite Program* (Fort Belvoir, Va.: Defense Technical Information Center, 1958).

20 *"Do you realize":* Tokaty, "Soviet Rocket Technology," 523.

21 *William Hermann Godel was born:* William Godel, compiled military service record, National Archives, St. Louis.

21 *At one point, the younger Godel:* Kathleen Godel-Gengenbach, e-mail correspondence with author.

21 *He was wounded twice:* Godel, compiled military service record, National Archives, St. Louis.

21 *The wound in his left leg:* Ibid.

22 *He made enough of a name:* Godel-Gengenbach, correspondence with author.

23 *In the United States, computer scientists:* See Redmond, *From Whirlwind to MITRE.*

23 *In 1947, President Truman:* McDougall, *Heavens and the Earth,* 97.

23 *when von Braun proposed research:* Ward, *Dr. Space,* 70.

23 *"professional gloom":* Ibid., 73.

24 *"It's a boy":* York, *Making Weapons, Talking Peace,* 69.

24 *"a blinding white fireball":* Rhodes, *Dark Sun,* 508.

24 *ignite the atmosphere:* The possibility of igniting the atmosphere was raised in an early technical report. E. J. Konopinski, C. Marvin, and E. Teller, "Ignition of the Atmosphere with Nuclear Bombs" (Los Alamos National Laboratory, LA-602, 1946). However, Teller later dismissed the possibility. Arthur Compton discussed the idea of the explosion creating a chain reaction in the ocean in his memoir, *Atomic Quest: A Personal Narrative* (New York: Oxford University Press, 1956).

24 *"World War II should have taught us":* Dwight D. Eisenhower, "I Shall Go to Korea Speech," Oct. 25, 1952.

24 *In the past two decades:* McDougall, *Heavens and the Earth,* 113.

25 *Rand, a newly established think tank:* That report was preceded by *Preliminary Design of an Experimental World-Circling Spaceship* (Santa Monica, Calif.: Rand, 1946).

25 *Back in Washington:* Godel-Gengenbach, correspondence with author; Godel, interview with Huff.

25 *Frustrated by the lack of coordination:* W. H. Lawrence, "Board to Conduct Psychology War: Gordon Gray Will Head Group to Direct Open and Covert Strategy of 'Cold War,'" *New York Times,* June 21, 1951.
25 *appointed Godel*: Frank Pace Jr. (secretary of the army) to Raymond B. Allen (director, Psychological Strategy Board), April 15, 1952, Harry S. Truman Library.
25 *Official correspondence:* Executive Assistant to the Director (name excised by the CIA), CIA Memorandum for the Record, Conversation with Mr. Godel, Department of Defense, Re: Mr. John Drew, June 25, 1956.
26 *But the infighting was bad enough*: Sarah-Jane Corke, *US Covert Operations and Cold War Strategy* (New York: Routledge, 2007), 202.
26 *In 1955, Donald Quarles:* Godel, unpublished interview with Huff.
26 *In a later unpublished interview:* Ibid.
26 *"Glad to know":* Wilson to Hoover, July 11, 1957, Department of Defense, Office of the Secretary of Defense, *Security Division Personal Interview, Mr. Raymond A. Loughton with William Godel,* Nov. 27, 1953. Freedom of Information Act Request 13-F-0963.
27 *Godel's role by then:* Donald Hess, interview with author; Godel, interview with Huff.
27 *"This is not a design contest"*: Quoted in Ward, *Dr. Space,* 96.
27 *the CIA and the NSA were monitoring Soviet launches:* McDougall, *Heavens and the Earth,* 117; Rand Araskog, interview with author.

CHAPTER 2: MAD MEN

28 *age of nuclear Armageddon:* Oliver M. Gale Papers, box 1, Washington Journal, vol. 1, July 1957 to Dec. 1958, Dwight D. Eisenhower Presidential Library.
29 *"Soap manufacturer Neil McElroy"*: United Press, "Ike to Name Ohioan as Chief of Defense," *Milwaukee Journal,* Aug. 7, 1957, 1.
29 *"vital activities in persuading housewives":* Associated Press, "Ike Names Neil McElroy, Head of Soap Firm, Defense Secretary," *Lewiston Morning Tribune,* Aug. 8, 1957, 1.
29 *At Strategic Air Command:* Gale, Washington Journal, Sept. 17, 1957, Eisenhower Library.
29 *"extremely able":* Ibid.
29 *"where horror is as much":* Ibid.
30 *"become a sort of czar"*: Medaris, *Countdown for Decision,* 153.
30 *who sported a black mustache:* Sheehan, *Fiery Peace in a Cold War,* 324.
30 *"salesman, promoter":* Gale, Washington Journal, Oct. 4, 1957, Eisenhower Library.
30 *"Von Braun was still wistful":* Ibid.
30 *Even in Huntsville, the Germans:* For the political struggles over the rocket launch, see McDougall, *Heavens and the Earth.*
30 *"scientific boondoggles":* Huff and Sharp, *Advanced Research Projects Agency,* II-1.
30 *When Wilson had visited Huntsville:* Medaris, *Countdown for Decision,* 155.
31 *"There was an instant":* Ibid.
31 *"Vanguard will never make it":* Ibid.
32 *The conversation did impart:* It is interesting to compare Medaris's and Gale's accounts of the cocktail party and dinner. Though factually similar, Medaris believed that he and von Braun had succeeded that evening in persuading McElroy

to let them proceed with a space launch, while Gale's account is clear that McElroy came away with no particular decision, or even urgency of decision.

32 *"Two generations after the event":* Roger D. Launius, "Sputnik and the Origins of the Space Age" (NASA, 1997), history.nasa.gov.

33 *In fact, the headline had nothing:* Cyrus F. Rice, "Today We Make History," *Milwaukee Sentinel,* Oct. 5, 1957, 1.

33 *"hysterical":* National Security Council, "Discussion at the 339th Meeting of the National Security Council, Thursday, October 10, 1957," Oct. 11, 1957, NSC Series, box 9, Eisenhower Papers, 1953–1961, Ann Whitman File, Eisenhower Library.

33 *"just a hunk of iron":* McDougall, *Heavens and the Earth,* 145.

34 *"In the West":* Lyndon Johnson, *The Vantage Point* (New York: Holt, Rinehart and Winston, 1971), 272.

34 *"Soon, they will be dropping bombs":* Dickson, *Sputnik,* 117.

34 *Though the Soviets were somewhat ahead:* McDougall, *Heavens and the Earth,* 184. McDougall writes that the Soviets were actually behind in all areas except for large boosters and space medicine.

34 *"Now, so far as the satellite":* Dwight D. Eisenhower, President's News Conference, Oct. 9, 1957.

34 *"Russians captured":* Ibid.

35 *"The same missile that launched":* Drew Pearson, "Space Talk Taboo for the Air Force," *Tuscaloosa News,* Oct. 20, 1957, 4.

35 *The launch sparked panic:* "No event since Pearl Harbor set off such repercussions in public life," wrote McDougall. *Heavens and the Earth,* 142.

36 *None of the suggestions:* Huff and Sharp, *Advanced Research Projects Agency,* II-2.

36 *Ernest Lawrence, the famed:* Ibid.

36 *"upstream research":* Procter & Gamble, "A Company History 1837–Today" (2006).

36 *"vast weapon systems":* U.S. House, *Department of Defense Ballistic Missile Programs,* 7.

37 *"no urgency on Mars":* Wang, *In Sputnik's Shadow,* 94.

37 *They eventually acquiesced:* James Killian is sometimes erroneously credited with coming up with the idea of ARPA, but all the available evidence indicated it was McElroy's idea, though he did discuss it with Killian. Huff and Sharp, *Advanced Research Projects Agency,* II-4.

37 *"when and if a civilian space agency":* Dwight D. Eisenhower, Memorandum for the Secretary of Defense, March 24, 1958, Eisenhower Library.

37 *"very great mistake":* Huff and Sharp, *Advanced Research Projects Agency,* II-13.

37 *When Sputnik launched:* "A New Realm of Flight," *Flight,* Oct. 18, 1957, 511.

38 *The Japanese newsmen:* "Setback for U.S. Prestige—the Satellite Effort That Failed," *New York Times,* Dec. 8, 1957, 1.

38 *The few dozen or so official viewers:* The description of the launch comes from the Associated Press, "Rocket Strains, Takeoff Fails," *Salt Lake Tribune,* Dec. 7, 1957, 1–2.

39 *"The Agency is authorized":* Department of Defense Directive, "Subject: Advanced Research Projects Agency," Feb. 7, 1958, No. 5105.15.

39 *"We must be forward looking":* Dwight D. Eisenhower, State of the Union address, Jan. 9, 1958.

CHAPTER 3: MAD SCIENTISTS

40 *"Let me dismiss my cab":* Gale, Washington Journal, Jan. 4, 1958, Eisenhower Library.

41 *"They're coming!":* York, interview with Williams/Gerard.

41 *"basically frantic":* York, interview with Finkbeiner.

41 *"the crazy Greek":* John W. Finney, "Atomic Inventor Was Held a Crank," *New York Times,* Feb. 14, 1958, 1, 8.

41 *"an Astrodome-like defensive shield":* York, *Making Weapons, Talking Peace,* 130.

42 *"His purpose was of epic proportions":* Ibid., 131.

42 *"nutty":* Robert Le Levier, interview with Finn Aaserud, American Institute of Physics.

43 *McElroy that month met with:* Gale, Washington Journal, Jan. 15, 1958, Eisenhower Library.

43 *After briefly considering:* Huff and Sharp, *Advanced Research Projects Agency,* II-25.

43 *It was an expedient solution:* Senior administration officials "never had in mind that it would survive," Godel later recounted. "It was a stop gap to get pressure off the [Pentagon] and the White House." Godel, interview with Huff.

43 *"urbane and handsome":* Ben Price, "ARPA Chief Is a Relatively Unknown Man," *Daytona Beach Morning Journal,* Aug. 3, 1958, 7A.

43 *"looked every inch":* Huff and Sharp, *Advanced Research Projects Agency,* II-21.

43 *"He'd show up with these gorgeous tans":* Huff, interview with author.

44 *"Eight years":* Gale, Washington Journal, Feb. 13, 1958, Eisenhower Library.

44 *"I went over there as chief scientist":* York, interview with Ann Finkbeiner.

44 *"personal consultant":* Roy Johnson, address to the ARPA-IDA Study Group, National War College, Washington, D.C., July 14, 1958, RG 330, National Archives, College Park.

44 *Johnson was the chief spokesman:* Araskog, interview with author.

45 *York, along with holding the title:* York, interview with Gerard/Williams.

45 *Robert Truax, a navy captain:* Robert L. Perry, *Management of the National Reconnaissance Program, 1960–1965* (Chantilly, Va., NRO History Office, 1961). Truax was really working as technical adviser to Richard Bissell, the CIA's deputy director for plans. Truax, described by Godel in his memoir as a "rocket nut," also went on to a colorful post-military career that included an ill-fated attempt to help Evel Knievel jump over Idaho's Snake River.

45 *"Roy Johnson set the pace":* Hess, interview with author.

46 *a signal emitted from a satellite:* William Guier and George Weiffenbach, "Genesis of Satellite Navigation," *Johns Hopkins APL Technical Digest* 19, no. 1 (1998): 14–71.

46 *"As gun powder":* Johnson, address to ARPA-IDA Study Group.

46 *In April 1958, shortly after arriving:* York, interview with Gerard/Williams.

46 *"ARPA is the only place":* York, interview with Finkbeiner.

47 *"the possibility that events":* Loper to Brigadier General A. D. Starbird, Philip Farley, and Spurgeon Keeny, memo, "Subject: ARGUS Experiment," April 21, 1958, Eisenhower Library.

47 *York made Argus his pet project:* York, *Making Weapons, Talking Peace,* 149.

47 *"was interesting science":* D'Antonio, *A Ball, a Dog, and a Monkey,* 207.

47 *Task Force 88:* This and other specific details of Argus come from a report prepared by the Defense Nuclear Agency, "Operation Argus 1958: United States Atmospheric Nuclear Weapons Tests Nuclear Test Personnel Review" (DNA 6039F, April 30, 1982).

48 *"historic experiment":* Killian to the president, memo, "Subject: Preliminary Results of the Argus Experiment," Nov. 3, 1958, Eisenhower Library.

48 *"greatest experiment":* Walter Sullivan, "Called 'Greatest Experiment,'" *New York Times,* March 19, 1959, 1.

48 *It was never clear who leaked:* Kistiakowsky, *Scientist in the White House,* 72. York claimed the leaker was a physicist for the navy. York, interview with Finkbeiner.

49 *"There could, however, be":* York, *Making Weapons, Talking Peace,* 131.

49 *"screwball":* U.S. Senate, *Investigation of Governmental Organization for Space Activities,* 125.

49 *"egg-shaped and the height":* Dyson, *Project Orion,* 2.

49 *"so the inhabitants are not killed":* U.S. Senate, *Investigation of Governmental Organization for Space Activities,* 135.

50 *Freeman Dyson recounted:* Dyson, *Project Orion,* 230.

50 *"appalled":* Ibid., 221.

50 *In ARPA's version of history:* Huff and Sharp, *Advanced Research Projects Agency,* III-39.

50 *"We were firm believers":* Von Braun, interview with Roger Bilstein and John Beltz. Glen E. Swanson, ed., *"Before This Decade Is Out—": Personal Reflections on the Apollo Program* (Washington, D.C.: National Aeronautics and Space Administration, NASA History Office, Office of Policy and Plans, 1999). The rocket scientist gives less credit to ARPA in an article he wrote for NASA, claiming that booster was his team's idea and the Pentagon was simply in "just the right mood" to fund it. Wernher von Braun, "Saturn the Giant," in *Apollo Expeditions to the Moon* (Washington, D.C.: NASA's Scientific and Technical Information Office, 1975).

51 *"a sorry string of failures":* Huff and Sharp, *Advanced Research Projects Agency,* III-9.

51 *"If the DOD decides":* Ibid., III-27.

51 *"Beset by enemies internally":* Ibid., II-20.

CHAPTER 4: SOCIETY FOR THE CORRECTION OF SOVIET EXCESSES

53 *According to heavily redacted:* From "Monthly Activity Digest," marked "top secret," and addressed to the director of the National Security Agency, Nov. 7, 1957. Released to the author under FOIA Case No. 75066A. Godel, according to the document, was the chairman of the Robertson Committee of Alternates.

53 *"radical funding reductions": 60 Years of Defending Our Nation* (Fort Meade, Md.: National Security Agency, 2012).

53 *Sputnik caught the NSA: Cryptologic Almanac 50th Anniversary Series* (Fort Meade, Md.: National Security Agency, 2000).

53 *"I don't know how he got there":* Hess, interview with author.

54 *He knew his name had been dropped:* Godel, interview with Huff.

54 *"Godel knew quite a bit":* Huff, interview with author.

54 *Godel said only:* Godel "became involved in ARPA, through the intelligence channel," recalled Rand Araskog, who worked for Godel in the Pentagon's Office of Special Operations and then followed his boss to ARPA. A fluent Russian speaker, Araskog was only twenty-six years old but had already spent two years at the NSA listening to intercepts from the Soviet Union's missile launches from Kapustin Yar. Shortly after ARPA was established, Araskog was asked to brief Roy Johnson on Soviet technology. Impressed by Araskog, Johnson asked him to transfer to ARPA, promising a good promotion. Araskog's main responsibility was to keep the agency apprised of the latest in Russian missile technology, although he also soon found himself appointed the chief speechwriter for Johnson. Araskog, interview with author.

54 *"Since the Soviets":* Buhl, *An Eye at the Keyhole.* Godel does not specify who assigned him.

54 *"it was love at first sight":* Godel, interview with Huff.

54 *Godel admired Johnson's vision:* Ibid.
55 *"one of the early production":* Buhl, *An Eye at the Keyhole.*
56 *he chalked his name:* "The Big Bird Orbits Words," *Life,* Jan. 5, 1959.
56 *"The stated objective":* Buhl, *An Eye at the Keyhole.*
56 *"It's a big empty shell":* Huff and Sharp, *Advanced Research Projects Agency,* III-24.
56 *"a publicity stunt":* Buhl, *An Eye at the Keyhole.*
56 *"felt that this young man":* York, interview with Martin Collins, American Institute of Physics. Though York and press articles attribute SCORE to Johnson, more reliable sources, including the Barber Associates history and Godel's own memoir, make it clear it was Godel's idea. Shortly after the launch, *Life* magazine featured a fourteen-page spread dedicated to Project SCORE, featuring exclusive photographs and a carefully staged insider's look into the launch. The magazine said that 150 reporters and photographers contributed to the report.
56 *"It must remain secret":* Buhl, *An Eye at the Keyhole.*
57 *To maintain the ruse:* Ibid. Godel says Sullivan asked to join ARPA because Hoover, the FBI director, wanted him to go to Cuba as a mole, an ill-advised mission that Sullivan thought best to avoid.
57 *"Only 88 people":* After the launch, the air force claimed only 35 people knew, a practical impossibility. In fact, according to Godel's account, the number grew to more than 200 by the day of launch.
57 *"They had recently proposed":* Buhl, *An Eye at the Keyhole.*
57 *but built by RCA:* Godel does not mention RCA, but Barbree's recollection is backed up by the firsthand account of an engineer on the program. M. Walter Maxwell, *Reflections: Transmission Lines and Antennas* (Newington, Conn.: American Radio Relay League, 1990). Maxwell writes that ARPA "contracted out the entire communications package that flew on the ATLAS to the RCA Laboratories in Princeton, N.J. for the design and fabrication, while SDRL engineers watched over our shoulders."
57 *"You know perfectly well":* Buhl, *An Eye at the Keyhole.*
57 *As a result, Wilber Brucker:* Yet another version of this story maintains that a member of the SCORE team at Fort Monmouth was the original voice on tape. Brigadier General Harold McD. Brown, "A Signals Corps Space Odyssey," *Army Communicator* (Winter 1982).
58 *"You are the project manager":* Godel, interview with Huff.
58 *"So, hoping that no radios":* Buhl, *An Eye at the Keyhole.*
59 *"good old RCA":* Barbree, *Live from Cape Canaveral,* 20.
59 *The official story:* It appears that at least one news outlet got advance word that the president's voice would be broadcast. United Press International, "Eisenhower's Voice May Be Beamed to Earth Stations from Outer Space," *Rome News-Tribune,* Dec. 17, 1958, 1.
59 *"the longest of his life":* Buhl, *An Eye at the Keyhole.*
60 *"indistinct garble":* Associated Press, "Atlas Voices Ike's Yule Wish," *Pittsburgh Post-Gazette,* Dec. 20, 1958, 1.
60 *"Biggest Moon":* "U.S. Orbits Biggest Moon," *Milwaukee Sentinel,* Dec. 19, 1958, 1–2.
60 *"Giant Size":* Associated Press, "Ours Is Giant Size!," *Daytona Beach Morning Journal,* Dec. 18, 1958, 1.
60 *"transcribing Christmas messages":* Godel, interview with Huff.
61 *"It was awkward":* York, interview with author.
61 *"We ought to consider":* U.S. Senate, *Investigation of Governmental Organization for Space Activities,* 249.

61 *"The guy who denied":* Huff and Sharp, *Advanced Research Projects Agency,* III-79.

61 *A May 27, 1959, article:* John W. Finney, "Pentagon Lacks Firm Space Plan," *New York Times,* May 27, 1959, 18.

61 *"had agreed that 'ARPA' ":* Johnson to Gale, note, Washington Journal, Eisenhower Library.

61 *York wrote to Johnson:* Bilstein, *Stages to Saturn,* 39.

62 *"The date of the transfer":* Huff and Sharp, *Advanced Research Projects Agency,* III-41.

62 *"single most important act":* York, *Making Weapons, Talking Peace,* 176.

62 *"Saturn was the biggest":* Godel, interview with Huff.

62 *There were already calls:* Ford Eastman, "Gen. Schriever Asks for ARPA Abolishment," *Aviation Week,* May 4, 1959, 28.

62 *Other critics called:* "Demise of ARPA Urged by Furnas," *Aviation Week,* June 1, 1959, 37.

63 *"You could call the environmental capsule":* Kenneth Owen, "U.S. Missile Tour," *Flight,* April 24, 1959, 579.

63 *code-named the Pied Piper:* Alfred Rockefeller Jr., "Historical Report, Weapon System 117L, 1 January–31 December 1956" (Western Development Division Headquarters, Air Research and Development Command), 1.

63 *a concept proposed by the Rand Corporation:* Perry, *A History of Satellite Reconnaissance,* XV.

63 *A satellite, on the other hand:* In fact, in 1960, the Soviets succeeded in shooting down a U-2. Its pilot, Gary Powers, was captured after parachuting to the ground in the Soviet Union.

63 *In one launch, the mice died prior to liftoff:* A second set of four mice did not fare much better. The launch was almost canceled when a humidity sensor inside the capsule went haywire. It turns out the sensor had been unfortunately placed under the mouse cage, and the mice promptly urinated on it. The launch proceeded, but the capsule with the live mice ended up in the ocean. National Reconnaissance Office, "Early 'Discoverer' History" (Oct. 20, 1966), 1.

64 *Killing mice in a program:* One version of the story is that engineers put in real mouse droppings. Edward Miller and George Christopher, "The Spitsbergen Incident," in *Intelligence Revolution 1960: Retrieving the Corona Imagery That Helped Win the Cold War,* ed. Ingard Clausen and Edward A. Miller (Chantilly, Va.: Center for the Study of National Reconnaissance, 2012), 97.

64 *The launch was a success:* Owen, "U.S. Missile Tour," 580.

64 *The plan was for the capsule:* Dwayne A. Day, "Has Anybody Seen Our Satellite?," *Space Review,* April 20, 2009.

64 *"two guys" in Longyearbyen:* Ibid.

64–65 *"a gold bucket and light colored shrouds":* Miller and Christopher, "The Spitsbergen Incident," 97.

65 *"The mission was foredoomed":* Buhl, *An Eye at the Keyhole.* Godel repeated the assertion that the Soviets never found the capsule in an unpublished interview with Lee Huff. In the years that have followed, no evidence has ever surfaced demonstrating that the Soviets found the capsule.

65 *grainy images of the Mys Shmidta air base:* See Day, Logsdon, and Latell, *Eye in the Sky.*

65 *According to an official history:* Dwayne Day argues that ARPA had no real power over Corona, but he bases that on an air force source. Day, correspondence with author.

66 *The air force was never happy:* Perry, *A History of Satellite Reconnaissance,* vol. 1, NRO, 264.

66 *"mad as hell":* Kistiakowsky, *Scientist in the White House,* 91.

66 *"professional artist":* United Press International, "Convair Research Chief Named ARPA Director," *St. Petersburg Times,* Nov. 5, 1959, 2A.

66 *After Congress raised concerns:* "Critchfield Declines Space Job Appointment," *Washington Star,* Nov. 15, 1959.

67 *"We mistreated [the scientists]":* Godel, interview with Huff.

67 *Just prior to his departure:* Johnson's final memo was revised and completed by William Godel and Lawrence Gise, a senior administrative official at the agency. Huff and Sharp, *Advanced Research Projects Agency,* III-71.

CHAPTER 5: WELCOME TO THE JUNGLE

68 *"Le Viet Minh":* Buhl, *An Eye at the Keyhole.*

68 *"a piece of blackmail":* Melby, oral history interview with Robert Accinelli, Ontario, Canada, Nov. 14, 1986, Harry S. Truman Library.

69 *"row of dominos":* President Eisenhower, News Conference, April 7, 1954.

69 *In Melby's view:* The year after the mission, an investigation was launched into John Melby, whose criticisms of the CIA had earned him the enmity of Bedell Smith, the notoriously bad-tempered director of the CIA under Eisenhower. Melby, oral history with Accinelli.

69 *Godel's memory:* The subsequent quotations also come from Godel's unpublished memoir. Buhl, *An Eye at the Keyhole.*

69 *The United States, he concluded:* Godel, interview with Huff.

69 *So close were Godel's relations:* Godel/Wylie trial transcript.

69 *Godel also worked closely:* Godel to Lansdale, Feb. 10, 1956, Hoover Institution Archive, Stanford. Godel and Lansdale were also friends from World War II, when they both worked in military intelligence. Godel-Gengenbach, e-mail correspondence with author.

70 *he persuaded Magsaysay's government:* Lansdale, *In the Midst of Wars,* 72.

70 *it cemented Lansdale's reputation:* The former CIA director Richard Colby named Lansdale among the top ten spies. Nashel, *Edward Lansdale's Cold War,* 16.

70 *printing of an almanac:* Lansdale, *In the Midst of Wars,* 226–27.

70 *helped Diem win a 1955 referendum:* Ibid., 333.

71 *"father of his country":* Ibid., 330.

71 *"Please be assured":* Godel to Diem, Feb. 10, 1956, Hoover Institution Archive.

71 *"You're going to be director":* Betts, interview with Williams/Gerard.

71 *"custodian":* York, interview with Williams/Gerard.

71 *"During my one year":* Betts, interview with Williams/Gerard.

71 *"ideas were a little wild":* Ibid.

72 *"We were trying to figure out":* Wilson, interview with author.

72 *"That's it, let's go home":* Ibid.

72 *Godel pitched York:* Godel/Wylie trial transcript.

73 *"that could assist in improving":* Godel, *Report on R&E Far East Survey, October–December, 1960,* 2.

73 *"devoted inadequate attention":* Ibid., 3.

73 *A report written by:* "The Kennedy Commitments and Programs, 1961," in Gravel, Chomsky, and Zinn, *Pentagon Papers,* 1–39.

73 *"This is the worst yet":* Schlesinger, *A Thousand Days,* 320.

73 *On January 28, 1961, Lansdale:* Langguth, *Our Vietnam,* 114–15.

74 *he even listed Fleming's book* From Russia, with Love: Hugh Sidey, "The President's Voracious Reading Habits," *Life,* March 17, 1961, 59.

74 *considerable skepticism:* Ironically, Lansdale's rise to fame as America's foremost counterinsurgent was largely linked to his purported portrayal as Alden Pyle in Graham Greene's *Quiet American,* published in 1955, which depicts an idealistic young CIA officer posing as an aid worker in Vietnam.

74 *"A Program of Action": Foreign Relations of the United States, 1961–1963,* vol. 1, *Vietnam, 1961,* doc. 52.

74 *gave presidential approval:* Captain Lawrence Savadkin to Godel, memo, "Subject: Project AGILE, a Joint Operation," Advanced Research Projects Agency, April 23, 1963, RG 330, National Archives, College Park. Savadkin cites "A Program of Action to Prevent Communist Domination in Vietnam" as the authority for starting Project AGILE.

74 *"to acquire directly":* Godel/Wylie trial transcript.

75 *"policy options":* Godel, interview with Huff.

75 *Reporting directly to Lansdale:* This statement is based on Bob Frosch, Don Hess, and Harold Brown, interviews with author. All three were in senior oversight positions, either at ARPA or in the Pentagon, and admitted they had little insight into Godel's programs.

75 *"I believe that this nation":* President Kennedy, Address to Congress on Urgent National Needs, May 25, 1961.

76 *regarded Godel with fear:* Deposition of Truong Quang Van, Godel/Wylie court records.

76 *"an airborne Volkswagen": Foreign Relations of the United States, Vietnam, 1961–1963,* doc. 96.

77 *In 1959, Diem had embarked:* Lawrence Grinter, "Population Control in South Viet Nam, the Strategic Hamlet Study," unpublished paper, May 1966, Pool Papers, MIT Archives. In Indochina, the French first dabbled in resettlement in 1951, when they moved some half a million peasants in the Kampot and Takeo area of Cambodia.

77 *The staff was still small:* William Godel, "Progress Report: Vietnam Combat Development and Test Center," Advanced Research Projects Agency, Sept. 13, 1961, Kennedy Library.

78 *"Godel has suggested":* Walt W. Rostow, "Guerrilla and Unconventional Warfare, 7/1/61–7/15/61," National Security Files, Kennedy Library.

78 *By early 1962, Diem agreed:* Gravel, Chomsky, and Zinn, *Pentagon Papers,* 128–59.

78 *The breathtaking goal:* Advanced Research Projects Agency, "Research and Development Effort in Support of the Vietnamese Rural Security Program" (prepared in Vietnam by the Rural Security Study Team under Project AGILE, Dec. 19, 1962), 11.

78 *The Defense Department's internal study:* "The Strategic Hamlet Program, 1961–1963," in Gravel, Chomsky, and Zinn, *Pentagon Papers,* 128–59.

79 *"clean up the Viet Cong threat":* The Kennedy Commitments and Programs, 1961, in Gravel, Chomsky, and Zinn, *Pentagon Papers,* 40–98.

79 *"wolf in sheep's clothing":* In court testimony, Godel says the name came from "German" Q boats, but he might have simply misspoken. It also may be that Godel, a German speaker, was more familiar with the German version.

79 *fake supply truck:* Godel/Wylie trial transcript.
80 *"in all truth General Taylor":* Buhl, *An Eye at the Keyhole.*
80 *"was at best quizzical":* Ibid.
80 *"Nothing that a good double":* Ibid.
80 *"American genius to work":* Lansdale, interview with Ted Gittinger, Sept. 15, 1981, National Archives, Lyndon B. Johnson Library.
80 *"Just stay behind":* Ibid.
80 *Kennedy approved Godel's plan:* Sorensen, *Kennedy,* 633.
81 *bought sporting pontoons:* Buhl, *An Eye at the Keyhole.*
82 *a small, jet-powered drone:* Ibid.
82 *"Means of Using the Montagnards":* William Godel to Brigadier General E. G. Lansdale, memo, Task Force Report on the Establishment of a Combat Development and Test Center, Vietnam, July 12, 1961, Project Agile, RG 330, National Archives, College Park.
82 *"fighting priest":* Nashel, *Edward Lansdale's Cold War,* 54–55. The "fighting priest" attracted international attention after Lansdale, at the behest of President Kennedy, published an anonymous report about the village in *The Saturday Evening Post* in May 1961.
82 *"All I want to know":* Appendix 8 to the Taylor Report, Research and Development, by George W. Rathjens and William H. Godel, JFKNSF-203-005: Research and Development in Vietnam and the Pacific Theater, NSF, box 203, Folder: Vietnam, Subjects: Taylor Report, 11/3/61, Rostow Working Copy, Tabs 6–8.
82 *"something the short":* Minutes of a meeting with Viet-Nam Task Force, in U.S. Department of State, *Foreign Relations of the United States, 1961–1963,* vol. 1, *Vietnam, 1961,* doc. 96.
82 *After an initial positive response: Hearings Before the Special Subcommittee on the M-16 Rifle Program,* House of Representatives, Committee on Armed Services, Special Subcommittee on the M-16 Rifle program, Washington, D.C., May 15, 1967.
82 *"Big Ears":* Buhl, *An Eye at the Keyhole.* These were not the only limitation. The audio had to be picked up on a radio by someone within range.
83 *"If the President":* Ibid.
83 *"ARPA advises the chemical":* Godel to Lansdale, memo, "Task Force Report on Establishment of a Combat Development and Test Center," July 12, 1961.
84 *"at the strong insistence":* NSF 194: Sept 20, 1961, memo forwarded by Robert H. Johnson to Walt Rostow containing a Sept. 13, 1961, report, Kennedy Library.
84 *"A covert ops guy":* Brown, interview with author.
84 *Warren Stark, who worked:* Stark, *Many Faces, Many Places,* 96.
84 *"Harold Brown sort of foisted":* Ruina, interview with author.
84 *"You people did the wrong thing":* Ibid.
85 *"I wasn't involved":* Ibid.
85 *"AGILE made Jack":* Godel, interview with Huff.

CHAPTER 6: ORDINARY GENIUS

86 *"He savaged me":* Kempe, *Berlin 1961,* 257.
86 *"Whichever figures are accurate":* Remarks of Senator John F. Kennedy in the U.S. Senate, National Defense, Feb. 29, 1960, Kennedy Library.

86 *Whether Kennedy really believed:* Kennedy was given an intelligence briefing by Allen Dulles, the director of the CIA, in July 1960, at a time when the CIA was beginning to get better intelligence through U-2 flights. Dwayne A. Day, "Of Myths and Missiles: The Truth About John F. Kennedy and the Missile Gap," *Space Review,* Jan. 3, 2006.

87 *A little more than two weeks:* Richard Reeves, "Missile Gaps and Other Broken Promises," *New York Times,* Feb. 10, 2009.

87 *"all the time we need":* Ruina, interview with Kai-Henrik Barth, American Institute of Physics.

88 *McNamara was not invited:* Ruina has repeated this story consistently in several interviews. Ruina, interview with author. A similar account of the meeting with Kennedy is described by Harold Brown in *Star Spangled Security,* 91.

88 *"The helicopter's waiting":* Ruina, interview with Williams/Gerard.

88 *nationwide program for fallout shelters:* Brown, *Star Spangled Security,* 92.

88 *"I don't think we should go ahead":* Ruina, interview with Finkbeiner. Nike Zeus continued on through 1962 as a research and development program but never proceeded to deployment. The army went on to pursue Nike-X, a system that employed better radar for tracking but was arguably also unworkable.

89 *The highest-ranking officer:* Ruina, interview with Williams/Gerard.

89 *"ARPA is not strong":* Huff and Sharp, *Advanced Research Projects Agency,* IV-42.

90 *"my whole life would have been different":* Ruina, interview with Finn Aaserud, American Institute of Physics.

90 *"The only two programs":* Ruina, interview with Barth, American Institute of Physics.

90 *"protect all the United States":* U.S. House, *Department of Defense Appropriations for 1962, Part 4: Research, Development, Test, and Evaluation,* 82.

90 *"anti-gravity, anti-matter":* Associated Press, "Defense Agency to Study Missile-Defense Measures," *Pacific Stars and Stripes,* March 2, 1959, 3.

91 *The agency had been inundated:* Huff and Sharp, *Advanced Research Projects Agency,* V-18.

91 *One proposal called for:* Bruno W. Augenstein, "Evolution of the U.S. Military Space Program, 1945–1960: Some Key Events in Study, Planning, and Program Development" (Rand Corp., Sept. 1982), 16.

91 *"mad scientist's dream":* York, *Race to Oblivion,* 131. York might have slammed the BB-laced spiderweb that was BAMBI, yet he had single-handedly orchestrated three high-altitude nuclear explosions to test the concept of a force field.

91 *"loony idea":* "Kill Bambi" became something of an insider's joke in the Pentagon. Brown, *Star Spangled Security,* 33.

91 *"Bambi brought in all the nuts":* Huff and Sharp, *Advanced Research Projects Agency,* V-19.

91 *"Isn't that a bit fantastic?":* U.S. House, *Department of Defense Appropriations for 1962, Part 4: Research, Development, Test, and Evaluation,* 82.

91 *Not only was it impractical:* Ibid.

92 *The scheme involved launching:* Jacques S. Gansler, *Ballistic Missile Defense: Past and Future* (Washington, D.C.: Center for Technology and National Security Policy, 2010), 38.

92 *"kind-of nutty":* Huff and Sharp, *Advanced Research Projects Agency,* V-19.

92 *Back in early 1958:* The National Advanced Research Projects Laboratory would be to the Defense Department, or ARPA, what the nuclear labs were to the Atomic Energy Commission, a sort of reservoir of scientific talent and idea generation. The

laboratory, in Wheeler's view, would be "much larger in size than Los Alamos or Livermore." See Aaserud, "Sputnik and the 'Princeton Three.'"

92 *ARPA was not interested:* York, *Making Weapons, Talking Peace,* 212.

92 *The proposals for that first meeting:* The idea was to alert nuclear-armed submarines of a possible nuclear launch order. It was initially called Project Bassoon and would require an eighty-five-hundred-foot-long antenna and most of the state of Wisconsin.

93 *"If by use of high speed":* John Wheeler, app. A-4, in *Study No. 1,* "Identification of Certain Current Defense Problems and Possible Means of Solution" (Institute for Defense Analyses, Advanced Research Projects Agency, 1958).

93 *Even that idea foundered:* "Wheeler judged that the immediate emergency had subsided, and that little emergencies were not the jobs of the academics." Aaserud, "Sputnik and the 'Princeton Three,'" 224.

93 *called Sunrise:* Finkbeiner, *Jasons,* 38.

93 *promptly changed its name to JASON:* Ibid., 40.

94 *"golden fleece":* Ibid., 39–40. Godel, in his interview with Huff, also alleged that the golden fleece joke came from the members' starting to promote their own projects to ARPA. There is certainly evidence for that allegation. The files of William Nierenberg, a prominent member of the group, intermix notes on JASON meetings with a proposal to ARPA for work on floating mid-ocean bases, which would be done at his home institution, the Scripps Institution of Oceanography. Nierenberg to Craig Fields, April 16, 1987, Scripps Institution of Oceanography Archives.

94 *As it turned out:* Finkbeiner, *Jasons,* 50.

94 *"kind of a truth squad":* Ruina, interview with Aaserud, American Institute of Physics, Aug. 8, 1991.

94 *"not good" idea:* Huff and Sharp, *Advanced Research Projects Agency,* IV-23.

94 *The particle beam:* Ibid., IX-31.

95 *"death radar sub-system":* Ibid., IV-23.

95 *"He was not ever frightened":* Ruina, interview with Finkbeiner.

95 *"aptly named, because it seesawed":* Rechtin, interview with Finn Aaserud, American Institute of Physics.

96 *"ARPA was generally of a mind":* Kresa, interview with author.

96 *"There's a better way to do it":* Ibid.

96 *"We're going to put":* Ibid. Indeed the JASONs recommended continuing the laser program. JASON, Project Seesaw (Alexandria, Va.: Institute for Defense Analyses, 1968), 3.

97 *Seesaw never fired a single shot:* If one counts the start of Seesaw in 1958, that means it was funded for thirteen years, through 1972, and would still likely hold the record for the longest-lived ARPA program.

97 *"There will not be any payoff":* Huff and Sharp, *Advanced Research Projects Agency,* V-19.

97 *The NSA wanted to use:* "Herzfeld told us in no uncertain terms that [Arecibo] had been funded as a wholly scientific and open facility, and would not be allowed to undertake classified studies, and that it was presumptuous of us to ask," Nate Gerson, an NSA cryptologist, later recounted. N. C. Gerson, "SIGINT in Space," *La Physique au Canada,* Nov./Dec. 1998, 357. (Previously published in *Studies in Intelligence* 28, no. 2.) This exchange is also cited in Bamford's *Puzzle Palace.*

97 *Herzfeld insisted Arecibo:* Ibid.

98 *"A nuclear detonation":* Ibid.

98 *"It is almost certain":* Edward Teller and Albert Latter, "The Compelling Need for Nuclear Testing," *Life,* Feb. 10, 1958, 70.

99 *The debate had become:* Lawyer, Bates, and Rice, *Geophysics in the Affairs of Mankind,* 132.

99 *ARPA got the work:* Charles Bates, interview with author. This view is also backed up by the Huff and Sharp ARPA history.

99 *Vela had three parts:* The two most significant parts of Vela ended up being Vela Hotel and Vela Uniform. Vela Sierra, which involved ground-based sensors to detect nuclear tests in space, was eventually folded into Vela Hotel. Some of the Vela work, it turns out, did not really require any exotic science. For example, detecting underwater explosions required little new research. ARPA conducted some underwater tests using conventional explosives under the code name CHASE, short for "cut holes and sink 'em." Huff and Sharp, *Advanced Research Projects Agency,* VII-15. "The ocean detection system was a nonproblem," Frosch said. Frosch, interview with author.

100 *"mildly crazy":* Frosch, interview with author.

100 *"When I was sitting there":* Ibid.

100 *"It was literally opened up":* Ibid.

101 *"a bunch of incompetents":* Huff and Sharp, *Advanced Research Projects Agency,* V-33.

101 *ARPA also encountered:* Lawyer, Bates, and Rice, *Geophysics in the Affairs of Mankind,* 122, 178.

101 *"You could do things a lot easier":* Peterson, interview with Kai-Henrik Barth, American Institute of Physics.

102 *"too secretive":* Ruina, interview with Barth, American Institute of Physics.

102 *"Romney never tried to mess":* Evernden, interview with Kai-Henrik Barth, American Institute of Physics.

102 *Now, with the Aardvark data:* Romney insisted the revisions were the result not of systemic errors but of getting more data. He had been relying on historical data of large Soviet nuclear tests and extrapolating down to make estimates about the detection of smaller tests, which might be confused with earthquakes. "The change came about as a result of additional information we got," Romney insisted. Romney, interview with author.

102 *During a July 3, 1962, meeting:* The President's Deputy Special Assistant for National Security Affairs to President Kennedy, memo, Washington, July 20, 1962, Kennedy Library.

102 *"withholding information":* Secretary of Defense McNamara to President Kennedy, memo, Washington, July 28, 1962, in *Foreign Relations of the United States, 1961–1963,* vol. 7, *Arms Control and Disarmament,* doc. 204.

102 *"honest mistake":* Ruina, interview with Barth, American Institute of Physics.

102 *"This is what can happen":* Ruina, interview with Barth, American Institute of Physics.

103 *"VELA seemed to indicate":* Seaborg, *Kennedy, Khrushchev, and the Test Ban,* 145.

103 *a spectacular success:* Huff and Sharp, *Advanced Research Projects Agency,* VII-29.

103 *"One that Mr. Kennedy wanted it":* Ibid., VI-8.

103 *Plate tectonics, which had previously:* Kai-Henrik Barth, "The Politics of Seismology: Nuclear Testing, Arms Control, and the Transformation of a Discipline," *Social Studies of Science* 33, no. 5 (Oct. 2003): 743–81.

104 *"Seismology and the New Global Tectonics":* Oliver, interview with Kai-Henrik Barth.

104 *"almost instantaneously transformed":* Lynn Sykes, "Seismology, Plate Tectonics,

and the Quest for a Comprehensive Test Ban Treaty, a Personal History of 40 Years at LDEO," in *International Handbook of Earthquake and Engineering Seismology* (Amsterstam: Academic Press, 2002), 1456.

104 *Following Kennedy's death:* As John Dumbrell points out in his book, *President Lyndon Johnson and Soviet Communism* (Manchester, U.K.: Manchester University Press, 2004), President Johnson approved the largest ever underground nuclear test—Operation Boxcar, a 1.3-megaton explosion—in the midst of negotiations over the Nuclear Nonproliferation Treaty.

105 *"not because we were geniuses":* Lukasik, interview with author.

CHAPTER 7: EXTRAORDINARY GENIUS

106 *"We have some big trouble":* Dobbs, *One Minute to Midnight,* 22.

106 *"Those sons of bitches":* Ibid.

106 *The CIA had used a massive:* Ibid., 16.

106 *"were recognized by more":* Department of Defense, *National Military Command and Control in the Cuban Crisis of 1962* (Washington, D.C.: DTIC, 1965), 50.

107 *There, he could work in peace:* Licklider, interview with J. William Aspray and Arthur L. Norberg, Charles Babbage Institute.

107 *"I do not want to be negative":* Licklider to General H. H. Wienecke (director for Remote Area Conflict), memo, April 16, 1964, Project AGILE, RG 330, National Archives, College Park.

108 *gun that shot micro-rockets:* It is likely Licklider was referring to the Gyrojet hand pistol, an experimental weapon that shot a small rocket. ARPA was involved in funding the weapon for use in village defense. Advanced Research Projects Agency, "Caliber .50 Gyrojet Hand Pistol," Oct. 25, 1962, Project AGILE, RG 330, National Archives, College Park.

108 *"These things got going":* Licklider, interview with Aspray and Norberg. The Lafayette Square reference is never explained in the interview, but it presumably means that Licklider's budget was being used to hide secret funds for other projects, which was a common tactic at ARPA.

108 *Among his colleagues:* Waldrop, *Dream Machine,* 2.

108 *"I knew that he didn't get into anyone's business":* Hess, interview with author.

109 *"messianic" vision:* Huff and Sharp, *Advanced Research Projects Agency,* V-52.

109 *By the 1990s, however:* News accounts directly linking ARPA's work to nuclear survivability first appear in the mid-1990s. This confusion shows up even more recently, however, in an otherwise insightful computer history by George Dyson. *Turing's Cathedral,* 330.

109 *This account sparked a counter-narrative:* "ARPANET and its progeny, the Internet, had nothing to do with supporting or surviving nuclear war—never did," wrote Katie Hafner, author of *Where Wizards Stay Up Late,* which chronicles the birth of computer networking. Hafner and Lyon, *Where Wizards Stay up Late,* 10. Unfortunately, this view is true only if one ignores the underlying reasons for the Pentagon's support of the ARPANET.

109 *"My son! My son!":* "POW Who Changed His Mind Meets Family in Capital," *Niagara Falls Gazette,* Nov. 21, 1952, 1.

110 *At his court-martial:* Fred L. Borch, "The Trial of a Korean War 'Turncoat': The Court-Martial of Corporal Edward S. Dickenson," *Army Lawyer,* Jan. 2013.

110 *"one in three American prisoners":* Consultation with Edward Hunter (author and

foreign correspondent), Committee on Un-American Activities, House of Representatives, 85th Cong., March 13, 1958 (Washington, D.C.: Government Printing Office, 1958).

111 *"In any future war":* Charles Bray, "Toward a Technology of Human Behavior for Defense Use," 538.

111 *"By the early 1960s the DOD":* Ellen Herman, *The Romance of American Psychology: Political Culture in the Age of Experts* (Berkeley: University of California Press, 1995), 128.

112 *"It will be obvious to you":* Bray to Licklider, May 24, 1961, Record Unit 179, Research Group in Psychology and the Social Sciences Records, 1957–1963, Smithsonian Institution.

113 *"The Truly SAGE System":* J. C. R. Licklider, "The Truly SAGE System; or, Toward a Man-Machine System for Thinking," manuscript, Aug. 20, 1957, Licklider Papers, MIT Libraries.

113 *"The fig tree is pollinated":* J. C. R. Licklider, "Man-Computer Symbiosis," *IRE Transactions on Human Factors in Electronics* (March 1960): 4–11.

115 *"We built a network like a fishnet":* Baran, interview with Stewart Brand, "Founding Father," *Wired,* March 2001.

115 *"The enemy could destroy 50, 60, 70 percent":* Ibid.

115 *"The early missile control systems":* Baran, interview with Judy O'Neill, Charles Babbage Institute.

116 *"research and* no *development":* Baran, interview with Brand, "Founding Father."

116 *"I pulled the plug":* Ibid.

116 *William Godel, then the deputy director:* Godel, interview with Huff.

116 *Brown was unhappy:* Brown, correspondence with author.

117 *One of his early program descriptions:* J. C. R. Licklider, Program Plan 93, Computer Network and Time-Sharing Research, April 5, 1963, RG 330, National Archives, College Park.

117 *"Who can direct":* Licklider, interview with Aspray and Norberg, Charles Babbage Institute.

117 *"I did realize":* Ibid.

117 *"asked to see me about something":* Ruina, interview with William Aspray, Charles Babbage Institute.

118 *"Tell me what has happened":* Ibid.

118 *Licklider ended up spending:* Under Licklider, the Behavioral Sciences Office was not involved in Southeast Asia. In 1964, Lee Huff, who had been ARPA's representative in Thailand, took over and began to fund work in that area. Huff, interview with author. See also Huff and Sharp, *Advanced Research Projects Agency,* VI-52-3.

118 *"There was a kind of a cloak and dagger":* Licklider, interview with Aspray and Norberg, Charles Babbage Institute.

118 *Licklider's immediate problem:* Huff and Sharp, *Advanced Research Projects Agency,* V-49.

119 *By 1960, the Pentagon's primary concern:* "NORAD/CONAD Historical Summary," Jan.–June 1960, Directorate of Command History, Office of Information Headquarters, NORAD/CONAD, 1.

119 *"great asset":* Licklider, interview with Aspray and Norberg.

119 *"At this extreme, the problem":* Licklider to Members and Affiliates of the Intergalactic Computer Network, memo, Advanced Research Projects Agency, April 25, 1963.

120 *After the heyday:* See Huff and Sharp, *Advanced Research Projects Agency,* Figure VII-I.

120 *"Okay, look, before you cancel":* Sproull, interview with Williams/Gerard.
120 *"almost killed the Internet":* Ibid.
121 *"I was in the army":* Sutherland, interview with author.
121 *"There came a moment":* Crocker, interview with author.
121 *"my major failure":* Sutherland, interview with William Aspray, Charles Babbage Institute.
122 *"I became a disciple":* Herzfeld, interview with author.
122 *"How much money":* Taylor, interview with William Aspray, Charles Babbage Institute.
123 *"Over the course of several hundred years":* Baran, interview with O'Neill, Charles Babbage Institute.
123 *"I did nothing":* Ruina, interview with Finkbeiner.
123 *Licklider was able to do:* Though it is often presumed in retrospect that the freedom ARPA enjoyed was a given, the agency was only a few years old, and there was little presumption of anything. It was a unique time for the agency, Ruina reminisced. "I had so much freedom; [if] I wanted to build a bridge from Chattanooga to Seattle, I could do it," he recalled in an interview almost fifty years later. "I look back at my years, and I didn't appreciate enough the kind of freedom I had to do as many good things that I could have done, and that would have been great for the country, and great for ARPA. I just didn't realize what special times they were." Ruina, interview with author.

CHAPTER 8: UP IN FLAMES

125 *"The growing U.S. military involvement":* Karnow, *Vietnam,* 270.
125 *"I don't see nothing":* Ibid., 270–71.
126 *"This program has been undertaken":* Memo forwarded by Robert H. Johnson to Walt Rostow, Sept. 20, 1961, containing a Sept. 13, 1961, report, Kennedy Library.
126 *That experiment appeared:* George Rathjens and William Godel, research and development app. of the Taylor report, Nov. 3, 1961, Kennedy Library.
126 *eliminate ground cover:* Ibid.
126 *That came on November 30, 1961:* Major William A. Buckingham Jr., "Operation Ranch Hand: Herbicides in Southeast Asia," *Air University Review,* July–Aug. 1983.
126 *"to be ignorant":* C. E. Minarek, Memorandum for the Record, "Subject: Meeting with Mr. William Godel on 4 December 1961," Defense Technical Information Center.
127 *To avoid detection:* Cecil, *Herbicidal Warfare,* 31.
127 *The security measures were not:* Ibid.
127 *The bands on the barrels:* Ibid., 32.
128 *"The development and use":* William H. Godel, Deputy Director, Vietnam Combat Development and Test Center, ARPA, Progress Report: Vietnam Combat Development and Test Center, Sept. 13, 1961, Kennedy Library.
128 *Another important distinction:* Department of Army pamphlet, *Area Handbook for South Vietnam,* April 1967.
129 *massive undertaking:* Quang Trach, the Vietnamese colonel who headed ARPA's Combat Development and Test Center, and Truong Quang Van, the intelligence chief's aide, both recalled in their court depositions that Diem was skeptical of the strategic hamlet program at that first meeting with Godel in June 1961; he wanted

a study done to see if the village leaders could be persuaded to participate, and that required "gifts," or more accurately, bribes, which could be used to persuade them to move the peasants.

129 *Diem's brother Nhu:* For all the literature that has been dedicated to the strategic hamlet program, there has been remarkably little discussion about what persuaded Diem to embark on a risky large-scale endeavor that had already failed on a small scale. Some accounts of the Vietnam War point to the CIA, although there is little documentation to back up that claim, while others have claimed it was Diem's brother. "While it was never explicitly stated, there does seem considerable reason to believe that the strategic hamlet scheme was the personal concept of President Diem's brother, Ngo-Dinh Nhu," wrote Milton E. Osborne in *Strategic Hamlets in South Vietnam: A Survey and a Comparison* (Ithaca: Cornell University Press, 1965), 26. Though the only evidence for his patrimony is a self-serving booklet published by the South Vietnamese government crediting Nhu with the idea.

129 *By the fall:* Gravel, Chomsky, and Zinn, *Pentagon Papers,* 128–59.

129 *"is a condition of the mind and heart":* ARPA, "Research and Development Effort in Support of the Vietnamese Rural Security Program" (Washington: ARPA, 1962), 12. Thomas Thayer Collection, U.S. Army Center of Military History, Washington, D.C.

129 *"machinery for formal government control":* Ibid.

130 *Farmers complained of forced labor:* John C. Donnell and Gerald C. Hickey, *The Vietnamese "Strategic Hamlets": A Preliminary Report* (Santa Monica, Calif.: Rand, 1962).

130 *Gripping a cigarette holder:* Hickey, *Window on a War,* 92.

130 *A marine general slammed his fist:* Ibid., 99. Harold Brown wrote that he could not recall Hickey's briefing, "but he is certainly right that I, along with most others in the government, was naïve about the nature and prospects of the Vietnam War." Brown, correspondence with author.

130 *In April 1962:* Stephen T. Hosmer and S. O. Crane, *Counterinsurgency: A Symposium, April 16–20, 1962* (Santa Monica, Calif.: Rand, 1962).

131 *"That it was the major facet":* Grinter, "Population Control in South Viet Nam, the Strategic Hamlet Study."

131 *Counterinsurgency experts like:* Lansdale admitted he did not realize until 1955 that Nhu was essentially running Diem's intelligence operations.

131 *The strategic hamlets, run by Nhu:* Sheehan, *Bright Shining Lie,* 265.

131 *Unable to adapt to the jungle:* Seymour Deitchman, interview with author. "Dogs were scroungers in Vietnam, never used as pets," Deitchman said. "And they were considered food."

131 *The spraying was supposed to be done:* Associated Press, "Attempt to Strip Jungles in Viet Nam a Flop So Far," *Tuscaloosa News,* March 30, 1962, 16.

131 *Empty barrels:* "The vapor from those barrels killed all the vegetation in the city for a mile around the [Research Development Field Unit] compound," Deitchman recalled. Deitchman, interview with author.

132 *At one point, Colonel Trach:* Ibid.

132 *"Well, do you know":* Ibid.

132 *"wanted to stick a finger":* Godel, interview with Huff.

132 *"For the type of conflict":* Advanced Research Projects Agency, Final Report, OSD/ARPA Research and Development Field Unit–Vietnam, Aug. 20, 1962, Defense Technical Information Center.

132 *"that the AR-15 is decidedly superior":* Reed, Van Atta, and Deitchman, *DARPA Technical Accomplishments,* vol. 1, 14-4.

133 *The army did not want to be told:* The lethality debate continued more than four decades later, with soldiers questioning the performance of the weapon in Iraq and Afghanistan. Anthony F. Milavic, "The Last 'Big Lie' of Vietnam Kills U.S. Soldiers in Iraq," *American Thinker,* Aug. 24, 2004.

133 *"It was a blatant, blithering failure":* Godel, interview with Huff.

134 *"regarded the ARPA field unit":* Cosmas, *MACV,* 51.

134 *"I do not feel well enough informed":* The phone call between Diem and Lodge is printed in Gravel, Chomsky, and Zinn, *Pentagon Papers.*

134 *ARPA personnel in the city:* Herzfeld, *Life at Full Speed,* 126–27.

135 *following the CIA's Bay of Pigs fiasco:* Charles Maechling, "Camelot, Robert Kennedy, and Counter-insurgency: A Memoir," *Virginia Quarterly* (Summer 1999).

135 *Wylie wanted to meet:* Godel/Wylie trial transcript.

135 *Assigned to ARPA as Godel's:* Ibid.

135 *"vultures":* Corson, *Betrayal,* 249.

135 *"This has about as much value":* "Memorandum for General Wienecke: Subject: Analysis of RAC Proposed Tasks by AGILE Program Managers," May 22, 1964, Project Agile, RG 330, National Archives, College Park.

135 *In the spy business, Class A agents:* Wolfgang W. E. Samuel, *American Raiders: The Race to Capture the Luftwaffe's Secrets* (Jackson: University Press of Mississippi, 2004). After World War II, for example, Class A agents were used to give cash to German scientists brought to the United States to keep them out of Soviet hands.

136 *McNamara suddenly demanded:* Godel/Wylie trial transcript.

136 *"You can be around people":* Ibid.

136 *"If you are in trouble":* Ibid.

137 *Believing a deal:* Karnow, *Vietnam,* 325–26.

137 *"rushed from the room":* Taylor, *Swords and Plowshares,* 301.

137 *In the Pentagon's Office of Special Operations:* Samuel Vaughan Wilson, interview with author.

138 *"in some way covert intelligence operations":* Godel/Wylie trial transcript.

138 *"The president's been shot":* Frosch, interview with author.

138 *It was late on a Friday evening:* All direct references to the Godel trial are taken from the Godel/Wylie trial transcript. John Loftis, another senior Pentagon official originally charged in the case, successfully petitioned to be tried separately and was cleared.

138 *"This is an important case":* Ibid.

139 *None of them had a change of clothes:* Ibid.

139 *"distant thunder that precedes":* "South Viet Nam: Forecast: Showers & a Showdown," *Time,* May 21, 1965.

139 *"to support what we called":* Godel/Wylie trial transcript.

140 *"I never met":* Ibid.

140 *"Bill Godel wouldn't bother":* The Landon Chronicles, an oral history of Dorothea Mortenson Landon and Kenneth Perry Landon, recorded by their son, Kenneth Perry Landon Jr. (1976–1989). Margaret and Kenneth P. Landon Papers, Wheaton College Archives.

140 *Wylie appeared nearly comatose:* Warren Stark, interview with author. Pretrial records also detail Wylie's efforts to claim mental illness. Godel/Wylie trial records.

140 *"It was a couple guys":* Cacheris, interview with author.

141 *Over the course of the Vietnam War:* See *Veterans and Agent Orange: Health Effects of Herbicides Used in Vietnam* (Washington, D.C.: National Academy Press, 1994).

141 *Agent Orange became synonymous:* The best technical assessment of the herbicide program is in Jeanne Mager Stellman et al., "The Extent and Patterns of Usage of Agent Orange and Other Herbicides in Vietnam," *Nature,* April 17, 2003, 681–87. The authors reconstruct the use of herbicides based on National Archives records. They concede the exact amount used is difficult to pin down, because in some cases only procurement amounts are recorded, which is not necessarily the same as what was actually sprayed.

141 *Of all the things Godel:* Today, ARPA jokes about its failures, like the mechanical elephant that would trudge through the jungles of Vietnam, but Agent Orange is never mentioned in any official materials. In an otherwise detailed account of the early years of the agency, the only reference to chemical defoliation is a brief note that chemicals were used to "clear roads and border areas and possibly disrupt Viet Cong food sources, culminating in a limited operational test which 'yielded generally inconclusive results.'" Huff and Sharp, *Advanced Research Projects Agency,* V-42.

141 *Godel left ARPA:* Godel even believed that the AR-15, later hailed as AGILE's greatest success, was, in fact, a failure for counterinsurgency. The debate over the U.S. Army's adoption of the AR-15 merely sidetracked plans to give the weapon to South Vietnamese troops, to help them fight in the jungles. Turning the AR-15 into a weapon for American forces defeated the entire point. "I neither know, nor give a damn, what the U.S. Army needs," Godel said. Godel, interview with Huff.

141 *"None":* Ibid.

142 *"stuff happens":* Herzfeld, interview with author.

CHAPTER 9: A WORLDWIDE LABORATORY

144 *"simulating the behavior":* U.S. House, *Department of Defense Appropriations for 1966. Part 5: Research, Development, Test, and Evaluation,* 565

144 *Herzfeld was born:* National Academy of Sciences, *Biographical Memoirs, Volume 80* (Washington, D.C.: National Academy Press, 2001), 162. It also appears that Charles Herzfeld's grandfather Karl Herzfeld converted and was alleged to be a "leading anti-Semite in the faculty of medicine." Walter Moore, *Schrödinger: Life and Thought* (Cambridge, U.K.: Cambridge University Press, 1992). See also Arthur Schnitzler, *My Youth in Vienna* (New York: Holt, Rinehart and Winston, 1970), 307.

144 *"I was no Mozart":* Herzfeld, *A Life at Full Speed,* 93.

144 *"towering intellect":* Herzfeld, interview with author.

144 *"I thought it was a declaration":* Ibid.

145 *Though he would not formally:* Ibid. This view is also supported by the Huff and Sharp history.

145 *"The only thing worth doing":* Ibid.

145 *"I think one could do":* Huff and Sharp, *Advanced Research Projects Agency,* VI-21.

145 *But interest in missile defense:* Ibid., VII-19. The history notes, "By the end of 1966 the drive for a comprehensive test ban had essentially ceased to exist as a matter of national priority."

145 *Almost from the start:* In a 1988 interview, when asked about his travels to look at networking, he recalled going to MIT a couple times, and also looking at military computers at Strategic Air Command, but it was clear the specifics of the program were beyond his daily purview. In his memoir, Herzfeld devotes fewer than two

pages to computer networking, an area in which he was deeply influential but that did not occupy much of his time.

145 *"I've been thinking about":* Herzfeld, interview with author.

145 *"more of a doer":* Ibid.

146 *"The question was, if you dig a tunnel":* Ibid.

146 *"We did experiments":* Ibid. Though Herzfeld did not expand on why Vietnam's trees were sick, it might have been the result of two years of chemical defoliation, initiated by ARPA.

146 *"We were expected to solve":* Herzfeld, interview with author.

146 *"I was involved in this":* Ibid.

147 *"I'm thinking of opening an office":* Stark, interview with author.

147 *"To be honest with you":* Ibid.

148 *The Pentagon's spending:* Circular 515-6, Department of Defense Research and Development Activities in U.S. Southern Command, July 13, 1965, Project AGILE, RG 330, National Archives, College Park.

148–49 *"Thailand was basically a laboratory":* Stark, interview with author.

149 *He paid for an hour:* Stark, *Many Faces, Many Places,* 100.

149 *Despite the admitted lack of knowledge:* R. H. Wienecke to Director, Program Management, memo, "Subject: Counterinsurgency Information Analysis Center," June 2, 1964, Project AGILE, RG 330, National Archives, College Park.

149 *The "people and politics" branch:* Advanced Research Projects Agency, Project AGILE, Semiannual Report, 1 July–31 Dec. 1963, Project AGILE, RG 330, National Archives, College Park.

149 *"With a proper approach": U.S. House, Department of Defense Appropriations for 1966. Part 5: Research, Development, Test, and Evaluation,* 137.

150 *"The need for ecological":* Stark to Director, Remote Area Conflict, memo, Feb. 18, 1963, Project AGILE, RG 330, National Archives, College Park.

150 *"We can communicate":* This quotation is repeated in many forms, in memoirs, in interviews with the author, and in the ARPA history. This specific version is from Stark's memoir, *Many Faces, Many Places,* 123.

151 *"had little potential":* Huff and Stark, *Advanced Research Projects Agency,* VI-40.

151 *"maintain a much more efficient":* Robert A. Kulinyi, "Program Review of the Southeast Asia Communications Research Project," in J. R. Wait et al., *Workshop on Radio Systems in Forested and/or Vegetated Environments* (Fort Huachuca, Az: Army Communications Command, 1974), 20.

151 *"While SEACORE may have had":* Huff and Stark, *Advanced Research Projects Agency,* VI-41.

152 *ARPA provided boats:* Stark and Deitchman, interviews with author.

152 *"All the South China Sea Junks":* Herzfeld, *A Life at Full Speed,* 137.

152 *As Stark walked down the street:* Stark, *Many Faces, Many Places,* 122.

152 *ARPA's Rural Security Systems Program:* Huff and Sharp, *Advanced Research Projects Agency,* VIII-47.

153 *The Thais attempted to mimic:* Ibid.

153 *"laughable":* Ibid., VIII-48.

153 *"I remember taking":* Stark, interview with author.

153 *Stark was beginning to think:* Ibid.

153 *Asked what could account:* Stark, *Many Faces, Many Places,* 110.

154 *two-hundred- to three-hundred-foot wire:* Christofilos to Foster, Aug. 29, 1966, Project AGILE, RG 330, National Archives, College Park. The letterhead indicated that

Christofilos was writing on behalf of the JASON group, managed by the Institute for Defense Analyses.

154 *"The approach appears feasible":* Herzfeld to Foster, memo, Nov. 23, 1966, Project AGILE, RG 330, National Archives, College Park.

154 *"a unique approach":* Foster to Christofilos, Nov. 22, 1966, Project AGILE, RG 330, National Archives, College Park.

154 *"the suspension of a wire":* Lieutenant Colonel Thomas F. Doeppner, memo, Nov. 8, 1966, Project AGILE, RG 330, National Archives, College Park.

154 *The Christofilos proposal:* Although the ARPA files do not say if the idea was pursued, in a later interview John Foster refers to an ARPA project that detected oxide using "a new frequency, a harmonic." Foster, interview with Williams/Gerard.

154 *The JASONs were mostly:* Stark, *Many Faces, Many Places,* 114.

154 *They did have ideas:* JASON Summer Study, meeting minutes, "The Thailand Study Group," June 26, 1967, Project AGILE, RG 330, National Archives, College Park.

154 *"ear cutting":* Finkbeiner, *Jasons,* 101–2.

155 *This speech was a watershed:* Daniel E. Harmon, *Ayatollah Ruhollah Khomeini* (Philadelphia: Chelsea House, 2005), 38.

155 *"You helpless creature":* Translation of Khomeini's speech, IRIB World Service, worldservice.irib.ir.

155 *In December 1963:* Norman H. Jones Jr., "Support Capabilities for Limited War in Iran," study, Rand, Dec. 1963.

156 *"Iran has been the primary":* John A. Reed Jr. to Colonel Jordan, memo, Aug. 11, 1967, RG 59, National Archives, College Park.

156 *"in the mold of the New Jersey":* George J. Wren, *Jersey Troopers II: The Next Thirty-Five Years (1971–2006)* (Bloomington, Ind.: iUniverse, 2009), 34.

156 *Schwarzkopf, whose son:* It came out years later that Schwarzkopf leveraged his success with the gendarmerie to help the CIA facilitate its successful coup. J. Dana Stuster, "The Craziest Detail About the CIA's 1953 Coup in Iran," *Passport* (blog), *Foreign Policy,* Aug. 20, 2013.

157 *"biggest challenge":* Herzfeld, interview with author.

157 *ARPA demonstrated how:* Ibid.

157 *"gotten too close":* Herzfeld, *Life at Full Speed,* 153.

157 *"trivial and floundering":* Gerald Sullivan, Memorandum for the Record, "Subject: Brief on RDFO(ME) Activities," Nov. 5, 1971, Project AGILE, RG 330, National Archives, College Park.

158 *"They gave us lots of insight":* Herzfeld, interview with author.

158 *"Can you detect":* Ibid.

158 *"We have recently completed":* Advanced Research Projects Agency, Report, "ARPA Research in Iran," April 26, 1970, Project AGILE, RG 330, National Archives, College Park. Though the report focuses on Iran, it also covers ARPA's "MEAFSA" program, or Middle East, Africa, and Southern Asia.

158 *Herzfeld spent several years:* In his memoir, Herzfeld recalls it was President Lyndon Johnson's concerns about Pakistan-India tensions that killed the agreement. The Project AGILE files in the National Archives make oblique references to a controversy over ARPA's India endeavor. It appears, based on other memos in the file, that the State Department was concerned about ARPA's involving itself in diplomatic issues.

158–59 *"We wanted to open another":* Hess, interview with author.

159 *"We did a study":* Stark, interview with author.

159 *"an impending insurgency":* Memorandum for Mr. Godel, "Subject: Insurgency-U.S. Style," June 8, 1964, Project AGILE files, National Archives, College Park.

160 *The closest AGILE ever got:* Harold Brown's Pentagon office issued a formal memo to ARPA on December 3, 1963.

160 *"Operation Barn Door":* Frosch, interview with author. In correspondence with the author, Brown wrote, "I may well have made that remark, but it did not reflect a lack of seriousness in the effort."

160 *"We couldn't possibly":* Sproull, interview with Williams/Gerard.

160 *"The Defense Department has":* Herzfeld to Director of Defense Research and Engineering, memo, "Congressional Query Regarding Star," Sept. 4, 1964, Project Star, RG 330, National Archives, College Park.

160 *"need to know" basis:* Although the code name "Star" was briefly acknowledged in a footnote in the 1975 history of ARPA, the records detailing ARPA's contribution to presidential security were not released by the National Archives until November 2013, in response to a request from the author.

161 *One proposal, for example:* ARPA officials grew frustrated with the Secret Service's lack of technical expertise, which often led them to propose weapons that sounded as if they were straight out of a Road Runner cartoon, like a nonlethal weapon that would immediately disable a potential assassin in a crowd. When ARPA officials did test some of the weapons the Treasury Department had in mind, like a ninja-inspired collapsible billy club with a stabbing end that projected out at high speed, they found it operationally useless. Either the club collapsed after several uses, or the users lost their grip.

161 *"There also exists a need":* H. Morris and G. J. Zissis to H. Tabor, memo, "An Examination of a Few Protective Concepts," March 9, 1964, Project Star, RG 330, National Archives, College Park.

161 *"move about when in the car":* Unsigned memo submitted by "John" to Alyce (presumably Alyce Pekors, a longtime secretary for Project AGILE) on April 16, 1964, describing various ideas submitted by ARPA staff. Project Star, RG 330, National Archives, College Park.

161 *The nonlethal weapon:* Other proposed weapons, like an aerosol gun that dispensed tear gas, risked disabling the agent using the weapon if he had not donned a protective mask ahead of time. The frustration went both ways, however. The Treasury Department accused ARPA of having its own James Bond–inspired suggestions, which were technically feasible, but operationally impractical, like electrifying chrome strips of the presidential limousine to prevent crowds from overturning the vehicle. An armored vehicle would be too heavy to overturn anyhow, the Treasury Department countered.

162 *The liquid squirt gun:* Colonel Harry Tabor to Dr. R. L. Sproull, memo, "Advice to Treasury Department on Non-lethal Weapons," Jan. 21, 1965, Project Star, RG 330, National Archives, College Park.

162 *Though the squirt guns:* Telegram sent by Harold Tabor in March 1967, Project Star, RG 330, National Archives, College Park.

162 *"The rationale was that a waving flag":* Brown, correspondence with author.

162 *common green bottle fly:* H. A. Ells and R. E. Kay, "Applicability of Olfactory Transducers to the Detection of Human Beings: Final Report, Advanced Research Projects Agency, Feb. 1, 1965–July 31, 1966," Project AGILE, RG330, National Archives, College Park.

162 *"nonlethal decay mechanisms":* Hughes Aircraft Company, "Proposal for a Study of

Non-lethal Decay Mechanisms, an Unsolicited Proposal to the Advanced Research Projects Agency," June 23, 1964, Project AGILE, RG 330, National Archives, College Park.

163 *the agency's involvement with Agent Orange:* In 1967, an agency directory still listed chemical defoliation as a program activity under AGILE.

163 *"as a large scale countermeasure":* ARPA, "Excerpts from Recent Trip Report: For Follow-Up Action Where Indicated," n.d., Project AGILE, RG 330, National Archives, College Park.

163 *"high priority":* Author unknown, "Dr. Herzfeld's Trip Actions," n.d., Project AGILE, RG 330, National Archives, College Park.

163 *"During my five years at ARPA":* Stark, *Many Faces, Many Places,* 93.

163 *Back in 1964, Godel had laid out:* ARPA, "Task Force 'Isolation in South Vietnam.'" The report itself is undated, but an accompanying transmission letter written by the director of ARPA's Remote Area Conflict program explains that it is a "study paper" written by ARPA for Harold Brown, the director of defense research and engineering. Major General R. H. Wienecke, "Memorandum for Brigadier General John Boles, director, JRATA (Subject: Border Security—S. Vietnam)," March 27, 1964, Project AGILE, RG 330, National Archives, College Park.

164 *A handwritten list:* Ibid. The handwritten list was attached to the task force report.

164 *The border-sealing proposal:* "I believe the cost has been far underestimated," Seymour Deitchman, Harold Brown's assistant for counterinsurgency, wrote after reviewing it. Deitchman, "Memorandum for the Director, ARPA," March 24, 1964, Project AGILE, RG 330, National Archives, College Park.

164 *"one of the most unusual":* Headquarters, Strategic Air Command, "History of Strategic Air Command, Jan.–June 1966," September 19, 1997, 118–19.

164 *That was because the purpose:* "Forest Fire as a Military Weapon, Final Report July 1970," U.S. Department of Agriculture, Forest Service, sponsored by the Advanced Research Projects Agency, Remote Area Conflict, Defense Technical Information Center.

165 *This time, ARPA claimed:* "History of Strategic Air Command, Jan.–June 1966," 118–19.

165 *"The country doesn't burn":* Deborah Shapely, "Technology in Vietnam: Fire Storm Project Fizzled Out," *Science,* July 21, 1972, 239–41.

165 *"This was clearly one":* Ibid.

166 *In 1966, Deitchman was working:* Finkbeiner, *Jasons,* 62–89.

166 *McNamara instead assigned:* Ibid., 77.

166 *"ARPA was cut out":* Lukasik, interview with author.

166 *"A lot of what we called 'dirty tricks'":* Tegnelia, interview with author.

167 *"in the desperation":* Buhl, *An Eye at the Keyhole.*

167 *"The barrier proved":* Tim Weiner, "Robert S. McNamara, Architect of a Futile War, Dies at 93," *New York Times,* July 6, 2009, A1.

167 *"network centric warfare":* Deitchman, "'Electronic Battlefield' in the Vietnam War."

CHAPTER 10: BLAME IT ON THE SORCERERS

168 *"Do you see anything":* Walter Slote, "Observations on Psychodynamic Structures in Vietnamese Personality: Initial Report on Psychological Study—Vietnam," 1966, Simulmatics Corporation, Pool Papers, MIT Archives.

169 *"The Viet Cong member":* Ibid.
169 *"It is my strong impression":* Ibid.
170 *"started to twist my arm":* Deitchman to Peter Hayes, e-mail, Feb. 23, 2002.
171 *With his growing expertise:* When McNamara was tapped to lead the Defense Department in 1961, he brought operations research back with him to the Pentagon, along with a cadre of his "whiz kids" determined to reform how the military did business. See Kaplan, *Wizards of Armageddon,* 256.
171 *"artificial moon":* Deitchman, interview with Bob Sheldon, in *Military Operations Research* 15, no. 2 (2010).
172 *"total solution of the problem":* Huff and Sharp, *Advanced Research Projects Agency,* VII-21.
172 *"mass polygraph for internal village security":* S. W. Upham to H. H. Hall, Aug. 25, 1965, Project AGILE, RG 330, National Archives, College Park.
173 *ARPA managed to stay out:* ARPA had typically stayed away from those sorts of proposals, sometimes because the ideas were impractical, sometimes because they were stupid, and often because Godel did not think they would work. At one point, Godel wrote that he preferred funding "a social science study on the mores and susceptibility" of Vietnamese to different "interrogation techniques" rather than the polygraph. Undated memo by Godel (attached to General Electric proposal on polygraph), Project AGILE, RG 330, National Archives, College Park.
173 *"a collection of $25 to $30 million worth":* Deitchman, interview with author.
173 *"Sy Deitchman was aghast":* Stark, interview with author.
173 *That included, in one report:* Herman Kahn and Garrett N. Scalera, *Basic Issues and Potential Lessons of Vietnam: A Final Report to the Advanced Research Projects Agency,* vol. 5, *A Summary of Economic Development Projects That Might Have or Might Still Be Helpful in Vietnam* (Croton-on-Hudson, N.Y.: Hudson Institute, 1970), 50A. The idea was to build canals that would be used as "barriers" to protect hamlets. Kahn's proposal, which seemed more appropriate to zombie defense than insurgent warfare, involved "using small dredges that could construct barrier canals around pacified hamlets in the Delta, cutting and blocking existing canals to create a perimeter enclosing hamlets and adjacent fields." Considering that the strategic hamlet program was, by this point, a notable failure, it is unclear what barricading them off would have done, even if it were practical.
173 *"Well, you could get Herman Kahn":* Lewis Sorley, *Vietnam Chronicles: The Abrams Tapes, 1968–1972* (Lubbock: Texas Tech University Press, 2004), 201.
173 *"What is Herman Kahn doing":* Deitchman, interview with author.
174 *"had to do with anger":* Morell, interview with author.
174 *"The new theme":* Elliott, *RAND in Southeast Asia,* 89.
175 *"When the Air Force is paying":* Ibid., 103.
175 *But Gouré's new slant:* Ibid., 125–26.
175 *"I got hold of McNamara's military assistant":* Deitchman, interview with author.
175 *Project Camelot:* The Camelot story is recounted in many places and is sometimes erroneously credited to ARPA. One contemporaneous account is by George E. Lowe, "The Camelot Affair," *Bulletin of Atomic Scientists,* May 1966.
176 *"could easily attract":* Deitchman, *Best-Laid Schemes,* 312.
176 *"As far as we in ARPA":* Ibid.
176 *"This is the A-bomb":* Thomas B. Morgan, "The People-Machine," *Harper's,* Jan. 1961, 53.
176 *"people machine":* The Simulmatics Corporation, *Human Behavior and the Electronic Computer,* information brochure, n.d., Pool Papers, MIT Archives.

177 *Pool originally suggested:* Pool to Warren Stark, draft letter, Advanced Research Projects Agency, Feb. 1, 1966, Pool Papers, MIT Archives.

177 *"My research work":* Hoc to Deitchman, Feb. 1, 1966, Pool Papers, MIT Archives.

177 *"It is possible to control":* Simulmatics Corporation, "Continuation of Psychological Warfare Weapons Project," Feb. 14, 1968, Pool Papers, MIT Archives.

178 *By unfortunate coincidence:* The Simulmatics report says the observers' reports were translated into English, subject in some cases to classification, and then "statistical tabulations made of the frequency of various kinds of reactions and remarks." No details of these results were given in the draft report.

178 *"sorcerer's project":* Joseph Hoc, "Testing New Psychological Warfare Weapons in Viet Nam," Simulmatics Corporation, 1968, Pool Papers, MIT Archives.

178 *"variables were contaminated":* Quinn to Pool, Oct. 2, 1968, Pool Papers, MIT Archives.

178 *"Curse You, Red Baron":* Hoc to Pool, Oct. 19, 1968, Pool Papers, MIT Archives.

178 *To head the Simulmatics office:* Alfred de Grazia, "A Brief Biography, 29 December 1919 to 31 August 2006," Alfred de Grazia personal archive, grazian-archive .com.

179 *"rank amateurishness":* Seymour Deitchman, Memorandum for the Record, "Subject: Simulmatics Corporation," Nov. 29, 1967, Project AGILE, RG 330, National Archives, College Park.

179 *"someone had taken a book":* L. A. Newberry, Research Development Field Unit–Vietnam, "Memorandum for the Record: Meeting at Simulmatics Village with Mr. Los, Mr. Nhon, and Dr. Melhado About TV Study," Dec. 17, 1967, Project AGILE, RG 330, National Archives, College Park.

179 *"Briefcase Directors":* Colonel William B. Arnold, "Memorandum for W. G. McMillan, Subject: Simulmatics Corporation," Nov. 21, 1967, Project AGILE, RG 330, National Archives, College Park.

179 *"Hate the U.S." banquet:* Ibid.

179 *"running around Vietnam":* Colonel John V. Patterson Jr. (director, ARPA Field Unit, Vietnam), "Memorandum for Garry L. Quinn, Program Manager. Subject: ARPA Contractors," n.d., Project AGILE, RG 330, National Archives, College Park.

179 *"Wonder what brand":* Ibid.

179 *"If ARPA is to perform":* De Grazia to Deitchman, June 27, 1967, de Grazia archive.

180 *"This Corporation has been operating":* McMillan to Deitchman, Dec. 9, 1967, "Termination of Simulmatics Research Activities in Republic of Vietnam," Project AGILE, RG 330, National Archives, College Park.

180 *"One aspect of it may be":* Deitchman to Colonel W. B. Arnold, Sept. 8, 1967, Project AGILE, RG 330, National Archives, College Park.

180 *Deitchman's reaction was identical:* Pool did use his political connections to appeal directly to the White House, lobbying officials there to set up a social science research center in Vietnam, ostensibly under the rubric of "Vietnamization." Naturally, Simulmatics would be the lead contractor for the center; a handwritten budget drawn up by Pool showed that most of the costs would go toward overhead. To help their cause, Simulmatics had Father Hoc, the Vietnamese priest, ghostwrite a letter to be signed by President Thieu's political adviser, requesting the center be set up. Even Greenfield, the company's president, realized that was going too far. "They'll see through the ruse because of the items which give away Hoc's authorship," Greenfield wrote to Pool. "It could backfire when they pick up the clues." The contract never came through. Greenfield to Pool, Aug. 15, 1968, Pool Papers, MIT Archives.

180 *gunrunning in Southeast Asia:* The Landon Chronicles.

180 *"Greenfield told me Godel":* Deitchman to George Tanham, personal, May 14, 1968, Project AGILE, RG 330, National Archives, College Park.

181 *If he refused a senior Thai official:* Ibid.

181 *"bureaucratic reasons":* Deitchman, *Best-Laid Schemes,* 319.

181 *From start to finish:* Rohde, "Last Stand of the Psychocultural Cold Warriors," 233. Rohde's article provides the most comprehensive account of Simulmatics' misadventures in Vietnam.

181 *"The fact and means of measurement":* Deitchman, *Best-Laid Schemes,* 447.

182 *"As we thought about that":* Deitchman, interview with author.

182 *"How does this get to be":* U.S. House, *Department of Defense Appropriations for 1968,* 168.

183 *Saigon's once bustling:* Hickey, *Window on a War,* 242.

184 *Only Thailand managed to avoid:* Herzfeld claims that ARPA "helped the Thais preserve the integrity of Thailand." Lee Huff, who helped establish the program there, believed the Thai government was largely responsible for keeping the insurgency in check. American officials had "a tendency to underestimate the people that they were working with," he said. Herzfeld and Huff, interviews with author.

184 *"As to the reason":* Deitchman, correspondence with author.

185 *"AGILE was an abysmal failure":* Huff and Sharp, *Advanced Research Projects Agency,* VIII-50. Herzfeld repeated the "glorious failure" in conversation with the author in 2013.

CHAPTER 11: MONKEY BUSINESS

186 *On October 22, 1964:* Adam Nossiter, "Are Mississippi Deaths Linked to N-Bomb Tests?," *Tuscaloosa News,* May 17, 1990, 7B.

186 *"bad roll":* Bates, interview with author.

186–87 *"These were poor people":* Ibid.

187 *"spoon into Jell-O":* U.S. House, *Department of Defense Appropriations for 1968. Part III,* 143.

187 *Its missile defense work:* The radar supported the Pacific Range Electromagnetic Signature Study.

187 *"first attempt to get in bed":* Lukasik, correspondence with author.

188 *"some special projects":* Herzfeld, *Life at Full Speed,* 140.

188 *"research in such fields":* Quoted in Huff and Sharp, *Advanced Research Projects Agency,* VIII-52.

188 *"Its operation, greatly complicated":* Ibid.

189 *Moscow Viral Study:* Associated Press, "Tracking the Microwave Bombardment," *Spartanburg Herald Tribune,* June 12, 1988, A20.

189 *At five microwatts per square centimeter:* U.S. Senate, *Radiation Health and Safety,* 269.

189 *The United States wanted to figure out:* Barton Reppert, "The Zapping of an Embassy: The Mystery Still Lingers," *Spartanburg Herald Tribune,* June 12, 1988, A19.

189 *The new ARPA office:* Koslov's professional biography, including the dates of his ARPA position, are included in U.S. Senate, *Joint Hearing Before the Subcommittees on Military Construction of the Committee on Appropriations and the Committee on Armed Services, Military Construction Appropriations for 1974* (Washington, D.C.: Government Printing Office, 1973).

189 *"He did his best":* Frosch, correspondence with author.

190 *"a master at maneuvers":* John L. Sloop, *Liquid Hydrogen as a Propulsion Fuel, 1945–1959,* Scientific and Technical Information Office, National Aeronautics and Space Administration, 1978.

190 *Colleagues at ARPA:* George Lawrence, interview with author.

190 *"Tell your nominal boss":* Lukasik, correspondence with author.

190 *"We were used as a cutout":* Kresa, interview with author.

191 *"to use for some of the space work":* Huff, interview with author.

191 *The only public information:* Robert Cooksey and Des Ball, "Pine Gap's Two Vital Functions," *Age,* July 2, 1969, 6.

191 *"When I visited a foreign country":* Lukasik, interview with author.

191 *"is to initiate a selective portion":* R. Cesaro, "Memorandum for the director, defense research and engineering, Subject: Project BIZARRE," Sept 26, 1967. This memo, in addition to *Operational Procedures for Project Pandora Test Facility,* was released as part of a Freedom of Information Act request (80-FOI-2208), filed by Michael Drosnin.

191 *Thus was born ARPA Program Plan 562:* Ibid.

192 *One theory that officials floated:* Lukasik, interview with author.

192 *In fact, ARPA-sponsored translations:* Lawrence, interview with author.

192 *MKULTRA:* According to the Church Committee report, the original MKULTRA charter did include radiation, though it is unclear if the agency ever did experiment with microwaves and behavior. U.S. Senate, *Select Committee to Study Governmental Operations with Respect to Intelligence Activities* (Washington, D.C.: Government Printing Office, 1976), bk. 1, 390.

193 *"much as embassy employees":* Jack Anderson, "The Strange Secret of 'Operation Pandora,'" *Florence Times Daily,* May 10, 1972, 4.

193 *The radiation results were so convincing:* In fact, the test results appeared much more ambiguous than Cesaro made out. As declassified documents would later show, the monkey Cesaro referenced worked ten hours a day, seven days a week. When its performance degraded on the twelfth day, Cesaro concluded that this was the result of radiation (rather than exhaustion). After the monkey stopped working, the radiation continued for two more days and then was turned off. After another three days without radiation, the monkey returned to normal work. Again, Cesaro said this result was from terminating the radiation (rather than the five days of rest, irrespective of the radiation). Cesaro noted that after the monkey resumed its normal tasks, it worked for five days without radiation, at which point the radiation resumed for an additional eight days, and then the monkey's performance again began to slip. Cesaro also attributed this work stoppage to the radiation. Simple math shows that Cesaro was making logical leaps of faith. In the first run of experiments, when the monkey was irradiated continuously, its performance degraded on the twelfth and thirteenth days. On the second run, the monkey worked for five days without radiation, and eight days with radiation, and on the thirteenth day, its performance slipped. The results, in other words, were roughly the same regardless of the radiation; the monkey's performance got worse around the twelfth or thirteenth day. The description of these tests can be found in U.S. Senate, *Radiation Health and Safety.*

193 *"potential weapon applications":* R. Cesaro, "Memorandum for the director, defense research and engineering, Subject: Project BIZARRE."

194 *"The extremely sensitive nature":* Ibid.

194 *In minutes from a May 12, 1969, meeting:* Institute for Defense Analyses, Minutes of

Pandora Meeting of May 12, 1969, in *Operational Procedures for Project Pandora Test Facility.*

194 *"gonadal protection be provided":* U.S. Senate, *Radiation Health and Safety,* 1199.

195 *One experimental protocol, called Big Boy:* Ibid.

195 *"The answer to that was no":* McIlwain, interview with author.

195 *"If there is an effect":* U.S. Senate, *Radiation Health and Safety,* 1189.

195 *"He was all over the place":* Lukasik, correspondence with author.

196 *"In brief, I am forced to conclude":* Koslov to Lukasik, "Review of Project Pandora Experiments," Nov. 4, 1969.

197 *"You can't just hit a target": Hearings on Research, Development, Test, and Evaluation Program for Fiscal Year 1971 Before Sub-committee of Committee on Armed Services,* House of Representatives, March 25, 1970 (Washington, D.C.: Government Printing Office, 1970), 8500.

198 *nuclear depth charge:* Peter Papadakos, "QH-50 Evolution," www.gyrodynehelicopters.com. That never happened, however, because the QH-50, developed at the behest of Admiral Arleigh Burke, the chief of naval operations and a rare drone enthusiast, retired. With him gone, enthusiasm for the QH-50 dissipated.

198 *initiated two drone projects:* Nite Panther was loaded with sensors, such as day and night television cameras, radar to track moving targets, and a laser designator. Later, ARPA even experimented with tethered balloons, also developed by the Advanced Sensors Office, to relay video from the QH-50 and extend its range. In all, ARPA, either on its own or in conjunction with the military services, conducted nine different configurations of the QH-50 under names like Desjez (short for "Destroyer Jezebel") and Blow Low, which used a classified electro-optical sensor. Michael J. Hirschber, "To Boldly Go Where No Unmanned Aircraft Has Gone Before: A Half-Century of DARPA's Contributions to Unmanned Aircraft," *48th AIAA Aerospace Sciences Meeting Including the New Horizons Forum and Aerospace Exposition, 4–7 January 2010, Orlando, Florida* (American Institute of Aeronautics and Astronautics, 2010).

198 *Egyptian Goose:* Reed, Van Atta, and Deitchman, *DARPA's Technical Accomplishments,* 1:17-1-2.

198 *a "checkered" history:* Huff and Sharp, *Advanced Research Projects Agency,* VIII-53.

199 *"unsuccessful":* Tietzel, "Summary of ARPA-ASO, TTO Aerial Platform Programs."

199 *While some of ARPA's QH-50s:* The exact fate of Nite Gazelle is unclear, but interviews with Stephen Lukasik and Peter Papadakos, among others, confirm that it crashed.

199 *Stephen Lukasik says he believed:* Lukasik, interview with author.

199 *Though Nite Gazelle was never used:* Tietzel, "Summary of ARPA-ASO, TTO Aerial Platform Programs."

199 *Deitchman recalled clashing:* Deitchman, interview with author.

199 *Though the six prototypes:* The first was called Camp Sentinel II, and the five additional radar were Camp Sentinel III. Reed, Van Atta, and Deitchman, *DARPA Technical Accomplishments,* 1:15-2-5.

199 *Fred Wikner, a physicist who served:* Wikner, interview with author. In fact, there were some half a dozen of the radar sent to Vietnam, so Wikner's comment, though pertinent, should be taken with a grain of salt.

200 *Jacobson achieved infamy:* Marlene Cimons, "Infertility Doctor Is Found Guilty of Fraud, Perjury," *Los Angeles Times,* March 5, 1992. One might argue these later misdeeds had no connection to his earlier State Department work, but the secrecy

surrounding the radiation investigation, and its lack of peer review, invited sloppy results. At the very least, Dr. Jacobson's later work calls into question his judgment.

200 *"I look at it as still a major":* Associated Press, "Tracking the Microwave Bombardment."

200 *Project Pandora was often cited:* In his book *Currents of Death* (New York: Simon & Schuster, 1989), the *New Yorker* writer Paul Brodeur describes as "mysterious" McIlwain's determination that the Moscow Signal showed no effect. There was nothing particularly mysterious in McIlwain's view; he just crunched the numbers.

200 *"The lesson learned":* Associated Press, "Tracking the Microwave Bombardment."

200 *"showy stunt":* Lukasik, correspondence with author.

201 *"The bird tumbled out of the sky":* Papadakos, interview with author.

201 *"I argued very hard":* Herzfeld, interview with author.

CHAPTER 12: BURY IT

202 *Cyrus Vance:* Huff and Sharp, *Advanced Research Projects Agency,* VI-7.

202 *"Would it be desirable":* U.S. House, *Hearings Before a Subcommittee of the Committee on Appropriations, Department of Defense Appropriation for 1965,* 154.

203 *"Congress hated AGILE":* Lukasik, interview with author.

203 *To protect its bureaucracy:* Lukasik, e-mail correspondence with author.

203 *On March 23:* Huff and Sharp, *Advanced Research Projects Agency,* IX-23.

203 *On its own, the new name:* A commonly repeated myth was that the *D* in "DARPA" was added to signify the agency's move to defense applications. This is wrong; interviews with contemporaneous officials, ARPA's institutional history, and records from the time all indicate it was an administrative change.

203 *"This was not a minor point":* Lukasik says the transition to "ARPA" as an acronym did not occur until George Heilmeier became director in 1975. Lukasik, interview with author.

204 *The agency's budget:* Huff and Sharp, *Advanced Research Projects Agency,* I-1.

204 *"the flow of 'Presidential assignments'":* Lukasik, e-mail correspondence with author.

204 *"It did not serve as a useful":* Ibid.

204 *"mechanical elephant":* Thomas O'Toole, "'Walking' Truck Is Drafted by U.S. Army," *New York Times,* March 30, 1966, 22.

204 *"damned fool":* Huff and Sharp, *Advanced Research Projects Agency,* VI-42.

205 *"open the door":* Bell Aerosystems, *Individual Mobility System,* undated pamphlet, Jet Belt, RG 330, National Archives, College Park.

205 *ARPA eventually stopped funding:* Steve Lehto, *The Great American Jet Pack: The Quest for the Ultimate Individual Lift Device* (Chicago: Chicago Review, 2013), 91.

205 *"The agency got rid of it":* James R. Chiles, "Air America's Black Helicopter," *Air & Space Magazine,* March 2008.

206 *"general dishonesty":* Lukasik, e-mail correspondence with author.

206 *"AGILE, counterinsurgency":* Lukasik, interview with author.

206 *In Washington, the key to killing:* In 2014, the Pentagon announced it was changing the name of the Joint Improvised Explosive Device Defeat Organization, the bomb-fighting agency that was created following the invasion of Iraq. This is a classic example of a first step in eliminating the organization.

206 *"Look, the Nixon Doctrine":* Lukasik, interview with author.

206 *"bury it someplace":* Lukasik, interview with Williams/Gerard.

206 *"We shifted from little guys":* Lukasik, interview with author.

207 *On the day after Christmas:* Lukasik to Secretary of Defense, memo, "Taking Stock," Dec. 26, 1972, Gerald R. Ford Library.

207 *ARPA was asked to step in:* ARPA Research in Iran, April 26, 1970, Project AGILE, RG 330, National Archives, College Park.

207 *"Iran's military leaders neither assign":* October 1969 Summary for Ambassador MacArthur, Project AGILE, RG 330, National Archives, College Park.

207 *"disappointing":* Delavan P. Evans (director) to Chief, ARMISH/MAAG, memo, "Subject: Continuation of ARPA Effort with CREC," March 16, 1970, RG 330, National Archives, College Park.

207 *That work pleased the shah:* Abbas Milani, *The Shah* (New York: Palgrave Macmillan, 2012), 36.

207 *In 1970, ARPA proposed a new approach:* "Technical Program Plan Military Systems Analysis," rcv. Aug. 26, 1970.

208 *"The reason is simple":* Large to Tachmindji, June 10, 1972, Project AGILE, RG 330, National Archives, College Park.

208 *"pompous":* Ibid.

208 *"near-religious faith":* Tim Weiner, "Robert Komer, 78, Figure in Vietnam, Dies," *New York Times,* April 12, 2000.

208 *"Toufanian was the top clerk":* Precht, interview with author.

208 *Not only did the shah:* "The Iranian Deals" (pt. 3 of the BAE Files), *Guardian,* June 8, 2007.

208 *Cordesman wanted to study:* Harold C. Kinne, Memorandum for the Record, Aug. 19, 1972, Project AGILE, RG 59, National Archives, College Park.

209 *"mask the analysis work":* John H. Rouse Jr. to Douglas Heck (minister-counselor, American embassy, Iran), Dec. 19, 1972, RG 59, National Archives, College Park.

209 *What other analysis was being masked:* Cordesman, phone conversation with the author. He declined to comment even after being sent declassified correspondence from the archives regarding his work.

209 *When the navy demonstrated:* Currie, interview with author.

209 *Iran became the only foreign country:* Dario Leone, "Thirty Minutes to Choose Your Fighter Jet: How the Shah of Iran Chose the F-14 Tomcat over the F-15 Eagle," *Aviationist,* Feb. 11, 2013.

210 *Fearful of the hijackers' finding:* Patricia Sullivan, "Robert Schwartz; Defense Official Was Hostage in Hijacking," *Washington Post,* June 18, 2007.

210 *Both men were eventually released:* Stark, interview with author.

210 *"This would be unfortunate":* Henry Precht, Memorandum, "Subject: ARPA," March 7, 1974, RG 59, National Archives, College Park.

210 *"there are limits":* J. G. Dunleavy, J. H. Ott, S. Goddard, and R. D. Minckler, "A Status Report on Equipment and Devices for Disposal of Improvised Explosive Devices in Urban Environments," sponsored by the Overseas Defense Research Office, Advanced Research Projects Agency, Sept. 1971, Project AGILE, RG 330, National Archives, College Park.

211 *After spending nearly $20 billion:* Sandra Erwin, "Technology Falls Short in the War Against IEDs," *National Defense,* Oct. 20, 2010.

211 *"was a neat idea":* Lukasik, interview with author.

211 *Lukasik took what was left.* There was a precedent for the Tactical Technology Office. When missile defense was taken away from DARPA just a few years prior, the pre-

vious director had created the Strategic Technology Office, which the ARPA history says worked on "truly exotic weapons concepts," like lasers and particle beams. At just under $70 million, the Strategic Technology Office was still, like Defender before it, the largest part of ARPA's budget at the time.

212 *"I took a really bad dish":* Lukasik, interview with author.

212 *It was not until 1972:* In his memoir, *A Life at Full Speed,* Charles Herzfeld says that ARPA's Advanced Sensors Office was involved in developing the laser-guided bombs that took down the bridge. However, there are no other ARPA officials, or air force officials, who can confirm this, nor is there any documentation to back up this claim. Without a doubt, ARPA explored laser-guided bombs during the Vietnam War, but it does not appear, based on available sources, that it was directly involved in the bombs used in this operation.

213 *By the time Wikner arrived:* Wikner says the number was more than 50 percent, though this is hard to confirm because the casualty statistics are typically reported by division. However, there were certainly divisions that counted mines and booby traps as contributing well over 50 percent of casualties.

213 *"That's the greatest accomplishment":* Wikner, interview with author.

213 *Office of Net Technical Assessment:* The office was eventually eliminated and its mission subsumed by the Office of Net Assessment, headed by Andrew Marshall, which was established in 1973.

213 *Soviet Union had increased:* Lewis Sorley, ed., *Press On! Selected Works of General Donn A. Starry* (Fort Leavenworth, Kans.: Combat Studies Institute Press, 2009), 1:23.

214 *"no more than speed bumps":* Donn Starry, "Opening Remarks," in *China's Revolution in Doctrinal Affairs: Emerging Trends in the Operational Art of the Chinese Liberation Army,* ed. James Mulvenon and David Finkelstein (Alexandria, Va.: CNA, 2005), 374.

214 *"Nuke 'em till they glow":* Wikner, interview with author.

214 *"What we want":* Ibid.

214 *"Steve, we have to do something":* Ibid.

215 *"If we had called this Project Smart Kill":* Lukasik, interview with author.

215 *"wrote as colorlessly":* Ghamari-Tabrizi, *Worlds of Herman Kahn,* 354.

215 *An unusual figure:* For a biographical portrait of Wohlstetter, see Abella, *Soldiers of Reason.*

216 *"To deter an attack":* Albert Wohlstetter, "The Delicate Balance of Terror," *Foreign Affairs,* Jan. 1959.

216 *The LRRDPP was approaching:* In addition to Wohlstetter, the study included Joseph Braddock and Don Hicks, two physicists who were part of a group of scientists who frequently advised the Pentagon on nuclear policy.

217 *"Look, we have something to accomplish":* Lukasik, interview with author.

217 *"near zero miss":* Defense Advanced Research Projects Agency, Defense Nuclear Agency, *Summary Report of the Long Range Research and Development Planning Program,* DNA-75-03055, Feb. 7, 1975. Declassified Dec. 31, 1983.

217 *Once there, it would release:* Van Atta and Lippitz, *Transformation and Transition, Vol. II,* IV-15.

217 *Assault Breaker:* Even when Assault Breaker was in full swing, the army was not necessarily ready to give up its tactical nukes. In 1983, it emerged that the army was looking at putting tactical nuclear weapons on the Assault Breaker missiles, essentially undermining the very purpose of the system, which was to lessen dependence

on nuclear weapons. "You can't tie your hands behind your back," *The Washington Post* quoted one army official as saying.

218 *"ARPA gets 60 to 70 percent":* Wikner, interview with author.

218 *"It all came from counterinsurgency":* Lukasik, interview with author.

CHAPTER 13: THE BUNNY, THE WITCH, AND THE WAR ROOM

219 *Vietcong flag:* Marshall Kilduff, "SDS to Hold Rally," *Stanford Daily,* Jan. 29, 1969, 1.

219 *That month, Stanford's board:* For a full account of the rift, see C. Stewart Gillmor, *Fred Terman at Stanford: Building a Discipline, a University, and Silicon Valley* (Stanford, Calif.: Stanford University Press, 2004).

220 *At 10:30 p.m.:* Leonard Kleinrock, "The Day the Infant Internet Uttered Its First Words," www.lk.cs.ucla.edu.

220 *Stephen Lukasik, like Charles Herzfeld:* One of the best explanations for the complex motives underlying the ARPANET's creation can be found in Stephen Lukasik's article "Why ARPANET Was Built," *IEEE Annals of the History of Computing,* July–Sept. 2011, 4–21.

220 *"ARPA's computing program continued":* Waldrop, *Dream Machine,* 278.

221 *"kill die factor":* University of Illinois Archives (RS41/66/969) from *A Byte of History: Computing at the University of Illinois* exhibit, March 1997.

221 *Concerned about the ability:* Hafner and Lyon, *Where Wizards Stay Up Late,* 228–29.

221 *"What exotic studies":* U.S. House, *Department of Defense Appropriations for 1972,* 323.

221 *"Al, I understand what this is about":* Lukasik, interview with author.

222 *Believing the work:* Ibid.

222 *"Friends and enemies alike":* Tim Weiner, "Sidney Gottlieb, 80, Dies; Took LSD to C.I.A.," *New York Times,* March 10, 1999.

223 *Rockefeller Commission:* Seymour Hersh, "Family Plans to Sue C.I.A. over Suicide in Drug Test," *New York Times,* July 10, 1975.

223 *Church Committee:* U.S. Cong., Select Committee on Intelligence, *Project MKULTRA, the CIA's Program of Research in Behavioral Modification: Joint Hearing Before the Select Committee on Intelligence and the Subcommittee on Health and Scientific Research of the Committee on Human Resources,* U.S. Senate, 95th Cong., 1st sess., Aug. 3, 1977 (Washington, D.C.: Government Printing Office, 1977).

223 *"They are, and I'll say this":* Lukasik, interview with author.

223 *"Steve, let me show you":* Ibid.

224 *"Major impetus behind the Soviet drive":* Sheila Ostrander and Lynn Schroeder, *Psychic Discoveries Behind the Iron Curtain* (Englewood Cliffs, N.J.: Prentice-Hall, 1970), 7.

225 *"quiet, low-profile classified investigation":* Hal Puthoff, correspondence with author.

225 *"I thought this was a lot":* Lukasik, interview with author.

225 *"Everyone pretty much felt":* Young, interview with author.

226 *Just as he was packing his bags:* Lawrence, interview with author. Despite the seeming overlap in subject and time, Lawrence said he had no knowledge of Pandora when it was going on, even though his work at Walter Reed and then ARPA coincided with the dates of the program and he later became involved with some of the researchers, like Ross Adey, who were sponsored by Pandora. This claim is believable, however, because Pandora was top secret.

226 *"to such places"*: "Request and Travel Authorization for TDY Travel of DoD Personnel," July 3, 1975, George H. Lawrence personal collection.

227 *"The image of electronic equipment"*: Donald Moss, ed., *Humanistic and Transpersonal Psychology: A Historical and Biographical Sourcebook* (Westport, Conn.: Greenwood, 1998), xix.

227 *Among the people he recruited:* Fields to Lawrence, March 21, 1970, Lawrence personal collection.

228 *"Biofeedback offers much less powerful":* These were the conclusions made by Lawrence, based on the ARPA work and presented at the NATO conference Dimensions and Stress, June 29–July 3, 1975. George Lawrence, "Use of Biofeedback for Performance Enhancement in Stress Environments," in *Stress and Anxiety,* vol. 3, ed. Irwin G. Sarason and Charles D. Spielberger (New York: Hemisphere/Wiley), 1976.

228 *On the flip side:* Ibid.

228 *A cover of* Time *magazine:* "Boom Times on the Psychic Frontier," *Time,* March 4, 1974.

229 *He liked the witches:* Lawrence, interview with author.

229 *According to Puthoff:* Puthoff, correspondence with author.

229 *"At the time we were concerned":* Ibid.

230 *"He and Hyman and I made this trip":* Lawrence, interview with author.

230 *Lawrence was drinking heavily:* Van de Castle, interview with author.

230 *"Okay, show me a fucking miracle":* Ibid.

230 *The Israeli performer dramatically covered:* Many of the details of the trip are recounted in Hyman to Charles Anderson (president of SRI), April 5, 1973, Lawrence personal collection.

231 *Geller soon emerged triumphant:* "The Magician and the Think Tank," *Time,* March 12, 1973.

231 *"Targ and Puthoff, from the way":* Hyman to Charles Anderson, Lawrence personal collection.

231 *Puthoff countered that the experiments:* Puthoff, correspondence with author.

231 *Lawrence, after stomping his foot:* The visit is also recounted in John L. Wilhelm, *Search for Superman* (New York: Pocket Books, 1976).

232 *"Colonel Mitchell, what do you think":* Lawrence, interview with author.

233 *"the intelligent yarmulke":* Lukasik, interview with author.

233 *"Can these observable electrical brain signals":* Jacques Vidal, "Toward Direct Brain-Computer Communication," *Annual Review of Biophysics and Bioengineering* 2 (June 1973): 157–80.

233 *Within a few years:* John Hebert, "Man/Machine Interface Utilizes Human Brain Waves," *Computerworld,* June 28, 1976, S/2.

233–34 *"Soon, for example, a computer monitoring":* George Lawrence, "Biocybernetics: Program Plan," Lawrence personal collection.

234 *"The discipline of biocybernetics":* George Lawrence, ARPA, Dec. 1975 program summary, Lawrence personal collection.

234 *"They certainly haven't been":* Lawrence, interview with author.

235 *Geller's advocates, who believed:* The eventual project that emerged from the SRI work ended up with army intelligence and the Defense Intelligence Agency, rather than the CIA.

235 *"At the very least":* Lawrence, interview with author.

235 *"wandering off the territory":* Currie, interview with author.

CHAPTER 14: INVISIBLE WAR

238–39 *"Oh my God, he's dead":* Brown, interview with author.

240 *They named it Paradise Ranch:* Johnson, *Kelly,* 122–23.

240 *"He was on a scaffold":* Allen Atkins, interview with author.

240 *"Oh, that was his gas mask"*: Ibid. Alan Brown recalls the events in a similar way, though he said he was not privy to the official cover story or what precisely was told to the hospital staff. He agrees that staff did not believe what they were told. Brown, interview and personal correspondence with author.

240 *"General Delivery, Las Vegas":* UPI, "Crash of Plane Admitted, but Other Details Lacking," *Eugene Register-Guard,* May 12, 1978, 16C. Press reports at the time erroneously reported the pilot's last name as Parks.

241 *The Pentagon a week later:* UPI, "Pentagon Plays Down Plane Crash," *Milwaukee Sentinel,* May 13, 1978, 2.

241 *he would reveal years later:* UPI, "Chief Stealth Fighter Pilot Tells of Initial Test Flights," *Lodi News-Sentinel,* Sept. 30, 1989, 5.

241 *"perform missions along":* The TR-1 and the high-altitude aircraft were likely the official cover story. UPI, "Did a Secret U.S. Spy Plane Crash?," *Montreal Gazette,* May 13, 1978, 12.

241 Flight International *reported:* "Stealth Airplane Lost in Nevada," *Flight International,* May 27, 1978, 1591.

241 *"to disestablish DARPA":* Tegnelia, interview with author.

241 *The stealth aircraft's journey:* Robert Moore and Chuck Myers, correspondence with author. A minor discrepancy in their recollections is over whether the meeting took place at DARPA or in the Pentagon.

242 *"Walk Stealthily":* Myers, e-mail correspondence with author.

242 *Israel reportedly lost:* The exact number—as well as the cause of the losses—is still disputed. But the principal reason for those losses is not typically contested. Simon Dunstan, *The Yom Kippur War: The Arab-Israeli War of 1973* (Oxford: Osprey, 2007), 30.

242 *"Signature is the most important":* Myers, correspondence with author.

243 *He handed Moore:* Moore and Myers, correspondence with author.

243 *"drastic operational techniques":* Moore, interview with author.

243 *The mini-drone was not invisible:* Atkins, interview with author.

244 *"real world":* Ibid.

244 *Moore was intrigued:* Based on author interviews with Moore and Atkins and correspondence with Myers.

244 *"the high stealth aircraft":* Moore says he adopted the term "stealth" after chats with a naval officer on DARPA's staff. Myers had also used the word "stealth" in some of his own papers on Harvey. Moore, interview with author.

244 *five companies:* Aronstein and Piccirillo, *Have Blue and the F-117A,* 26.

246 *"clicked":* Currie, interview with author. Lukasik, on hearing that he was about to be replaced, resigned at the beginning of 1975, and Heilmeier came in a few months later. Lukasik, interview with author.

246 *"He wanted more technology":* Heilmeier, interview with author.

246 *DARPA orders:* "ARPA orders are not a good source of information—they are written by smart people to fool dumb people, meaning all the bureaucrats who have to sign off, or just not raise a question," Lukasik insisted. "They are blatant sales pitches." Lukasik, interview with author.

246 *Heilmeier, on the other hand:* Heilmeier, interview with author.

246 *"That's pure bullshit":* Heilmeier, interview with Aspray, Charles Babbage Institute.

246 *Licklider left DARPA:* "I think that we are at a watershed in the history of ARPA-IPTO," Licklider wrote, recapping his conversations with Heilmeier, shortly after the new director came over. J. C. R. Licklider to Allen Newell, e-mail, "Subject: Request for Advice," April 1, 1975, Carnegie Mellon University Libraries Digital Collections.

247 *"First what are you trying to do":* Heilmeier, interview with author. The Heilmeier catechism has several different wordings, though it is always constructed from some version of these seven questions. This version is verbatim from the author's interview with Heilmeier.

247 *"Why don't you try":* Heilmeier, interview with Williams/Gerard.

247 *"He went through all my programs":* Moore, correspondence with author.

248 *Moore's idea for the high stealth aircraft:* Moore, interview with author. The confusion over the stealth concept persisted even in the name. Though some accounts reference that initial study as "Harvey," the name does not appear to have ever been formally used by DARPA. Myers believed the DARPA study was going to explore Harvey, while DARPA from the outset was going in a different direction.

248 *"We needed to penetrate":* Currie, interview with author.

248 *Daniel reported back to Ben Rich:* Stevenson, *$5 Billion Misunderstanding,* 19.

249 *"Ben, we are getting the shaft":* Rich and Janos, *Skunk Works,* 23. There are some factual errors in Rich's chronology of the early days of the stealth competition, so it is difficult to know who first alerted him to the stealth competition. The DARPA study program at that point was not classified, so it is likely word was getting out from multiple sources.

249 *"This was exactly the kind of project":* Ibid.

249 *Heilmeier warned Rich:* Ibid., 24.

250 *Perko agreed to the offer:* Atkins, interview with author. Rich, in his book, *Skunk Works,* says the $1 offer came from George Heilmeier, and he actually turned it down. There is likely some truth in all the accounts. Heilmeier would surely not have offered Lockheed a contract without consulting Perko, who was in charge of the program. Likewise, Perko would not have allowed Lockheed's unorthodox entry into the program without Heilmeier's approval.

250 *"I will support your lightweight fighter":* Currie, interview with author. The offer to Jones was something of a bluff. Currie said he would have supported the F-16 regardless, because it was the defense secretary's favored project.

251 *"The air force ought to support":* This conversation is reconstructed from interviews and correspondence with Heilmeier, Moore, Currie, and Myers. The "motherhood" quotation is taken from Myers's account published in Stevenson, *$5 Billion Misunderstanding,* 21.

251 *"You have to admire a man":* Heilmeier, interview with author.

251 *"The difference for the Soviet guys":* Brown, interview with author.

252 *As Kelly Johnson:* Quoted in Aronstein and Piccirillo, *Have Blue and the F-117A,* 9.

253 *"It wasn't like the Russians":* Brown, interview with author. The debate over how much Ufimtsev contributed to stealth continues today. Overholser has said that Ufimtsev's theories were incorporated into the Echo computer program after the initial Have Blue design was completed. Aronstein and Piccirillo, *Have Blue and the F-117A,* 72. But it is clear that Ufimtsev's theory was at least informing Overholser's thinking as early as 1974 or 1975. Also see Stevenson, *$5 Billion Misunderstanding,* 17.

253 *"Well, that's stupid":* Brown, interview with author.

253 *"Grown men cried that day":* Stevenson, *$5 Billion Misunderstanding,* 22.
254 *As the aircraft lifted off:* Heilmeier, interview with author.
254 *"It's an empty bottle":* Ibid.
255 *"No self-respecting":* Brown, interview with author.
255 *"I'm still convinced":* Myers, interview with author.
255 *"I am announcing today":* Secretary of Defense Harold Brown, "Statement on Stealth Technology," Defense Department News Conference, Washington, D.C., Aug. 22, 1980.
256 *"can't afford to fail":* Tegnelia, interview with author.

CHAPTER 15: TOP SECRET FLYING MACHINES

257 *"already in an arms race":* Ronald Reagan's Speech at the VFW Convention, Chicago, Aug. 18, 1980.
258 *"Sure, Dick, yeah":* Cooper, interview with Alex Roland.
258 *"give an enema":* Ibid.
259 *"Cap, I've heard that song":* Ibid.
259 *"The solution is well":* Ronald Reagan's televised address announcing the Strategic Defense Initiative on March 23, 1983.
259 *"What other President":* FitzGerald, *Way Out There in the Blue,* 15.
260 *Eighth Card:* Beason, *E-Bomb,* 97.
260 *The Eighth Card study:* Hans Mark, "The Airborne Laser from Theory to Reality: An Insider's Account," *Defense Horizons,* April 2002, 2.
260 *By the time Reagan announced:* The Teller plan was one of several pitches Reagan had received in the weeks and months leading up to his announcement. FitzGerald, *Way Out There in the Blue,* 206.
260 *Everybody including DeLauer and myself:* Cooper, interview with Roland.
260 *"Although we appreciate":* FitzGerald, *Way Out There in the Blue,* 142–43.
260 *"I spent the following ten days":* Cooper, interview with Roland.
260–61 *"When the President made his announcement":* Cooper, interview with Williams/Gerard.
261 *"He was not a happy camper":* Kahn, interview with author.
261 *Officials in DARPA's Directed Energy Office:* To explain the view of officials in the Directed Energy Office, Kahn cited the parable of the "chicken and the pig" discussing a breakfast of ham and eggs. "The chicken is involved, but the pig is committed," Kahn explained. "The Directed Energy Office was committed. They were going to be involved no matter what because it was their technology. They were the pig." Ibid.
261 *In the end, Cooper decided:* In all likelihood, the retreat to West Virginia was more of Cooper's attempt to gather information. "If office directors thought there was democracy in the agency, that was a result of lots of discussion but never management by vote," said Larry Lynn, Cooper's deputy and later a director of DARPA. Lynn, correspondence with author.
262 *"We were spending money":* Cooper, interview with Williams/Gerard.
263 *Cost estimates for Aquila:* General Accounting Office, *Aquila Remotely Piloted Vehicle: Recent Developments and Alternatives* (Washington, D.C.: General Accounting Office, 1986).
263 *Compared with the United States:* Figures from Israeli Air Force website: http://www.iaf.org.il/4968-33518-en/IAF.aspx.

263 *"Karem wasn't really pushing":* Atkins, interview with author.

263 *Teal Rain:* Van Atta and Lippitz, *Transformation and Transition,* vol. 2. DARPA officials still will not discuss some of the specific aircraft developed under Teal Rain, but two officials did indicate the goal was to eventually replace aircraft like the U-2 and the SR-71.

263 *Even thirty years later:* Like the manned aircraft programs, some unclassified projects were really covers for classified ones. For example, DARPA was openly funding development of a massive Boeing-built drone called Condor, a high-altitude, long-endurance unmanned aircraft whose two-hundred-foot wingspan matched that of a Boeing 747. But only one prototype was built, and the aircraft appears to have been a cover for a classified military mission, which was canceled.

264 *"Gentlemen, everything I see":* Richard Whittle, "The Man Who Invented the Predator," *Air & Space Magazine,* April 2013.

264 *As an initial step:* Ibid.

264 *When that design proved successful:* Chuck Heber, interview with author.

264 *"The navy was in it":* Atkins, interview with author.

264 *"changed the world":* Whittle, *Predator,* 310.

265 *"Looking at it from the side":* Atkins, interview with author.

265 *"Most people didn't know":* Ibid.

265 *"We built some full-scale models":* Ibid.

265 *The "black" program involved:* Ibid.

265 *Those shifts are difficult to mask:* According to Atkins, one lesson was the importance of using stiff blades, because slapping blades were a dead giveaway. Ibid.

266 *"If you could ever get beyond":* Ibid.

266 *"No, that's too ugly":* Ibid. A Russian version of a forward-swept wing, the Sukhoi S-37, similarly failed to make it beyond the prototype stage.

266 *"swallowing the agency's budget":* Tegnelia, interview with author.

266 *"I had one program that lost":* Atkins, interview with author.

267 *"It was Ambrose facing me":* Ibid.

267 *Ambrose was known as a fierce protector:* John Cushman Jr., "In Budget War, Some Fall Amid Din and Others Go in Silence," *New York Times,* Feb. 24, 1988.

267 *"There are over half a dozen":* Ibid. Atkins, interview with author.

268 *"Well, Jim, we're going to do the program":* Ibid.

268 *"This is something DARPA":* Tony duPont, interview with author.

269 *The route duPont took:* Heppenheimer, *Facing the Heat Barrier,* 215.

269 *Tether, a science fiction fan:* Ibid.

269 *Space planes were an ambition:* Back in 1958, the Maneuverable Recoverable Space Vehicle, or MRS-V, never got beyond the concept stage before DARPA lost its space programs. An air force effort from the same time period, called the X-20 Dyna-Soar, was also eventually canceled.

269 *"not unlike keeping a candle lit":* David Schneider, "A Burning Question," *American Scientist,* Nov.–Dec. 2002.

270 *"Can you do this mission?":* Tony duPont, interview with author.

270 *"I'll study it":* Ibid.

270 *That was the beginning of Copper Canyon:* Ibid.

271 *The secret mission would require two pilots:* Ibid.

271 *"globe-girdling reconnaissance system":* U.S. House, *Department of Defense Authorization of Appropriations for Fiscal Year 1986,* 661.

271 *"Over the past year":* Ibid.

271 *"Let's do it":* Heppenheimer, *Facing the Heat Barrier,* 217.
271 *"Interesting":* Ibid., 218.
272 *"You can't say that":* Colladay, interview with Williams/Gerard.
272 *"going forward with research":* Ronald Reagan, State of the Union address, Feb. 4, 1986.
273 *"It's better to have them pissing":* Carol duPont, interview with author.
273 *"There's where Bob and I":* Atkins, interview with author.
273 *Williams did almost the exact opposite:* Tony duPont, interview with author. This version of events is also backed up by several histories of the National Aerospace Plane.
273 *In the meantime, the costs:* U.S. General Accounting Office, *National Aero-space Plane: Restructuring Future Research and Development Efforts* (Washington, D.C.: USGAO, 1992).
273 *"If we stuck with":* Tony duPont, interview with author.
274 *In fact, it could not maneuver:* Heppenheimer, *Facing the Heat Barrier,* 219.
274 *Furious, Robert Duncan:* Larry Schweikart, "The Quest for the Orbital Jet: The National Aero-Space Plane Program (1983–1995)," *The Hypersonic Revolution: Case Studies in the History of Hypersonic Technology, Vol. 3* (Bolling Air Force Base, Washington, D.C.: Air Force History and Museums Program, 1998), 155.
274 *"When I saw that happen":* Cooper, interview with Alex Roland.
274 *Almost $2 billion was spent:* U.S. General Accounting Office, *National Aero-space Plane: Restructuring Future Research and Development Efforts.*
274 *It would spend $30 billion:* James A. Abrahamson and Henry F. Cooper, "What Did We Get for Our $30-Billion Investment in SDI/BMD?" (Washington, D.C.: National Institute for Public Policy, 1999).
275 *American Cold War defense spending:* Office of the Assistant Secretary of Defense (Comptroller), National Defense Budget Estimates FY1986, March 1985.
275 *never even took flight:* Atkins says the Space Shuttle disaster of January 28, 1986, which killed all seven crew members, spelled the end of DARPA's X-Wing stealth helicopter. He argues that in the aftermath of *Challenger,* the space agency did not want to embark on any risky flight tests, let alone a project that was really a cover for a secret military aircraft. Atkins, interview with author. Ray Colladay, on the other hand, says the aircraft's aerodynamics were flawed. Colladay, correspondence with the author.

CHAPTER 16: SYNTHETIC WAR

276 *In the mid-1980s:* Counting tanks was a Cold War obsession of military analysts, and those figures were contested. However, no one disputed that the Warsaw Pact had a quantitative advantage. Jack Mendelsohn and Thomas Halverson, "The Conventional Balance: A TKO for NATO?" *Bulletin of the Atomic Scientists* 45, no. 2 (1989): 31.
276 *DARPA sent four new simulators:* Thorpe, interview with author.
277 *"Significant breakthroughs":* Captain Jack Thorpe, "Future Views: Aircraft Training, 1980–2000," Air Force Office of Scientific Research (Sept. 15, 1978), Bolling Air Force Base, Washington, D.C.
277 *"Well, that's kind of a neat idea":* Thorpe, interview with author.
278 *That year, DARPA:* John Rhea, "Planet SIMNET," *Air Force Magazine,* Aug. 1989.

278 *That same year, the first networked:* U.S. Cong., Office of Technology Assessment, *Distributed Interactive Simulation of Combat* (Washington, D.C.: Government Printing Office, 1995).

279 *In July 1989, Craig Fields:* Fields's lengthy tenure was not without critics. "That is a crime," Allan Blue, a scientist who had headed DARPA's Information Processing Techniques Office in the 1970s, remarked on Fields's lengthy tenure. Blue, interview with William Aspray, Charles Babbage Institute.

279 *"Abrasive" was typically the second:* Michael Schrage, "Will Craig Fields Take Technology to the Marketplace via Washington?," *Washington Post,* Dec. 11, 1992, D3.

279 *"Moving out of the Pentagon":* Fields, interview with Williams/Gerard.

280 *"We trundled out":* Cooper, interview with Roland.

280 *Strategic Computing Initiative:* The head of the program was Robert Kahn, a cousin of the famed futurist Herman Kahn, who with Vint Cerf had co-developed the communication protocols that would become the underpinnings of the modern Internet. He returned to DARPA in 1979 to head up the Information Processing Techniques Office, hoping to revive Licklider's vision of basic research that could revolutionize computer science.

280 *"I came back":* Cooper, interview with Roland.

280 *The Pilot's Associate:* Roland, *Strategic Computing Initiative,* 283–84.

280 *"giving up its work":* Ray Colladay, the next DARPA director, had almost no interactions with the Strategic Computing Initiative, which was already petering out when he arrived. Yet he called it a "great success." It was unclear why. Andrew Pollack, "Pentagon Wanted a Smart Truck; What It Got Was Something Else," *New York Times,* May 30, 1989.

280 *"It's over in my mind":* Ibid. Not surprisingly, the Japanese threat that Cooper had hyped back in 1983 in order to get congressional support also turned out to be nothing more than a mirage.

281 *nearly $50 billion:* Figures from U.S. Census Bureau, trade in goods with Japan, 1989, www.census.gov.

281 *"There is a basic conflict":* James Fallows, "Containing Japan," *Atlantic Monthly,* May 1989.

281 *The agency was funding:* Sematech, which was funded by member dues and eventually about half a billion dollars from DARPA, was credited with helping save the U.S. chip-manufacturing base.

281 *"At a time when":* Andrew Pollack, "America's Answer to Japan's MITI," *New York Times,* March 5, 1989.

282 *Consumer electronics:* The critics, however, pointed out the numbers did not back him up. The electronics market was a small fraction of U.S. semiconductor output—about 5 percent by some reports—and HDTV was only 1 percent. Even using more optimistic projections, it was difficult to see how having the Pentagon subsidize the consumer market was really going to help the semiconductor industry. Marc Busch, *Trade Warriors: States, Firms, and Strategic-Trade Policy in High-Technology Competition* (Cambridge, U.K.: Cambridge University Press, 2001), 104.

282 *"had never run":* Colladay, interview with author.

282 *Gallium arsenide chips:* George Whitesides, "Gallium Arsenide: Key to Faster, Better Computing," *Scientist,* Oct. 31, 1988.

283 *"other transactions":* NASA had used "other transaction authority" in the past, but the Pentagon had not. Dunn, interview with author.

283 *But Fields wanted to make Gazelle:* Ibid.

283 *"She operated in a completely different manner":* Ibid.
283 *On April 9, 1990, DARPA issued a press release:* Department of Defense press release.
284 *"Gazelle is typical":* Andrew Pollack, "Pentagon Investment Made in Chip Company," *New York Times,* April 10, 1990, D1.
284 *"The Defense Department":* Ibid.
284 *"Bring me the agreement quick":* Dunn, interview with author.
284 *"was somewhat displeased":* Richard Dunn, "A History of the Defense Advanced Research Projects Agency," 2000, unpublished history. Courtesy of Richard Dunn.
284 *Pentagon officials initially claimed: Defense Daily,* April 26, 1990, 150.
285 *"Well, I think you should check":* Fields, interview with Williams/Gerard.
285 *Donald Hicks, who held the position:* John Ronald Fox, *Defense Acquisition Reform, 1960–2009: An Elusive Goal* (Washington, D.C.: Government Printing Office, 2012), 140.
285 *"Did you hear Craig got fired?":* Herzfeld, *Life at Full Speed,* 231.
286 *Eight AH-64 army Apache helicopters:* There are a number of good accounts of this mission. Richard Mackenzie, "Apache Attack," *Air Force Magazine,* Oct. 1991.
286 *"There is no way":* McBride, interview with author.
286 *"We were instantiating":* Ibid.
286 *The F-117's first bomb:* James P. Coyne, "A Strike by Stealth," *Air Force Magazine,* March 1992.
286 *"highway of death":* Paul Eng, "High-Tech Radar Plane for Gulf Ground War," *ABC News,* March 20, 2003.
286 *"one of the more unlikely heroes":* Peter Grier, "Joint STARS Does Its Stuff," *Air Force Magazine,* June 1991.
287 *"Look, we don't need":* Reis, interview with author.
287 *When DARPA was told:* McBride, interview with author.
287 *"He had sort of been put there":* Reis, interview with author.
288 *"It would be like a living history":* Thorpe, interview with author.
289 *"There were still the tread marks":* Ibid.
289 *"You would be able":* Ibid.
289 *Reis, the director, showed the video:* Reis, interview with author.
289 *"Gee, if we had this earlier":* Ibid.
289 *Thorpe's SIMNET:* The extent of SIMNET's contribution to online gaming is open to debate, because the technologies were being developed in parallel. For example, Rtime, a Seattle-based software company, patented its technology for distributed gaming based on its SIMNET work. Teresa Riordan, "Patents: A Dangerous Monopoly?," *New York Times,* Feb. 1, 1999.
289 *In technology circles: Wired* magazine featured SIMNET and Jack Thorpe in two articles, including its second issue of the magazine. Bruce Sterling, "War Is Virtual Hell," *Wired,* March/April 1993; Frank Hapgood, "SIMNET," *Wired,* April 1997.
290 *"SIMNET was irrelevant":* Macgregor says the victory at 73 Easting was the result of investments made in people, not technology. He specifically cited live-fire practice with real ammunition in Germany and seven weeks of intensive training in Saudi Arabia. Macgregor, correspondence with author.
290 *"Elegant effort":* Gorman, interview with author.
291 *"I don't know if we're":* McBride, interview with author.
291 *difficult to find:* "DARPA chief spells out new initiatives," *Defense Daily,* March 20, 1992, 475.
292 *DARPA contracted to have a simulation facility:* Neyland, *Virtual Combat,* 58–59.

See also Denise Okuda, quoted in "High Tech Comes to Life," *Science Friday,* NPR, Aug. 27, 2010.

292 *"You could figure everything":* Murphy, interview with author.

292 *Almost all mention of it:* "Warbreaker Tabbed for About $600 Million in FY '94–99," *Defense Daily,* Oct. 6, 1992, 28.

292 *"It's easy to simulate":* Cosby, interview with author.

293 *In 2000, the same year:* Harris, *Watchers,* 179.

293 *"asymmetric threat is physically small":* Tom Armour, "Asymmetric Threat Initiative" (presentation at DARPATech 2000, Sept. 8, 2000).

293 *The DARPA director that year:* Frank Fernandez, interview with author.

CHAPTER 17: VANILLA WORLD

294 *The vice president wanted:* Tether, interview with Williams/Gerard.

294 *In the summer of 2001:* Ibid.

294 *DarkStar, a stealthy drone:* It would take two years before it would fly again, and at that point DARPA was getting ready to turn it over to the air force. In 1999, the Pentagon canceled DarkStar. "It was not a successful design, let's put it that way," DARPA's former director Gary Denman said. Denman, interview with author.

294 *"DARPA had become a backwater":* Tether, interview with Williams/Gerard.

295 *"was a person who liked":* Ibid.

295 *The headlining act:* Goldblatt, interview with author.

295 *At DARPA, however, Goldblatt:* Ibid. Goldblatt's interest in the subject was personal. His daughter suffered from cerebral palsy, caused by damage to the developing brain, and he was frustrated after a friend at Harvard Medical School told him it would never be cured.

295 *Goldblatt was inspired by:* DARPA's brain-computer interface might have thought it was inspired by science fiction like *Firefox,* but more likely *Firefox* was actually inspired by the DARPA program of the 1970s (the book was published in 1977, not long after reports of the Pentagon's secret stealth aircraft were leaking to the press and when George Lawrence's biocybernetics program and the concept of "thought controlled weapons" were making the rounds).

295 *The electrodes would read:* See Nicolelis, *Beyond Boundaries.* Goldblatt's office was funding Miguel Nicolelis, a neurobiologist at Duke University, who implanted electrodes in a video-game-loving monkey named Aurora. In 2003, Nicolelis's team announced it had succeeded in getting Aurora to move a robot arm just by thinking about moving it, as if it were her own appendage (and rewarding her with juice). Aurora was taught to use a joystick that manipulated a cursor on a computer screen, as well as a robot arm in another room. After a while, the joystick was removed, but Aurora, who had come to associate the joystick with moving the cursor, continued to manipulate the cursor—and the robot arm—just by thinking about the movement.

295 *Goldblatt's other ongoing research:* Noah Shachtman, "Be More Than You Can Be," *Wired,* March 2007.

296 *"It was fantastic":* Ibid. Tether, interview with Williams/Gerard.

296 *Even as threat reports on al-Qaeda surged: 9/11 Commission Report,* 266–73.

296 *"to keep someone from taking a plane":* Ibid., 396.

297 *"Hey, looks like another plane":* Tether, interview with Williams/Gerard.

298 *starting to watch television coverage:* Based on author's personal observation inside the Pentagon on September 11, 2001.

298 *DARPA had already begun exploring:* John Poindexter, interview with author. Some details are also drawn from Harris, *Watchers,* 93.

298 *The "laboratory" was all smoke*: Poindexter, interview with author.

299 *Within months of 9/11:* Ibid. Poindexter's recollection was that the office's budget for 2002 was $75 million, and $150 million in 2003.

299 *"We really were armor-proofed":* Tether, interview with Williams/Gerard.

300 *He then pursued a doctorate:* Poindexter, interview with author.

300 *Poindexter also brought PROFS:* Ibid.

300 *Poindexter was later indicted:* David Johnston, "Poindexter Is Found Guilty of All 5 Criminal Charges for Iran-Contra Cover-Up," *New York Times,* April 8, 1990.

300 *It was an interest dating back:* Harris, *Watchers,* 20–24.

301 *Soon, Poindexter was working:* Scientific Engineering Technical Assistance workers, whose numbers had exploded in the Pentagon in recent years, provide everything from low-level administrative assistance to high-level technical advice, as was the case for Poindexter.

301 *"The idea with the technology":* Poindexter, interview with author.

301 *"where we would run exercises":* Ibid.

301 *"Admittedly, we were looking":* Ibid.

302 *"A Manhattan Project for Combating Terrorism":* Ibid.

302 *Poindexter's idea was to create:* In an interview with the author, Poindexter insisted that what he was proposing was pattern analysis, not data mining, which has become a pejorative term. Nonetheless, contemporaneous briefing materials from this period, including Poindexter's, refer to the DARPA-sponsored work as data mining, so it appears appropriate to use in this case.

302 *"in a compound with barbed wire":* Poindexter, interview with author.

302 *Manhattan Project Terrorism:* From Poindexter's slide presentation, "A Manhattan Project for Combating Terrorism," Oct. 2001. A copy of the presentation was provided to the author by Poindexter.

302 *Sharkey was earning good money:* Poindexter, interview with author. Sharkey's reluctance to reenter DARPA is also described in O'Harrow, *No Place to Hide,* 183.

303 *"never really was convicted":* Belfiore, *Department of Mad Scientists,* 192.

303 *At first, Tether's belief:* John Markoff, "Chief Takes Over New Agency to Thwart Attacks on U.S.," *New York Times,* Feb. 13, 2002.

304 *Even though the idea:* Poindexter, interview with author.

304 *Poindexter had the Latin phrase:* Harris, *Watchers,* 197.

304 *"I know six good ways":* Lukasik, interview with author.

304 *Vanilla World:* Poindexter and Lukasik, interviews with author.

305 *"What this technology means":* Quoted in *Surveillance Technology: Joint Hearings Before the Subcommittee on Constitutional Rights of the Committee on the Judiciary and the Special Subcommittee on Science, Technology, and Commerce of the Committee on Commerce,* 5.

305 *"I don't want to affect":* Eric Horvitz, interview with author.

305 *Unlike the JASONs:* Until 2002, DARPA provided the bulk of JASONs funding; the group existed independently and could pursue work with other agencies. ISAT, on the other hand, is run by DARPA and works only for DARPA.

306 *"The idea basically was":* Horvitz, interview with the author.

306 *Horvitz described his concept:* What Horvitz was describing is actually more like a

panopticon, and yet a "hall of mirrors," which is about confusion and distortion of images, is somehow appropriate given the misunderstandings about privacy concerns.

307 *"If your search for that pattern":* Poindexter, interview with author.

307 *"It was a Big Brother":* Ibid.

307 *Rotenberg said he understood:* Ibid.

308 *"Perfect surveillance":* Rotenberg, correspondence with author.

308 *Over the first half of 2002:* Harris, *Watchers,* 176.

308 *The NSA, not surprisingly:* Ibid., 217.

308 *"We have had a lot of fun":* Tony Tether, welcoming speech at DARPATech 2002.

309 *"One of the significant new data sources":* Poindexter, speech at DARPATech 2002. DARPA has removed it from its website, but it is archived on several locations on the Web, including on the Federation of American Scientists website.

309 *"robot race":* Scott Burnell, "DARPA to Fund All-Terrain Robot Race," UPI, Aug. 2, 2002.

310 *"a vast electronic dragnet":* John Markoff, "Pentagon Plans a Computer System That Would Peek at Personal Data of Americans," *New York Times,* Nov. 9, 2002.

310 *"Every purchase you make":* William Safire, "You Are a Suspect," *New York Times,* Nov. 14, 2002.

311 *"I am reading these things":* Tether, interview with the Defense Writers Group, Oct. 22, 2003.

311 *"Tony didn't understand":* Poindexter, interview with author.

311–12 *DARPA had awarded contracts:* Poindexter pointed out that another aspect of the program was to find some reward other than money to be used by government employees (because it was believed that it would not be acceptable for government employees to use real money). Poindexter, correspondence with author.

312 *"colorful examples":* Robin Hanson, The Policy Analysis Market (and FutureMAP) Archive, mason.gmu.edu.

312 *"The idea of a federal betting parlor":* "Amid Furor, Pentagon Kills Terrorism Futures Market," CNN.com, July 30, 2003.

312 *"The congressmen and senators":* Poindexter, interview with author.

312 *"Congress claimed they had closed":* If there is some irony to Poindexter's pursuit of Total Information Awareness, it is that he was actually one of the first people to be convicted of a crime based on evidence gathered from an early version of electronic communications. When the investigation into Iran-contra started in November 1986, Poindexter wiped out more than five thousand electronic messages. The messages were recovered from a two-week backup system and became key evidence against him at trial. See Lawrence E. Walsh, *Final Report of the Independent Counsel for Iran/Contra Matters,* vol. 1, *Investigations and Prosecutions* (Washington, D.C.: Government Printing Office, 1993).

313 *"would take too long":* Poindexter, correspondence with author.

313 *The Advanced Research and Development Activity:* Poindexter, interview with author. The transfer to the NSA is also detailed in Harris, *Watchers,* 246.

313 *"ring-knocking master of deceit":* Safire, "You Are a Suspect."

313 *Poindexter envisioned Cherry Vanilla World:* Poindexter, interview with author.

313 *Total Information Awareness, Poindexter insisted:* Quantifying the impact of Poindexter is difficult. The most apt description comes from Harris, who credits him as "a source of philosophical gravity" for those who pioneered today's data collection systems. Harris, *Watchers,* 363.

314 *"Some government officials":* Aftergood, correspondence with author.
314 *"They thought we were making":* Goldblatt, interview with author.
315 *"We got rid of that crazy":* Garreau, *Radical Evolution*, 270.
315 *"I came over to tell you":* Tether, interview with Williams/Gerard. In his interview, Tether says that Rumsfeld communicated with him through intermediaries, like Gingrich. In general, Rumsfeld's lack of interest in DARPA is reflected in the dearth of material in his online archive of documents from his government career. That view is similarly confirmed by George Heilmeier, whose time at DARPA spanned part of Rumsfeld's first term as defense secretary.
315 *"greatest strategic thrust":* Tether, interview with Williams/Gerard.
315 *Futuristic aircraft were fine:* In the years following September 11, DARPA funded a series of quick-reaction projects to assist with deployed forces, but most involved specific tactical technologies, like a sniper detection system.

CHAPTER 18: FANTASY WORLD

316 *"I believe strongly":* Tether, correspondence with the author, via DARPA public affairs, 2007.
316 *"Imagine 25 years from now":* Tether, DARPATech 2002.
317 *"Welcome to our world!":* Tony Tether, opening speech, DARPATech 2004.
318 *"polymer ice":* "We drew a picture of an adversary slipping down the stairs in a building," said Mitchell Zakin, the DARPA program manager who came up with the idea. Tether approved it immediately. Zakin, interview with author.
318 *"hafnium bomb":* Sharon Weinberger, "Scary Things Come in Small Packages," *Washington Post Magazine,* March 26, 2004.
318 *"Gee, probably all the people":* Tether, interview with Williams/Gerard.
319 *His favorite phrase:* Ibid. In the interview, Tether used the expression "holy cow" twelve times.
319 *"People, especially people":* Ibid. Considering that a *New York Times* op-ed kicked off the debate, it is hard to countenance Tether's view that privacy concerns over Total Information Awareness were solely a product of West Coast liberals out of touch with the realities of war.
320 *Shortly before Dunn retired:* Prizes for scientific or technical achievements were not new; the British government in the nineteenth century had offered rewards for ways to calculate geographic longitude, to help ships navigate. Among the winners was John Harrison, who developed the chronometer. But there was no precedent for a U.S. government prize.
320 *Robotics was one of the early suggestions:* Dunn, interview with author.
320 *He envisioned a 250-mile:* Tether, interview with Williams/Gerard.
320 *In the end, however:* Ibid.
321 *"belly straddling the outer edge":* Joseph Hooper, "From DARPA Grand Challenge 2004: DARPA's Debacle in the Desert," *Popular Science,* April 6, 2004.
321 *"I know you guys":* Tether, interview with Williams/Gerard.
321 *by 2004, it had dropped:* John Markoff, "Pentagon Redirects Its Research Dollars," *New York Times,* April 2, 2005.
322 *"The message of the complaints":* U.S. House, *The Future of Computer Science Research in the U.S.,* 41.
322 *In 2002, he suddenly ended:* The JASONs, who had started out as mostly physicists,

had diversified over the years, adding members from the disciplines of biology, chemistry, and computer science, among others. But the younger members were often busier—and less visible—than the older members, and the perception was that the JASONs were an aging group of physicists out of touch with disciplines of interest to the Defense Department. Their reputation was still for brilliance, but also arrogance.

322 *The JASONs, who had long prided themselves:* Fernandez, interview with author.

323 *"We actually thought it was interesting":* Horvitz, interview with author.

323 *Horvitz and Schmorrow did not realize:* Lawrence recalled running into Tether at a retirement party, and when Lawrence brought up biocybernetics, Tether was surprised to hear about it. Lawrence said Tether invited him to DARPA to talk about the program. "Later on, I called," Lawrence said. "I was met by a very patronizing response as though I was a prospective contractor trying to sneak in the side door. 'What exactly did Dr. Tether say to you? What do you mean? How long ago was this? He's busy.'" Lawrence, interview with author.

324 *"When I came to see Dylan":* Donchin, interview with author.

324 *scientists still disagreed:* Mary Cummings, interview with author.

324 *To illustrate its vision:* During the Cold War, the military had often created informational films about ambitious projects, featuring real footage of military tests, narrated by stiff military officers. In more recent years, DARPA, and the military, have hired professional PR firms to craft slick marketing videos featuring real actors, animation, and special effects.

325 *"I'm a data guy":* Gevins, interview with author.

325 *"It clearly was fake":* Ibid.

326 *In a published critique:* M. L. Cummings, "Technology Impedances to Augmented Cognition," *Ergonomics in Design* 18, no. 2 (Spring 2010): 25.

326 *"What didn't I see?":* Cummings, interview with author.

326 *"When I asked them":* Ibid.

327 *"There would be a dozen guys":* Hughes, interview with author.

327 *"Where DARPA started to fall":* Cummings, interview with author.

327 *The allure of science fiction:* Perhaps the strangest part of DARPA's Augmented Cognition program was that in the very same building, Michael Goldblatt's Defense Science Office was sponsoring research on an entirely different approach to brain-computer interface, this time working with actual sensors implanted in the brain. When asked about whether there was any connection or cooperation between the two seemingly related programs, Goldblatt said simply, "No. I'll leave it at that." Goldblatt, interview with author.

328 *Instead of a windshield:* Roland, *Strategic Computing,* 226.

328 *The smart truck did this badly:* Pollack, "Pentagon Wanted a Smart Truck."

328 *When Carnegie Mellon researchers:* Roland, *Strategic Computing,* 230.

328 *One of the first things he did:* Jackel, interview with author.

329 *Even Congress was possessed:* Section 220 of the Defense Authorization Act for Fiscal Year 2001 (H.R. 4205/P.L. 106-398), Oct. 30, 2000.

330 *"It was meant to be able":* Jackel, interview with author.

331 *"I never said anything":* Tether, interview with Williams/Gerard.

331 *Thrun, along with several other competitors:* Jackel, interview with author. Thrun dropped out of LAGR after the first phase to focus on the Grand Challenge.

332 *"It was our secret weapon":* Burkhard Bilger, "Auto Correct: Has the Self-Driving Car at Last Arrived?," *New Yorker,* Nov. 25, 2013.

332 *A dark horse candidate:* Lee Gomes, "Team of Amateurs Cuts Ahead of Experts in Computer-Car Race," *Wall Street Journal,* Oct. 19, 2005.

332 *"Holy cow, we did it":* Tether, interview with Williams/Gerard.

332 *But it is impossible to ignore:* Carnegie Mellon built the LAGR vehicles, and Stanford's Thrun had been one of the funded research teams.

332 *"At some place":* Jackel, interview with author.

333 *At one point, competing vehicles:* Michael Belfiore, "Slow-Motion Train Wreck at Auto-Bot Race," *Wired News,* March 3, 2007.

334 *"Imagine if we had convoys":* Gerry J. Gilmore, "Research Agency Showcases Robot-Driven Vehicles at Pentagon," American Forces Press Service, April 11, 2008.

334 *"Total Information Awareness got to the point":* Tether, interview with Williams/Gerard.

334 *"George Orwell America":* Cynthia L. Webb, "Someone to Watch Over Us," *Washington Post,* Nov. 21, 2002.

334 *When Tether took over in 2001: Department of Defense Fiscal Year (FY) 2005 Budget Estimates February 2004, Research, Development, Test, and Evaluation, Defense-Wide,* vol. 1, *Defense Advanced Research Projects Agency.* DARPA's budget was linked to the overall Pentagon budget, so as the military's budget started to grow with the war on terror, DARPA rode on its coattails.

335 *No one in the Pentagon:* The Rumsfeld library, Donald Rumsfeld's online repository of correspondence from his government career, rarely mentions DARPA.

335 *If during the Vietnam War:* The Pentagon's post-9/11 program to send anthropologists to Iraq and Afghanistan, called the Human Terrain System, was reminiscent of DARPA's Vietnam-era work but had nothing to do with the modern agency. "DARPA, as you know, has a particular organizational culture, which under Tony Tether was very anti–social science," wrote Montgomery McFate, the creator of the program, when asked why DARPA was never involved in the work. "DARPA, under program manager Bob Popp, was running a social science program called [Pre-conflict Anticipation and Shaping], which was involved in developing models to predict political instability. Tony Tether was not exactly supportive of it." McFate, correspondence with author.

335 *public face of DARPA's war effort:* Tony Tether, DARPATech 2004 speech.

336 *A DARPA program manager working:* "Transforming the Defense Industrial Base: A Roadmap," Office of the Deputy Undersecretary of Defense, Industrial Policy, 2003, B-125.

336 *"It sucked":* Zemach, interview with author.

336 *The Afghan smiled:* Ibid.

337 *Even in the rarefied atmosphere:* Kevin Geib and Laurie Marshall, "Voice Recognition Evaluation Report" (prepared for Office of Science and Technology National Institute of Justice, Washington, D.C., Oct. 7, 2003), 44.

337 *"Took too long to translate":* James D. Walrath, "Phraselator Questionnaire Responses, Army Research Laboratory," ARL-TN-0350, May 2009.

338 *When it came to modern threats:* When the war in Iraq turned into a full-blown insurgency, and roadside bombs became the number one killer of U.S. forces, DARPA "could have, but did not have a major role" in solving the IED problem, argued Robert Moore, a former deputy director of DARPA. Instead, the army formed its own team, the Rapid Equipping Force, and then the Office of the Secretary of Defense created the Joint Improvised Explosive Device Defeat Organization, a dedicated bomb-fighting agency. Moore, e-mail correspondence with author.

340 *Luke Skywalker–inspired prosthetic:* Michael Chorost, "A True Bionic Limb Remains Far Out of Reach," *Wired*, March 20, 2012.

341 *In his final year of congressional testimony:* Statement by Tony Tether, director, Defense Advanced Research Projects Agency, Submitted to the Subcommittee on Terrorism, Unconventional Threats, and Capabilities, House Armed Services Committee, U.S. House of Representatives, March 13, 2008.

CHAPTER 19: RETURN OF VOLDEMORT

342 *"If you supply data"*: Anonymous scientist, interview with author.

343 *"greatest influence on Petraeus's thinking":* Kaplan, *Insurgents,* 17.

343 *the French officer:* The book that Petraeus read, according to Kaplan, was Galula's *Counterinsurgency: Theory and Practice,* which was not part of Project AGILE. However, *Counterinsurgency* was preceded by Galula's more substantial work, the DARPA-supported *Pacification in Algeria, 1956–1958.* Galula's writing was also heavily influenced by his attendance at a 1962 ARPA-sponsored counterinsurgency conference. Ann Marlow, *David Galula: His Life and Intellectual Context* (Carlisle, Pa.: Strategic Studies Institute, 2010), 7–9, 48.

343 *"By 2009 [counterinsurgency] was":* Greg Jaffe, review of *The Insurgents,* by Fred Kaplan, and *My Share of the Task,* by Stanley A. McChrystal, *Washington Post,* Jan. 6, 2013.

344 *When she started making:* Dugan was described as "swaggering and stylish, with a helmet of thick, dark hair, piercing eyes, and a penchant for jeans, leather jackets, and scarves." Miguel Helft, "Google Goes DARPA," *Fortune,* Aug. 14, 2014.

344 *Dugan's financial ties:* Noah Shachtman and Spencer Ackerman, "DARPA Chief Owns Stock in DARPA Contractor," *Wired News,* March 7, 2011.

344 *"There is a time and a place":* Statement by Dr. Regina E. Dugan, director, Defense Advanced Research Projects Agency, Submitted to the Subcommittee on Emerging Threats and Capabilities, U.S. House of Representatives, March 1, 2011. Ironically, however, Dugan did support a whimsical DARPA contract for a "100 Year Starship" designed to travel to the stars.

345 *The Information Processing Techniques Office:* Lee, interview with author. This view was also expressed in other author interviews.

345 *"You know, Peter":* Ibid.

345 *"pure expression of what DARPA":* Ibid.

345 *"Oh my God":* Ibid.

346 *"program that shall not be named":* Dugan, interview with author.

346 *"very top secret":* "Transcript: Bob Woodward Talks to ABC's Diane Sawyer About 'Obama's Wars,'" *ABC News,* Sept. 27, 2010.

346 *It was designed:* Siobhan Gorman, Adam Entous, and Andrew Dowell, "Technology Emboldened the NSA: Advances in Computer, Software Paved Way for Government's Data Dragnet," *Wall Street Journal,* June 9, 2013.

347 *"If you're smart enough":* Rustan, interview with Ben Iannotta, "Change Agent," *Defense News,* Oct. 8, 2010.

347 *"a steady stream":* Lee, interview with author.

347 *"That idea might be stupid"*: Ibid.

348 *"trivial":* Andy Greenberg, "Mining Human Behavior at MIT," *Forbes,* Aug. 12, 2010.

348 *"social physics":* Pentland, interview with author.

348 *"If you look at the models":* Ibid.

349 *At the NSA, Garrett had been a key official:* J. Nicholas Hoover, "NSA Using Cloud Model for Intelligence Sharing," *Information Week,* July 20, 2009.

349 *"essentially every kind of data":* Gorman, Entous, and Dowell, "Technology Emboldened the NSA."

349 *Afghanistan, after ten years:* In 2014, it was revealed, as a result of documents leaked by Edward Snowden, that the NSA was intercepting, recording, and storing almost every cell phone call made in Afghanistan. "WikiLeaks Statement on the Mass Recording of Afghan Telephone Calls by the NSA," May 23, 2014, wikileaks.org.

349 *"Someone made the observation":* Lee, interview with author.

349 *"For example, we were trying":* Ibid.

350 *"You're from DARPA":* Dugan, interview with author.

350 *"BIG BREAKTHROUGH":* Department of Defense Inspector General, "Report of the Investigation: Doctor Regina E. Dugan, Former Director, Defense Advanced Research Projects Agency," April 9, 2013.

350 *One, called More Noses:* More Noses, for reasons not publicly explained, never got beyond the drawing board. Richard Weiss (DARPA public affairs officer), conversation with author, Nov. 14, 2014. "It didn't get past the idea stage," Weiss said.

351 *More Eyes, together with More Noses:* Department of Defense Inspector General, "Report of the Investigation: Doctor Regina E. Dugan, Former Director, Defense Advanced Research Projects Agency."

351 *"When Nexus 7 started up":* Pentland, interview with author.

352 *"actually build this big data":* Anonymous social scientist, interview with author.

352 *The program was a direct carryover:* Ibid.

352 *"For the military, social networks": Department of Defense Fiscal Year (FY) 2014 President's Budget Submission, DARPA, Justification Book,* vol. 1, April 2013.

352 *"how can we take all that data":* Cosby, interview with author.

353 *Dugan believed she could avoid:* Dugan also set up a privacy panel. Dugan, interview with author.

353 *"We had to have cover stories":* Lee, interview with author.

353 *price of exotic vegetables:* Kilcullen, *Counterinsurgency,* 60.

354 *"Go and go faster":* Dugan, interview with author.

354 *"I should have been with them":* Lee, interview with author.

354 *"It was the first operational deployment":* Dugan, interview with author.

354 *"a 90-day Skunk Works activity":* Statement by Regina E. Dugan, director, Defense Advanced Research Projects Agency, Submitted to the Subcommittee on Emerging Threats and Capabilities, U.S. House of Representatives, March 1, 2011.

355 *But much of the data:* Noah Shachtman, "Inside DARPA's Secret Afghan Spy Machine," *Wired News,* July 21, 2011.

355 *DARPA was touting Nexus 7's successes:* Ibid.

355 *The Synergy Strike Force was always:* Sean Gorman, interview with author.

355 *There were a few "burners":* Brian Calvert, "The Merry Pranksters Who Hacked the Afghan War," *Pacific Standard,* July 1, 2013.

356 *"the tiki bar at the edge of the universe":* McCarthy, interview with Vinay Gupta on Taj Beer for Data, youtube.com.

356 *"super-powered geeks":* Matthew Borgatti, "The Synergy Strike Force at STAR-TIDES," Oct. 10, 2011, GWOB.org.

356 *In 2010, around the time:* Huffman, interview with author.

356 *Ushahidi:* Anand Giridharadas, "Africa's Gift to Silicon Valley: How to Track a Crisis," *New York Times,* March 13, 2010.

356 *Soon, Ryan Paterson:* Gorman, interview with author.

356 *"We were referred to":* More Eyes regional coordinator, interview with author.
357 *Afghan participants, often drawn:* Ibid.
357 *"Generally speaking, U.S. forces":* Jeffrey E. Marshall, "All Source Intelligence and Operational Fusion: Fusing Crowd Sourcing and Operations to Strengthen Stability and Security Operations" (unpublished report for Thermopylae Sciences and Technology, n.d.).
357 *"Afghanistan Atmospherics":* Ryan Paterson et al., "Getting 'More Eyes' in Afghanistan: Experiments in Promoting Indigenous Self-Reporting of Local Conditions and Sentiment," May 11, 2011 (unpublished DARPA white paper provided to the author by an anonymous source).
357 *"to catalyze the local population":* Ibid.
358 *The members of the Synergy Strike Force:* One person working on the More Eyes contract explained that it was difficult to "get ahead" in the world of aid and development if his or her name was associated with military contracting. Anonymous, interview with author.
358 *"Afghanistan's DIY Internet":* Sebastian Anthony, *Extreme Tech,* June 22, 2011.
358 *"Was the More Eyes program":* Anonymous More Eyes scientist, interview with author.
359 *"The More Eyes Team quickly learned":* Paterson et al., "Getting 'More Eyes' in Afghanistan."
359 *On August 11, 2012:* Peretz Partensky, "Basketball Diaries, Afghanistan," *N+1,* Dec. 5, 2012. It is unclear why Saraj was killed, but the Taliban often target anyone who works for foreigners, and the subtleties of U.S. government contracting, whether civilian or military, likely would not matter.
360 *"There are no models":* Shachtman, "Inside DARPA's Secret Afghan Spy Machine."
360 *"It's the ultimate correlation tool":* Gorman, Entous, and Dowell, "Technology Emboldened the NSA." Without public data or published results, this statement should be viewed with caution.
360 *One thing was clear:* A Freedom of Information Act request filed with the Defense Department by the author in 2012 for Nexus 7 documents has so far not been completed. In a 2013 interview with the author, Arati Prabhakar declined to talk about the program in detail, citing security issues.
360 *The inspector general's report:* Because Dugan had already left government, the inspector general recommended that no action be taken based on the report's conclusions.

EPILOGUE: GLORIOUS FAILURE, INGLORIOUS SUCCESS

363 *"The intelligence aspects":* Buhl, *An Eye at the Keyhole.*
364 *That came to a halt in the 1980s:* Larry Lynn, interview with author.
365 *"I think it is unchanged":* Prabhakar, interview with author.
365 *"Facing that kind of existential threat":* Ibid.
365 *"gem of the Pentagon":* William Perry, interview with author.
365 *When asked about the episode:* It was a sensitive subject, because her job just prior to entering DARPA was at U.S. Venture Partners, a Silicon Valley firm whose portfolio included Solyndra, a solar power start-up that went bankrupt, after taking half a billion dollars in government-backed loans.
365 *In 2003, when the American military:* Gates, *Duty,* 147.

367 *"By the time you get"*: Reis, interview with author. Parkinson wrote that "perfection of planned layout is achieved only by institutions on the point of collapse." His argument was that "during a period of exciting discovery or progress there is no time to plan the perfect headquarters. The time for that comes later, when all the important work is done. Perfection, we know, is finality; and finality is death." Cyril Northcote Parkinson, *Parkinson's Law, and Other Studies in Administration* (Boston: Houghton Mifflin, 1957), 60–61.

367 *The attempts to replicate DARPA:* In 2013, Regina Dugan described her work to create a DARPA-like organization at Google. Her "DARPA model" was distilled to just three elements: "ambitious goals"; "temporary project teams"; and "independence." She omitted, however, a fundamental part of the DARPA model: having a "customer" willing to adopt and field a technology with no immediate commercial prospects. Without that key ingredient, the examples she cites, like stealth aircraft and satellite navigation, would never have made it beyond the prototype stage. Even Google, with its deep pockets, is unlikely to invest in a technology whose commercial application is decades away. Regina Dugan, "'Special Forces' Innovation: How DARPA Attacks Problems," *Harvard Business Review,* Oct. 2013.

368 *"140 program managers":* This statement has been cited by DARPA officials dating back at least to the 1980s. More recently, it was used by Tony Tether during a 2002 interview. William New, "Defense Research Agency Seeks Return to 'Swashbuckling' Days," *Government Executive,* May 13, 2002.

369 *"Long Range Research Development Plan":* Amaani Lyle, "DoD Seeks Future Technology via Development Plan," *DoD News,* Defense Media Activity (U.S. Department of Defense), Feb. 3, 2014.

370 *"This may be more like an entropy process":* Lukasik, interview with author.

370 *"transformed into intrusive government policies"*: Lukasik, "Advanced Research Projects Agency."

371 *"on restoring the injured":* Sanchez, interview with author.

Sources

A NOTE ON SOURCES

When I first embarked on this project in 2011, several people associated with DARPA questioned how I would write a history of the agency without access to materials on its classified projects, which form a substantial part of its legacy. My stock answer was that histories have been written based on declassified records and interviews of other, far more secretive agencies, including the National Security Agency and the Central Intelligence Agency, and the same could and would be done for DARPA. My bigger concern at the time was that DARPA officials would choose not to participate in a history that was not commissioned by the agency. Both concerns turned out to be wrong.

In the end, this book proves a history can be done without full access to classified records, and even without cooperation from the agency. Though no history is ever complete, the research for this book encompasses extensive archival materials in addition to more than three hundred hours of interviews, primarily with former DARPA officials. Those interviews spanned from those who worked in the agency during its first days in 1958 all the way up to its current director. Almost all of the former DARPA and Pentagon officials I contacted agreed to be interviewed, often at length.

As it turned out, while the concern about access to classified materials was not baseless, neither was it insurmountable. Much of the material in this book for the period prior to 1973 is drawn from the National Archives and Records Administration in College Park, Maryland, particularly Record Group 330, which covers the Office of the Secretary of Defense. The archivists warned me at the outset that a large portion of Record Group 330, including many of DARPA's records, is still classified. They were correct, but dozens of boxes of agency records have been declassified in recent years and are now finally accessible to researchers. The National Archives' collection of DARPA records provided a detailed portrait of the agency's first fifteen years and allowed me to cross-reference and fact-check interviews with those former DARPA officials who did their best to recall events from forty and, in some cases, fifty years prior.

It is, however, somewhat disconcerting that a number of documents previously made available have been withdrawn under the National Archives' post-9/11 review of "records of concern." The frequent withdrawals, usually of specific documents that are part of larger files, were rarely of any great historical concern. For example, an unclassified 1965 report to DARPA titled "State of the Art Study of Anti-metabolites" was typical of the many documents withdrawn in 2002. The study in question was part of DARPA's investigation into incapacitating chemicals that might be used in Vietnam. Presumably, the report was withdrawn in case potential terrorists might use the information to do harm (it is worth noting that the accompanying notes in the file indicate that antimetabolites would not make for useful weapons). The absurdity of believing that terrorists might delve into a half-century-old unclassified report in order to build weapons is itself a testament to the way secrecy has obscured common sense. It would be comical, were it not for the enormous effort required to process archival materials to make them available to the public. I cannot speak for all records, but in the case of DARPA's files the withdrawals appear to be a shameful waste of resources, while offering little protection of national security.

In addition to the records at College Park, this book draws on other National Archives facilities, including the Kennedy Presidential Library, the Eisenhower Presidential Library, and the Nixon Presidential Library. I also relied on the Godel trial and accompanying records, held by the National Archives (the trial records are normally held in Philadelphia but because of renovations were temporarily moved to Atlanta when I accessed them). These records, which included the trial transcript and an incomplete set of associated material, provided the best insight into some of Godel's Vietnam activities. In particular, depositions with two South Vietnamese officials shed light on Godel's mission in the country, providing a rare Vietnamese perspective on DARPA's efforts there.

Other archives I used included the Smithsonian Institution Archives, Washington, D.C.; the U.S. Army Center of Military History, Fort McNair, Washington, D.C.; the U.S. Army Military History Institute at Carlisle Barracks, Pennsylvania; the MIT Institute Archives & Special Collections; and the Mandeville Special Collections Library at the University of California, San Diego.

Materials from this book covering the mid-1970s and after, where official records are not yet available through the National Archives, are based largely on interviews I conducted with former and current officials, as well as unclassified materials that were made available to me, either by individuals or through other archives. I also relied on interviews with former DARPA officials conducted by historians, including those held at the Charles Babbage Institute and the American Institute of Physics. Several scholars and writers generously shared their source materials with me. Ann Finkbeiner provided me with transcripts of DARPA interviews she conducted in conjunction with her book on the JASONs. Alex Roland of Duke University shared interviews he conducted for his book on DARPA's Strategic Computing Initiative. And L. Douglas Keeney, who wrote an insightful history of Strategic Air Command, shared the command histories he obtained under the Freedom of Information Act. All three are a reminder that scholarship is about sharing information, and I am grateful to them for their intellectual generosity.

A significant set of materials was obtained from the Defense Department using the Freedom of Information Act, which yielded, among other treasures, an unpublished interview with William H. Godel on DARPA's early history. After three years, and finally litigation under the Freedom of Information Act, the Defense Department also released to me a full, unredacted set of interviews conducted with former DARPA directors as part of the agency's forty-fifth and fiftieth anniversary celebrations. (DARPA declined to release two interviews, with Herbert York and John S. Foster Jr., which were part of this set, claiming that they were outside the scope of my request. Stephen Lukasik graciously provided me with a copy of those two missing transcripts from his personal collection.)

While I had already interviewed most of the former living directors, these additional interviews sometimes reflected a more candid viewpoint, because they were conducted at the behest of DARPA. For those two former directors who declined to speak with me for this book—Tony Tether and Craig Fields—their interviews provided me with a window into their views of DARPA.

I also benefited greatly from private materials provided by a number of individuals, including sections of William H. Godel's unpublished memoir, shared by his daughter, Dr. Kathleen Godel-Gengenbach. At her request, the citation for this memoir uses H. A. H. Buhl Jr., her father's birth name. I do not cite page numbers for this manuscript, because it is in the process of being edited, and thus the page numbers will likely not correspond to any published or archival version made available in the future.

Richard Dunn, DARPA's former chief counsel, also provided me with a copy of his own unpublished history of the agency, and Stephen Lukasik graciously sent me a personal memoir of his time at DARPA, written for his family, and provided other materials from his private collection.

Because this book is part journalism and part history, I have elected to use a citation convention that blends aspects of both professions. Given the complexity of integrating material obtained from multiple archives, my interviews, interviews conducted by other writers and scholars, private archival collections, and documents obtained under the Freedom of Information Act and occasionally from unnamed sources, I've tried to provide the reader a precise understanding of where the material came from and enough information to find the original source documents, where publicly available. However, because the location of many archival documents changes over time, I do not always provide box numbers and file names. I also have not, in most cases, and at the recommendation of the publisher, included URLs for online sources. There are a few cases where I have noted web addresses, simply because there is no other way of indicating where the information came from.

The greatest challenge to documenting DARPA's history is DARPA itself, which has failed to record its past in any meaningful way since 1975. Unlike other national security agencies, which have compiled and declassified many volumes dealing with different chapters of their histories, DARPA's only institutional history is the Barber Associates study, written by Lee Huff and Richard Sharp and referenced extensively in this book. Their insightful and candid writing was an invaluable reference. Institutional histories, even well-done ones like the Barber Associates history, have limits. That history was written as an unclassified study, and this book, by having access to newly declassified documents on the Vietnam War period, fills in additional blanks. And while the Barber Associates history was appropriately critical of many aspects of DARPA's activities, it avoided some of the darker aspects of Project AGILE, particularly the agency's role in chemical defoliation. I have attempted to rectify that omission here.

The Institute for Defense Analyses has also done several excellent studies led by Richard Van Atta detailing DARPA's programmatic work, including the 1991 report *DARPA Technical Accomplishments*. The reports, while valuable, focus on programs and not on DARPA as an institution. Sadly, other efforts by DARPA to document its own history have resulted in nothing more than glossy PR materials.

DARPA's history, in the meantime, is literally dying off. Robert Cooper passed away in 2007. Seymour Deitchman, who tried to solve counterinsurgency through social science, died in 2013, at the age of ninety. George Heilmeier passed away in 2014. A number of other former directors and key officials interviewed in this book have either died since I began writing or slipped away into a haze of dementia.

Finally, whatever the limitations on source materials, an outside history of government agencies benefits greatly from the ability to be independent. What I lacked in access to classified materials I gained in the ability to write freely about DARPA's projects. There are classified projects from DARPA's more recent history that will likely come to light and provide additional insights into the agency's contributions. For example, the exact mission of Copper Canyon, the hypersonic space plane, might someday be confirmed, as might the various secret unmanned aircraft sponsored under Teal Rain. There, again, I doubt those revelations will change the portrait of the agency I have painted here. The truth is that classified projects that fail in development tend to remain classified, because there is usually no reason to reveal them, while successful projects, like Have Blue and Tacit Blue, eventually become public when they are used in military operations and can no longer be kept hidden. Nor is there necessarily a connection between classification and historical importance. It is slightly amusing to note, for example, that almost an entire box of materials on DARPA's sponsorship in the 1960s of a jet belt remains classified. Certainly not because the jet belt was successful. It was not, at least not as a military technology that would allow soldiers to fly around the battlefield.

Declassification and release of records would, however, clear up lingering questions related to William Godel's role at DARPA during the Vietnam War. Several people familiar with Godel believe his work at DARPA was part of a highly classified intelligence operation, for which Project AGILE was in part a cover. If that was the case, then it was so covert that apparently even Harold Brown, who oversaw DARPA and later went on to become secretary of defense, was ignorant of its true purpose. More than likely, the covert element of Godel's work, as described in this book, was his role as an envoy for Edward Lansdale and other highly placed American counterinsurgents who sought to influence and assist President Diem. The one mystery that could be solved by the full declassification and release of government documents relates to the events leading up to the investigation and trial of Godel, as well as rumors that circulated after his death.

Shortly after Godel's death in 2000, Joseph Trento, an independent journalist, contacted the Godel family with allegations that Godel had been a Soviet mole. Trento and his wife in 1989 had published a controversial book on Soviet moles in the CIA. Their co-author on that book was William Corson, the marine officer assigned to DARPA, who later testified against Godel in the 1965 fraud trial. According to Trento, Corson was secretly working on behalf of the CIA to investigate Godel. Proof for the allegation, other than the purported words of Corson, who died the same year as Godel, was nonexistent. Nonetheless, Trento in 2001 published the account in his book *The Secret History of the CIA*.

Did the CIA believe Godel was a Soviet mole? That is impossible to say based on the available documents, though Corson was allegedly a confidant of James Angleton, the CIA's famous Soviet mole hunter. The historical record does show that Godel had enemies in the CIA, and the investigation and his subsequent trial came amid a slew of Angleton's obsessive counterintelligence investigations—a few deserved, and some driven by paranoia and political revenge. Likewise, Corson's connections to the CIA are well established, though his exact assignments for the spy agency remain unclear.

What role suspicions of espionage actually played in Godel's ouster remains mired in secret records that as of today the FBI and CIA have been unwilling to divulge, despite long-standing Freedom of Information Act requests. Those records, if eventually released, may ultimately have little bearing on DARPA's place in history, but they would help shed light on Godel's downfall and bring closure to his family. For those who have never entered the Kafkaesque world of public access to government records, it is difficult to articulate the immense frustration one feels trying to access documents that the government wants to withhold, or the incredible satisfaction that comes from obtaining them. Sometimes the importance of obtaining records is more sentimental than substantive.

Shortly before her father's death, Godel's oldest daughter, Kathleen Godel-Gengenbach, was able to obtain at least one document related to her father's personal history: his original birth certificate, which had been sealed and bore his birth name, Hermann Adolph Buhl. She gave the birth certificate to her father shortly before he died.

"He cried," she wrote.

ARCHIVES

Army War College, U.S. Army Heritage and Education Center, Carlisle, Pennsylvania
Dwight D. Eisenhower Presidential Library, Abilene, Kansas
George H. Lawrence personal collection
George H. W. Bush Presidential Library, College Station, Texas
Gerald Ford Presidential Library, Ann Arbor, Michigan
Herbert F. York Papers, Mandeville Special Collections Library, University of California, San Diego
Hoover Institution Archives, Stanford, California
Ithiel de Sola Pool Papers, MC440, MIT Archives, Cambridge, Massachusetts
John F. Kennedy Presidential Library, Boston, Massachusetts

National Archives and Records Administration, Atlanta, Georgia
National Archives and Records Administration, College Park, Maryland
National Archives and Records Administration, St. Louis, Missouri
National Archives and Records Administration, Washington, D.C.
Research Group in Psychology and the Social Sciences Records, 1957–1963 (Record Unit 179), Smithsonian Institution, Washington, D.C.
Stephen J. Lukasik personal collection
Thomas Thayer Collection, U.S. Army Center of Military History, Washington, D.C.
William Nierenberg, Mandeville Special Collections Library, University of Southern California, San Diego

INTERVIEWS

Author Interviews and Correspondence (2007–2014)

Stephen J. Andriole, office director, DARPA
Rand Araskog, special assistant, DARPA
Ronald Arkin, professor, Georgia Institute of Technology
Allen Atkins, director, Aerospace Technology Office, DARPA
Natalie Atkins, executive assistant, DARPA
Charles Bates, program manager, DARPA
Robert Brodkey, program manager, DARPA
Alan Brown, director of engineering, Lockheed Martin
Harold Brown, Secretary of Defense
Eric Cartwright, program manager, DARPA
William Casebeer, program manager, DARPA
Vincent Cerf, program manager, DARPA
Ray Colladay, director, DARPA
L. Neale Cosby, consultant, DARPA
Steve Crocker, program manager, DARPA
Mary Cummings, professor, Duke University
Malcolm Currie, director of defense research and engineering, Department of Defense
Larry Davis, professor, University of Maryland
Larry Dubois, office director, DARPA
Seymour Deitchman, director of AGILE, DARPA
Gary Denman, director, DARPA
Emanuel Donchin, professor, University of South Florida
Regina Dugan, director, DARPA
Richard Dunn, chief counsel, DARPA
Carol duPont, vice president, duPont Aerospace Company
Tony duPont, president, duPont Aerospace Company
Frank Fernandez, director, DARPA
Robert Fossum, director, DARPA
John S. Foster, director of defense research and engineering, Department of Defense
Robert Frosch, deputy director, DARPA
Ken Gabriel, deputy director, DARPA
Alan Gevins, neuroscientist
Kathleen Godel-Gengenbach (correspondence only), daughter of William Godel
Michael Goldblatt, director of Defense Sciences Office, DARPA

Paul Gorman, general (retired), U.S. Army
Sean Gorman, data scientist
Richard Hallion, chief scientist, U.S. Air Force
Charles "Chuck" Heber, program manager, DARPA
George Heilmeier, director, DARPA
Charles Herzfeld, director, DARPA
Donald Hess, director of administration, DARPA
Eric Horvitz, computer scientist, Microsoft
Lee Huff, director, Behavioral Sciences Office, DARPA
Todd Huffman, Synergy Strike Force
Todd Hughes, program manager, DARPA
Michiaki Ikeda, Nagasaki atomic bomb survivor
Larry Jackel, program manager, DARPA
Robert Kahn, director, Information Processing Techniques Office, DARPA
Deepak Khosla, scientist, HRL Laboratories
Kent Kresa, director, Strategic Technology Office, DARPA
George Lawrence, program manager, DARPA
Peter Lee, director, Transformational Convergence Technology Office, DARPA
Geoffrey Ling, director, Biological Technologies Office, DARPA
Stephen J. Lukasik, director, DARPA
Christian Macedonia, program manager, DARPA
Hans Mark, director of defense research and engineering, Department of Defense
Dennis McBride, program manager, DARPA
Robert Moore, deputy director, DARPA
David Morell, program manager, DARPA
Walter Munk, professor, Scripps Institution of Oceanography, La Jolla, California
Ronald Murphy, program manager, DARPA
Charles "Chuck" Myers (correspondence only), director for air warfare, Office of the Secretary of Defense
David Neyland, office director, DARPA
Sean O'Brien, program manager, DARPA
Ward Page, program manager, DARPA
Peter P. Papadakos, executive director, Gyrodyne Helicopter Historical Foundation
Dennis Papadopoulos, professor, University of Maryland
Constantine "Jack" Pappas, naval officer
Sandy Pentland, professor, MIT
John Perry, program manager, DARPA
William Perry, Secretary of Defense
John Poindexter, director, Information Awareness Office, DARPA
Arati Prabhakar, director, DARPA
Hal Puthoff (correspondence only), scientist, SRI International
George Rathjens, deputy director, DARPA
Victor Reis, director, DARPA
Carl Romney, deputy director, DARPA (and of the Air Force Technical Applications Center)
Sven Roosild, program manager, DARPA
Jack Ruina, director, DARPA
Justin Sanchez, director, Biological Technologies Office, DARPA
Warren Stark, program manager, DARPA
Ivan Sutherland, program manager, DARPA

James Tegnelia, acting director, DARPA
Anthony J. Tether (correspondence only), director, DARPA
Jack Thorpe, program manager, DARPA
Robert Van de Castle, professor, University of Virginia
Amy Vanderbilt, program manager, DARPA
Fred Wikner, director, Office of Net Technical Assessments
Samuel V. Wilson, lieutenant general (retired), deputy assistant for Special Operations
Stuart Wolf, program manager, DARPA
Herbert York, chief scientist, DARPA
Stephen Young, program manager, DARPA
Mitch Zakin, program manager, DARPA
Ken Zemach, consultant, Rapid Equipping Force

*Note: The preceding list includes only those who agreed to be interviewed on the record. I have not included specific dates, because many of those listed above were interviewed multiple times, with follow-up phone calls and frequent e-mail correspondence. I have given their title as it most closely related to their work at or with DARPA.

American Institute of Physics, College Park, Maryland

Richard Blankenbecler, May 5, 1987
Robert S. Cooper, Sept. 3, 1993
Freeman Dyson, Dec. 17, 1986
Jack Evernden, June 18, 1998
John S. Foster, Dec. 3, 15, 1968, Jan. 7, 1969
Edward A. Frieman, June 26, 1986, Dec. 4 and 5, 2006
Robert Frosch, July 10, 23, Aug. 19, Sept. 15, and Oct. 6, 1981, May 28, 1998
Charles Herzfeld, July 28, 1991
Donald Le Vine, July 29, 1991
Stephen J. Lukasik, April 21, 1987
Gordon J. F. MacDonald, March 21, 1994
Jon Peterson, Oct. 21, 1997
Eberhardt Rechtin, April 24, 1987
Carl Romney, Jan. 20 and 28, 1998
Jack Ruina, May 29, 1998, Aug. 8, 1991
Robert Sproull, July 11, 1983
Carlisle Martin Stickley, Sept. 22, 1984
Alexander J. Tachmindji, Aug. 7, 1991, March 24, 1993
Charles H. Townes, Jan. 28 and 31, 1984, May 20 and 21, 1987
Kenneth M. Watson, Feb. 10, 1986
Stuart Wolf, March 23, 2006
Herbert F. York, Sept. 24, 1980, Feb. 7, 1986, April 24, 2008

Charles Babbage Institute, University of Minnesota, Minneapolis

Paul Baran, March 5, 1990
George Heilmeier, March 27, 1991

Charles Herzfeld, Aug. 6, 1990
J. C. R. Licklider, Oct. 28, 1988
Stephen J. Lukasik, Oct. 17, 1991
Jack Ruina, April 20, 1989

Ann Finkbeiner Interviews

Frank Fernandez, Nov. 5, 2004
Stephen J. Lukasik, June 25, 2005
Jack Ruina, Feb. 12, 2002
Herbert F. York, June 16, 2002

Alex Roland Interviews

Lynn Conway, Jan. 12 and March 7, 1994
Robert Cooper, May 12, 1994
Robert Duncan, May 12, 1994
Robert Kahn, Aug. 2, 1993, Nov. 29, 1994
Steven Squires, June 17, July 12, and Dec. 21, 1994

Interviews Commissioned by DARPA

Austin Betts, Dec. 23, 2003
Ray Colladay, Jan. 16, 2007
Robert Cooper, Feb. 23, 2007
Gary Denman, Jan. 17, 2006
Frank Fernandez, Jan. 4, 2007
Craig Fields, March 7, 2007
Robert Fossum, March 14, 2007
John S. Foster, April 20, 2007
William Godel, June 17, 1975 (conducted by Lee Huff)
George Heilmeier, Jan. 16, 2007
Charles Herzfeld, Feb. 23, 2007
Stephen J. Lukasik, Jan. 17, 2007
Larry Lynn, Dec. 8, 2006
Victor Reis, Jan. 17, 2007
Jack Ruina, Jan. 11, 2007
Robert Sproull, Dec. 7, 2006
Anthony J. Tether, May 1, 2007, Feb. 13, 2009
Herbert York, Jan. 5, 2007

*Note: The preceding interviews, with the exception of William Godel, were conducted by Williams/Gerard on behalf of DARPA.

Selected Bibliography

BOOKS AND JOURNAL ARTICLES

Aaserud, Finn. "Sputnik and the 'Princeton Three': The National Security Laboratory That Was Not to Be." *Historical Studies in the Physical and Biological Sciences* 25, no. 2 (1995): 185–239.

Abella, Alex. *Soldiers of Reason: The Rand Corporation and the Rise of the American Empire.* Orlando, Fla.: Harcourt, 2008.

Aronstein, David C., and Albert C. Piccirillo. *Have Blue and the F-117A: Evolution of the "Stealth Fighter."* Reston, Va.: American Institute of Aeronautics and Astronautics, 1997.

Bamford, James. *The Puzzle Palace: A Report on America's Most Secret Agency.* Boston: Houghton Mifflin, 1982.

Barbree, Jay. *"Live from Cape Canaveral": Covering the Space Race, from Sputnik to Today.* New York: HarperCollins, 2007.

Beason, Doug. *The E-bomb: How America's New Directed Energy Weapons Will Change the Way Future Wars Will Be Fought.* Cambridge, Mass.: Da Capo Press, 2005.

Belfiore, Michael P. *The Department of Mad Scientists: How DARPA Is Remaking Our World, from the Internet to Artificial Limbs.* Washington, D.C.: Smithsonian Books, 2009.

Bilstein, Roger E. *Stages to Saturn: A Technological History of the Apollo/Saturn Launch Vehicles.* Washington, D.C.: Scientific and Technical Information Branch, National Aeronautics and Space Administration, 1980.

Boot, Max. *Invisible Armies: An Epic History of Guerrilla Warfare from Ancient Times to the Present.* New York: Liveright, 2013.

Bray, Charles W. "Toward a Technology of Human Behavior for Defense Use." *American Psychologist* 17, no. 8 (1962): 527–41.

Brown, Harold. *Star Spangled Security: Applying Lessons Learned over Six Decades Safeguarding America.* With Joyce Winslow. Washington, D.C.: Brookings Institution Press, 2012.

Buckingham, William A. *Operation Ranch Hand: The Air Force and Herbicides in Southeast Asia, 1961–1971.* Washington, D.C.: Office of Air Force History, U.S. Air Force, 1982.

Buhl, H. A. H., Jr. *An Eye at the Keyhole.* Unpublished manuscript (courtesy of K. Godel-Gengenbach), 1976.

Burke, David Allen. *Atomic Testing in Mississippi: Project Dribble and the Quest for Nuclear*

Weapons Treaty Verification in the Cold War Era. Baton Rouge: Louisiana State University Press, 2012.
Cecil, Paul Frederick. *Herbicidal Warfare: The Ranch Hand Project in Vietnam*. New York: Praeger, 1986.
Chertok, B. E. *Rockets and People*. Vol. 2. Washington, D.C.: National Aeronautics and Space Administration, NASA History Office, Office of External Affairs, 2005.
Christofilos, N. C. "The Argus Experiment." *Journal of Geophysical Research* 64, no. 8 (1959): 869–75.
Coleman, Elisheva R., Samuel A. Cohen, and Michael S. Mahoney. "Greek Fire: Nicholas Christofilos and the Astron Project in America's Early Fusion Program." *Journal of Fusion Energy* 30, no. 3 (2011): 238–56.
Corson, William R. *The Betrayal*. New York: W. W. Norton, 1968.
Cosmas, Graham A. *MACV: The Joint Command in the Years of Escalation, 1962–1967*. Washington, D.C.: Center of Military History, U.S. Army, 2006.
D'Antonio, Michael. *A Ball, a Dog, and a Monkey: 1957, the Space Race Begins*. New York: Simon & Schuster, 2007.
DARPA: 50 Years of Bridging the Gap. Tampa: Faircount, 2008.
Day, Dwayne A., John M. Logsdon, and Brian Latell, eds. *Eye in the Sky: The Story of the Corona Spy Satellites*. Washington, D.C.: Smithsonian Institution Press, 1998.
Deitchman, Seymour J. *The Best-Laid Schemes: A Tale of Social Research and Bureaucracy*. Cambridge, Mass.: MIT Press, 1976.
———. "The 'Electronic Battlefield' in the Vietnam War." *Journal of Military History* 72, no. 3 (July 2008): 869–87.
———. "A Lanchester Model of Guerrilla Warfare." *Operations Research* 10, no. 6 (1962): 818–27.
Dickson, Paul. *Sputnik: The Shock of the Century*. New York: Walker Publishing, 2001.
Dobbs, Michael. *One Minute to Midnight: Kennedy, Khrushchev, and Castro on the Brink of Nuclear War*. London: Arrow, 2009.
Dyson, George. *Project Orion: The True Story of the Atomic Spaceship*. New York: Henry Holt, 2002.
———. *Turing's Cathedral: The Origins of the Digital Universe*. New York: Pantheon Books, 2012.
Finkbeiner, Ann K. *The Jasons: The Secret History of Science's Postwar Elite*. New York: Viking, 2006.
FitzGerald, Frances. *Way Out There in the Blue: Reagan, Star Wars, and the End of the Cold War*. New York: Simon & Schuster, 2000.
Garreau, Joel. *Radical Evolution: The Promise and Peril of Enhancing Our Minds, Our Bodies—and What It Means to Be Human*. New York: Doubleday, 2005.
Gates, Robert. *Duty: Memoirs of a Secretary at War*. New York: Alfred A. Knopf, 2014.
Gertner, Jon. *The Idea Factory: Bell Labs and the Great Age of American Innovation*. New York: Penguin, 2012.
Ghamari-Tabrizi, Sharon. *The Worlds of Herman Kahn: The Intuitive Science of Thermonuclear War*. Cambridge, Mass.: Harvard University Press, 2005.
Gravel, Mike, Noam Chomsky, and Howard Zinn. *The Pentagon Papers: The Senator Gravel Edition*. Boston: Beacon, 1971.
Hafner, Katie, and Matthew Lyon. *Where Wizards Stay Up Late: The Origins of the Internet*. New York: Simon & Schuster, 1996.
Harris, Shane. *The Watchers: The Rise of America's Surveillance State*. New York: Penguin, 2010.

Heppenheimer, T. A. *Facing the Heat Barrier: A History of Hypersonics*. Washington, D.C.: NASA, 2006.

Herken, Gregg. *Cardinal Choices: Presidential Science Advising from the Atomic Bomb to SDI*. Stanford, Calif.: Stanford University Press, 2000.

Herzfeld, Charles M. *A Life at Full Speed: A Journey of Struggle and Discovery*. Arlington, Va.: Potomac Institute Press, 2014.

Hickey, Gerald Cannon. *Window on a War: An Anthropologist in the Vietnam Conflict*. Lubbock: Texas Tech University Press, 2002.

Johnson, Clarence L. *Kelly: More Than My Share of It All*. With Maggie Smith. Washington, D.C.: Smithsonian Institution Press, 1985.

Kaplan, Fred M. *The Insurgents: David Petraeus and the Plot to Change the American Way of War*. New York: Simon & Schuster, 2013.

———. *The Wizards of Armageddon*. New York: Simon & Schuster, 1983.

Karnow, Stanley. *Vietnam, a History*. New York: Viking, 1983.

Kempe, Frederick. *Berlin 1961: Kennedy, Khrushchev, and the Most Dangerous Place on Earth*. New York: G. P. Putnam's Sons, 2011.

Kilcullen, David. *Counterinsurgency*. Oxford: Oxford University Press, 2010.

Kistiakowsky, George B. *A Scientist at the White House: The Private Diary of President Eisenhower's Special Assistant for Science and Technology*. Cambridge, Mass.: Harvard University Press, 1976.

Langguth, A. J. *Our Vietnam: The War, 1954–1975*. New York: Simon & Schuster, 2000.

Lansdale, Edward Geary. *In the Midst of Wars: An American's Mission to Southeast Asia*. New York: Fordham University Press, 1991.

Lawyer, L. C., Charles C. Bates, and Robert B. Rice. *Geophysics in the Affairs of Mankind: A Personalized History of Exploration Geophysics*. Tulsa: Society of Exploration Geophysicists, 2001.

Lukasik, Stephen J. *Advanced Research Projects Agency, 1966–1974*. Personal memoir, 2014.

McDougall, Walter A. *The Heavens and the Earth: A Political History of the Space Age*. New York: Basic Books, 1985.

Medaris, John B. *Countdown for Decision*. New York: Putnam, 1960.

Nashel, Jonathan. *Edward Lansdale's Cold War*. Amherst: University of Massachusetts Press, 2005.

Neyland, David L. *Virtual Combat: A Guide to Distributed Interactive Simulation*. Mechanicsburg, Pa.: Stackpole, 1997.

Nicolelis, Miguel. *Beyond Boundaries: The New Neuroscience of Connecting Brains with Machines—and How It Will Change Our Lives*. New York: Times/Henry Holt, 2011.

Norberg, Arthur L. *A History of the Information Processing Techniques Office of the Defense Advanced Research Projects Agency*. Minneapolis: Charles Babbage Institute, 1992.

O'Harrow, Robert. *No Place to Hide*. New York: Free Press, 2005.

Prehmus, Drew. *General Sam: A Biography of Lieutenant General Samuel Vaughan Wilson*. CreateSpace Independent Publishing Platform, 2013.

Redmond, Kent C., and Thomas M. Smith. *From Whirlwind to MITRE: The R&D Story of the SAGE Air Defense Computer*. Cambridge, Mass.: MIT Press, 2000.

Rhodes, Richard. *Dark Sun: The Making of the Hydrogen Bomb*. New York: Simon & Schuster, 1995.

Rich, Ben R., and Leo Janos. *Skunk Works: A Personal Memoir of My Years at Lockheed*. Boston: Little, Brown, 1994.

Richelson, Jeffrey. *The Wizards of Langley: Inside the CIA's Directorate of Science and Technology*. Boulder, Colo.: Westview Press, 2001.

Rohde, Joy. *Armed with Expertise: The Militarization of American Social Research During the Cold War*. Ithaca, N.Y.: Cornell University Press, 2013.

———. "The Last Stand of the Psychocultural Cold Warriors: Military Contract Research in Vietnam." *Journal of the History of the Behavioral Sciences* 47, no. 3 (2011): 232–50.

Roland, Alex. *Strategic Computing: DARPA and the Quest for Machine Intelligence, 1983–1993*. With Philip Shiman. Cambridge, Mass.: MIT Press, 2002.

Romney, Carl. *Detecting the Bomb: The Role of Seismology in the Cold War*. Washington, D.C.: New Academia, 2009.

Schlesinger, Arthur M. *A Thousand Days: John F. Kennedy in the White House*. Boston: Houghton Mifflin, 1965.

Seaborg, Glenn T., and Benjamin S. Loeb. *Kennedy, Khrushchev, and the Test Ban*. Berkeley: University of California Press, 1981.

Sheehan, Neil. *A Bright Shining Lie: John Paul Vann and America in Vietnam*. New York: Random House, 1988.

———. *A Fiery Peace in a Cold War: Bernard Schriever and the Ultimate Weapon*. New York: Random House, 2009.

Simon, Jack S. [Seymour Deitchman]. *War and Marriage: A 1960s Love Story*. CreateSpace Independent Platform, 2008.

Singer, Peter W. *Wired for War: The Robotics Revolution and Conflict in the Twenty-First Century*. New York: Penguin, 2010.

Sorensen, Theodore C. *Kennedy*. New York: Harper & Row, 1965.

Stark, Warren. *Many Faces Many Places: My Life Story*. BookSurge, 2001.

Stevenson, James P. *The $5 Billion Misunderstanding: The Collapse of the Navy's A-12 Stealth Bomber Program*. Annapolis, Md.: Naval Institute Press, 2001.

Sweetman, Bill, and James C. Goodall. *Lockheed F-117A: Operation and Development of the Stealth Fighter*. Osceola, Wis.: Motor International, 1990.

Taylor, Maxwell D. *Swords and Plowshares*. New York: W. W. Norton, 1972.

Thayer, Thomas C. *War Without Fronts: The American Experience in Vietnam*. Boulder, Colo.: Westview, 1985.

Townes, Charles H. *How the Laser Happened: Adventures of a Scientist*. New York: Oxford University Press, 1999.

Volmar, Axel. "Listening to the Cold War: The Nuclear Test Ban Negotiations, Seismology, and Psychoacoustics, 1958–1963." *Osiris* 28, no. 1 (2013): 80–102.

Waldrop, M. Mitchell. *The Dream Machine: J. C. R. Licklider and the Revolution That Made Computing Personal*. New York: Viking, 2001.

Wang, Zuoyue. *In Sputnik's Shadow: The President's Science Advisory Committee and Cold War America*. New Brunswick, N.J.: Rutgers University Press, 2009.

Ward, Bob. *Dr. Space: The Life of Wernher von Braun*. Annapolis, Md.: Naval Institute Press, 2005.

Weinberger, Sharon. "Defence Research: Still in the Lead?" *Nature* 451, no. 7177 (2008): 390–93.

———. *Imaginary Weapons: A Journey Through the Pentagon's Scientific Underworld*. New York: Nation Books, 2006.

Weiner, Tim. *Legacy of Ashes: The History of the CIA*. New York: Doubleday, 2007.

Whittle, Richard. *Predator: The Secret Origins of the Drone Revolution*. New York: Henry Holt, 2014.

York, Herbert F. *Arms and the Physicist*. New York: AIP Press, 1995.

———. *Making Weapons, Talking Peace: A Physicist's Odyssey from Hiroshima to Geneva.* New York: Basic Books, 1987.

———. *Race to Oblivion: A Participant's View of the Arms Race.* New York: Simon and Schuster, 1970.

CONGRESSIONAL HEARINGS

U.S. Congress. House. Committee on Appropriations. Subcommittee on Department of Defense Appropriations. *Department of Defense Ballistic Missile Programs.* 85th Cong., 1st sess., Nov. 20, 1957.

U.S. Congress. House. Committee on Astronautics and Space Exploration, Select. *Astronautics and Space Exploration.* 85th Cong., 2nd sess., April 15–18, 21–25, 28–30, May 1, 5, 7, 8, 18, 1958.

U.S. Congress. House. Subcommittee on Department of Defense Appropriations, Committee on Appropriations. *Department of Defense Appropriations for 1959. Part 6.* 85th Cong., 2nd sess., April 3, 14, 16–18, 21–25, 28–30, 1958.

U.S. Congress. Joint. Committee on Atomic Energy. *Developments in the Field of Detection and Identification of Nuclear Explosions (Project Vela) and Relationship to Test Ban Negotiations.* 87th Cong., 1st sess., July 25–27, 1961.

U.S. Congress. Senate. Committee on Space and Astronautics. *National Aeronautics and Space Act. Part 1.* 85th Cong., 2nd sess., May 6–8, 1958.

U.S. Congress. House. Committee on Appropriations. Subcommittee on Department of Defense Appropriations. *Military Construction Appropriations for 1959. Overall Program, Department of the Air Force.* 85th Cong., 2nd sess., June 16, July 3, 4, 8–11, 14–16, 1958.

U.S. Congress. Senate. Committee on Appropriations. *Department of Defense Appropriations for 1959.* 85th Cong., 2nd sess., June 6, 9–13, 16–18, 20, 23, 25–27, 30, July 2, 3, 7–9, 15, 16, 1958.

U.S. Congress. Senate. Committee on Space and Astronautics. *Nominations.* 85th Cong., 2nd sess., Aug. 14, 1958.

U.S. Congress. Senate. Committee on Aeronautical and Space Sciences. Subcommittee Investigating Preparedness, Committee on Armed Services. *Missile and Space Activities.* 86th Cong., 1st sess., Jan. 29, 30, 1959.

U.S. Congress. House. Committee on Science and Astronautics. *Missile Development and Space Sciences.* 86th Cong., 1st sess., Feb. 2–5, 9, 10, 17, 18, 24, March 2, 12, 1959.

U.S. Congress. House. Committee on Science and Astronautics. *Satellites for World Communication.* 86th Cong., 1st sess., March 3, 4, 1959.

U.S. Congress. House. Committee on Science and Astronautics. *Space Propulsion.* 86th Cong., 1st sess., March 16–20, 23, 1959.

U.S. Congress. Senate. Committee on Aeronautical and Space Sciences. Subcommittee on Government Organization for Space Activities. *Investigation of Governmental Organization for Space Activities.* 86th Cong., 1st sess., March 24, 26, April 14, 15, 22–24, 29, May 7, 1959.

U.S. Congress. House. Committee on Appropriations. Subcommittee on Department of Defense Appropriations. *Department of Defense Appropriations for 1960. Part 6.* 86th Cong., 1st sess., April 14–16, 20–22, 24, 27, 1959.

U.S. Congress. House. Committee on Science and Astronautics. *Review of the Space Program. Part 1.* 86th Cong., 2nd sess., Jan. 20, 22, 25–29, Feb. 1–5, 1960.

U.S. Congress. House. Committee on Science and Astronautics. *To Amend the National Aeronautics and Space Act of 1958.* 86th Cong., 2nd sess., March 8–10, 14–17, 21, 22, 24, 28–31, April 4, 1960.

U.S. Congress. House. Committee on Appropriations. Subcommittee on Department of Defense Appropriations. *Department of Defense Appropriations for 1961. Part 6: Research, Development, Test, and Evaluation.* 86th Cong., 2nd sess., March 9, 11, 14–18, 1960.

U.S. Congress. Senate. Committee on Appropriations. *Department of Defense Appropriations for 1961. Part 2.* 86th Cong., 2nd sess., March 21–25, 28–30, April 1, 4, 12, May 18, 24–26, 1960.

U.S. Congress. Joint. Committee on Atomic Energy. Special Joint Subcommittee on Radiation and Subcommittee on Research and Development. *Technical Aspects of Detection and Inspection Controls of a Nuclear Weapons Test Ban. Part 1 of 2 Parts.* 86th Cong., 2nd sess., April 19–22, 1960.

U.S. Congress. House. Committee on Science and Astronautics. *Research and Development for Defense.* 87th Cong., 1st sess., Feb. 16, 17, 20, 22, 1961.

U.S. Congress. House. Committee on Appropriations. Subcommittee on Department of Defense Appropriations. *Department of Defense Appropriations for 1962. Part 4: Research, Development, Test, and Evaluation.* 87th Cong., 1st sess., March 23, April 20, 21, 24–26, May 1, 2, 1961.

U.S. Congress. Senate. Committee on Appropriations. *Department of Defense Appropriations for 1962.* 87th Cong., 1st sess., April 18–21, 26, 27, May 2, 3, 5, 9, 10, June 7–9, 13, July 10, 11, 18, 19, 26, 1961.

U.S. Congress. House. Committee on Appropriations. Subcommittee on Department of Defense Appropriations. *Department of Defense Appropriations for 1963.* 87th Cong., 2nd sess., March 19–23, 1962.

U.S. Congress. Senate. Committee on Appropriations. *Department of Defense Appropriations for 1963.* 87th Cong., 2nd sess., Feb. 14–16, 27, 28, March 1, 2, 5, 13, 28–30, April 2–6, May 15–18, 22–24, 1962.

U.S. Congress. Joint. Committee on Atomic Energy. *Developments in Technical Capabilities for Detecting and Identifying Nuclear Weapons Tests.* 88th Cong., 1st sess., March 5–8, 11, 12, 1963.

U.S. Congress. House. Committee on Appropriations. Subcommittee on Department of Defense Appropriations. *Department of Defense Appropriations for 1964. Part 6.* 88th Cong., 1st sess., March 29, April 3, May 2, 3, 6–10, 13–17, 20, 1963.

U.S. Congress. House. Committee on Appropriations. *Department of Defense Appropriations for 1965. Part 5: Research, Development, Test, and Evaluation, Appropriation Language, Testimony of Members of Congress, Organizations, and Interested Individuals.* 88th Cong., 2nd sess., March 6, 11–13, 16–19, 1964.

U.S. Congress. Senate. Committee on Appropriations. *Department of Defense Appropriations for 1966. Part 5: Research, Development, Test, and Evaluation.* 89th Cong., 1st sess., March 30, 31, April 5, 7–9, 13, 1965.

U.S. Congress. House. Committee on Armed Services. *Hearings on FY67 Defense Research, Development, Test, and Evaluation Program.* 89th Cong., 2nd sess., Jan. 24–26, Feb. 21, 23–25, 28, March 2, 3, 1966.

U.S. Congress. House. Committee on Science and Astronautics. *Report of the Committee on Science and Astronautics.* 89th Cong., 2nd sess., Jan. 24, 1966.

U.S. Congress. House. Committee on Science and Astronautics. *Report of the Committee on Science and Astronautics.* 89th Cong., 2nd sess., Jan. 24, 1966.

U.S. Congress. Committee on Foreign Affairs. Subcommittee on International Organizations and Movements. *Behavioral Sciences and National Security.* 89th Cong., 2nd sess., Jan. 25, 1966.

U.S. Congress. Joint. Committee on Atomic Energy. *Annual Report to Congress of the Atomic Energy Commission for 1965.* 89th Cong., 2nd sess., Jan. 31, 1966.

U.S. Congress. House. Committee on Science and Astronautics. *Message from the President of the United States Transmitting Report to Congress on Aeronautics and Space Activities, 1965.* 89th Cong., 2nd sess., Jan. 31, 1966.

U.S. Congress. House. Committee on Appropriations. Subcommittee on Independent Offices Appropriations. *Independent Offices Appropriations for 1967.* 89th Cong., 2nd sess., Feb. 1, 2, March 8, 9, 16, 17, 21, 28, 30, 31, April 4, 5, 20, 1966.

U.S. Congress. House. Committee on Appropriations. Subcommittee on Department of Defense Appropriations. *Department of Defense Appropriations for 1967. Part 1.* 89th Cong., 2nd sess., Feb. 14, 17, 28, 1966.

U.S. Congress. Senate. Committee on Appropriations. Subcommittee on Defense Appropriations. *Military Procurement Authorizations for FY67.* 89th Cong., 2nd sess., Feb. 23, 25, 28, March 8–10, 24, 25, 29–31, 1966.

U.S. Congress. House. Committee on Science and Astronautics. Subcommittee on Advanced Research and Technology. *1967 NASA Authorization.* 89th Cong., 2nd sess., Feb. 23–25, March 1–3, 7, 8, 1966.

U.S. Congress. Senate. Committee on Appropriations. Subcommittee on Department of Defense Appropriations. *Department of Defense Appropriations for FY67.* 89th Cong., 2nd sess., Feb. 23, 25, 28, March 8–10, 24, 25, 29–31, 1966.

U.S. Congress. Senate. Committee on Aeronautical and Space Sciences. *NASA Authorization for FY67.* 89th Cong., 2nd sess., Feb. 28, March 1–4, 1966.

U.S. Congress. House. Committee on Appropriations. Subcommittee on Department of Defense Appropriations. *Department of Defense Appropriations for 1967.* 89th Cong., 2nd sess., April 5, 6, 18–21, 26, 27, 1966.

U.S. Congress. House. Committee on Armed Services. *United States Defense Policies in 1965.* 89th Cong., 2nd sess., April 6, 1966.

U.S. Congress. House. Committee on Armed Forces. *Authorizing Defense Procurement.* 89th Cong., 2nd sess., May 16, 1966.

U.S. Congress. Senate. Committee on Aeronautical and Space Sciences. *Staff Report Prepared for the Use of the Committee on Aeronautical and Space Sciences.* 89th Cong., 2nd sess., May 19, 1966.

U.S. Congress. House. Committee on Appropriations. Subcommittee on Department of Defense Appropriations. *Department of Defense Appropriations for 1968. Part 1.* 90th Cong., 1st sess., Feb. 27, 28, March 1, 2, 1967.

U.S. Congress. House. Committee on Appropriations. Subcommittee on Department of Defense Appropriations. *Department of Defense Appropriations for 1968. Part II.* 90th Cong., 1st sess., March 6, 9, 15, 1967.

U.S. Congress. House. Committee on Appropriations. Subcommittee on Department of Defense Appropriations. *Department of Defense Appropriations for 1968. Part III.* 90th Cong., 1st sess., March 20, 1967.

U.S. Congress. House. Committee on Armed Services. *Hearings on FY68 Defense Research, Development, Test, and Evaluation Program Hearing.* 90th Cong., 1st sess., April 17–20, 24, 1967.

U.S. Congress. Senate. Committee on Aeronautical and Space Sciences. *NASA Authorization for FY68. Part 2.* 90th Cong., 1st sess., April 21, 25–27, 1967.

U.S. Congress. House. Committee on Appropriations. Subcommittee on Defense Appropriations. *Department of Defense Appropriations for 1968.* 90th Cong., 1st sess., April 26, 1967.

U.S. Congress. House. Committee on Armed Services. Subcommittee on the M-16 Rifle Program. *Hearings Before the Special Subcommittee on the M-16 Rifle Program.* 90th Cong., 1st sess., May 15, 16, 31, June 21, July 25–27, Aug. 8, 9, 22, 1967.

U.S. Congress. Senate. Committee on Appropriations. *Department of Defense Appropriations for FY68. Part 2.* 90th Cong., 1st sess., July 12, 13, 17–19, 1967.

U.S. Congress. House. Committee on the Armed Services. Subcommittee on the M-16 Rifle Program. *Full Committee Consideration of H.R. 12910, H.R. 11767, H.R. 4903, S. 223, H.R. 4772, H.R. 5943, and H.R. 9796, to Authorize the Extension of Certain Naval Vessel Loans Now in Existence Hearing.* 90th Cong., 1st sess., Sept. 26, 1967.

U.S. Congress. Senate. Committee on Armed Services. *Authorization for Military Procurement, Research, and Development, FY69, and Reserve Strength.* 90th Cong., 1st sess., Feb. 1, 2, 5, 7, 15, 16, 20, 21, 27–29, March 4, 14, 1968.

U.S. Congress. House. Committee on Appropriations. Subcommittee on Department of Defense Appropriations. *Department of Defense Appropriations for 1968. Part 1.* 90th Cong., 2nd sess., Feb. 14, 19–21, 26, 28, 29, March 1, 1968.

U.S. Congress. House. Committee on Appropriations. Subcommittee on Department of Defense Appropriations. *Department of Defense Appropriations for 1968.* 90th Cong., 2nd sess., March 4–7, 12, 1968.

U.S. Congress. Senate. Committee on Foreign Relations. *Defense Department Sponsored Foreign Affairs Research.* 90th Cong., 2nd sess., May 9, 1968.

U.S. Congress. Senate. Committee on Appropriations. *Department of Defense Appropriations for FY69. Part 1.* 90th Cong., 2nd sess., May 6–8, 17, 1968.

U.S. Congress. Senate. Committee on Appropriations. *Department of Defense Appropriations for FY69. Part 4.* 90th Cong., 2nd sess., May 23, June 11, July 10, 12, 15, 1968.

U.S. Congress. House. Committee on Science and Astronautics. Subcommittee on Science, Research, and Development. *1970 National Science Foundation Authorizations.* 91st Cong., 1st sess., Jan. 1, 1969.

U.S. Congress. House. Committee on Science and Astronautics. Subcommittee on Science, Research, and Development. *Institutional Grants Bill (H.R. 35).* 91st Cong., 1st sess., Feb. 18–20, 25–27, 1969.

U.S. Congress. Senate. Committee on Foreign Relations. Subcommittee on International Organization and Disarmament Affairs. *Strategic and Foreign Policy Implications of ABM Systems. Part 1.* 91st Cong., 1st sess., March 6, 11, 13, 21, 26, 28, 1969.

U.S. Congress. House. Committee on Appropriations. Subcommittee on Department of Defense Appropriations. *Department of Defense Appropriations for 1970. Part 1.* 91st Cong., 1st sess., March 11, 12, 17–20, 24, 25, 1969.

U.S. Congress. House. Committee on Science and Astronautics. Subcommittee on Science, Research, and Development. *National Science Foundation Authorization Volume 1.* 91st Cong., 1st sess., March 17, 18, 20, 24–28, April 1, 1969.

U.S. Congress. House. Committee on Education and Labor. Subcommittee on Education. *National Science Research Data Processing and Information Retrieval System.* 91st Cong., 1st sess., April 29, 30, 1969.

U.S. Congress. Senate. Committee on Armed Services. *Authorization for Military Procurement, Research, and Development.* 91st Cong., 1st sess., April 22, 23, 29, 30, May 13–15, June 3, 4, 1969.

U.S. Congress. Senate. Committee on Aeronautical and Space Sciences. *Authorization for FY70. Part 2.* 91st Cong., 1st sess., May 1, 6, 9, 1969.

U.S. Congress. House. Committee on Appropriations. Subcommittee on Department of Defense Appropriations. *Department of Defense Appropriations for 1970. Part 6.* 91st Cong., 1st sess., April 30, June 9, 16–19, 23, 24, 1969.

U.S. Congress. Senate. Committee on Appropriations. *Department of Defense Appropriations for FY70.* 91st Cong., 1st sess., June 10, 12, 13, 16, Sept. 15–17, 25, 1969.

U.S. Congress. Senate. Committee on Appropriations. *Department of Defense Appropriations for FY70. Part 2.* 91st Cong., 1st sess., June 17–20, 1969.

U.S. Congress. House. Committee on Appropriations. Subcommittee on Department of Defense Appropriations. *Department of Defense Appropriations for 1970. Part 5: Research, Development, Test, and Evaluation.* 91st Cong., 1st sess., July 1, 2, 7–11, 14, 17, 1969.

U.S. Congress. House. Committee on Armed Services. Subcommittee on Air Defense of Southeastern U.S. *Hearings on Air Defense of Southeastern U.S.* 91st Cong., 1st sess., Nov. 5–7, 12–14, 28, Dec. 5, 1969.

U.S. Congress. House. Committee on Appropriations. Subcommittee on Department of Defense Appropriations. *Department of Defense Appropriations for 1971. Part 5.* 91st Cong., 2nd sess., Feb. 10, March 19, April 6–8, 13, 15–16, 20–22, 1970.

U.S. Congress. Senate. Committee on Appropriations. Subcommittee on Department of Defense Appropriations. *Department of Defense Appropriations for FY71. Part 1.* 91st Cong., 2nd sess., April 13, 14, 15, May 11–13, 15, 20, 1970.

U.S. Congress, Senate. Committee on the Judiciary. Subcommittee on Constitutional Rights. *Surveillance Technology: Joint Hearings Before the Subcommittee on Constitutional Rights of the Committee on the Judiciary and the Special Subcommittee on Science, Technology, and Commerce of the Committee on Commerce,* 94th Cong., 1st Sess., June 23, Sept. 9, 10, 1975.

U.S. Congress. Senate. *Radiation Health and Safety: Hearings Before the Committee on Commerce, Science, and Transportation.* 95th Cong., 1st sess., June 16, 17, 27, 28, 29, 1977.

U.S. Congress. House. Committee on Armed Services. *Department of Defense Authorization of Appropriations for Fiscal Year 1986 and Oversight of Previously Authorized Programs, Part 4 of 7. Research, Development, Test, and Evaluation-Title II.* 99th Cong., 1st sess., March 6, 7, 19, 20, 21, 26, 27, 28, 29, April 2, 3, 4, 16, 1985.

U.S. Congress. House. Committee on Science, *The Future of Computer Science Research in the U.S.* 109th Cong., 1st sess., May 12, 2005.

GOVERNMENT PUBLICATIONS AND REPORTS

Note: DARPA over its history has supported thousands of reports, and countless more publications relate to its projects. I have elected to list below only those government publications that have the most relevance to the subjects covered in this book or that provide unique information about the agency.

Battelle Memorial Institute. *Advanced Research Projects Project Agile Quarterly Report. April 1–30 June 1963.* Alexandria, Va.: Defense Technical Information Service, 1963.

———. *Advanced Research Projects Project Agile Semiannual Report. July 1–Dec. 31, 1963.* Alexandria, Va.: Defense Technical Information Service, 1963.

Betts, R. R., and Frank H. Denton. *An Evaluation of Chemical Crop Destruction in Vietnam.* Santa Monica, Calif.: Rand, 1967.

Buckingham, William A. *Operation Ranch Hand: The Air Force and Herbicides in Southeast Asia.* Washington, D.C.: Office of Air Force History, U.S. Air Force, 1981.

Byron, E. V. *Operational Procedure for the Project Pandora Microwave Test Facility*. Silver Spring, Md.: Johns Hopkins University Applied Physics Laboratory, Oct. 1966.

Chandler, Craig C., and Jay R. Bentley. *Forest Fire as a Military Weapon*. Washington, D.C.: U.S. Department of Agriculture, Forest Service, 1970.

CIRADS III Proceedings. Washington, D.C.: Advanced Research Projects Agency, 1968.

Combat Development and Test Center Viet Nam, Monthly Report. June 1963.

Cosby, L. Neale. *SIMNET: An Insider's Perspective*. Alexandria, Va.: Institute for Defense Analyses, 1995.

Darrow, Robert Arthur, George B. Truchelut, and Charles M. Bartlett. *Oconus Defoliation Test Program*. Frederick, Md.: U.S. Army Biological Center, 1966.

Davison, W. Phillips. *User's Guide to the Rand Interviews in Vietnam*. Santa Monica, Calif.: Rand, 1972.

Delmore, Fred J. *Review and Evaluation of ARPA/OSD "Defoliation" Program in South Vietnam. Research Phase: 15 July 1961–12 January 1962*. Alexandria, Va.: Defense Technical Information Center, 1962.

Donnell, John C., and Gerald Cannon Hickey. *The Vietnamese "Strategic Hamlets": A Preliminary Report*. Santa Monica, Calif.: Rand, 1962.

Donnell, John C., Guy J. Pauker, and Joseph Jermiah Zasloff. *Viet Cong Motivation and Morale in 1964: A Preliminary Report*. Santa Monica, Calif.: Rand, 1965.

Elliott, Duong Van Mai. *RAND in Southeast Asia: A History of the Vietnam War Era*. Santa Monica, Calif.: Rand, 2010.

Godel, William H. *Report on R&E Far East Survey, October–December, 1960*. Washington, D.C.: Advanced Research Projects Agency, 1960.

Gorman, Paul. *SuperTroop via I-Port: Distributed Simulation Technology for Combat Development and Training Development*. Alexandria, Va.: Institute for Defense Analyses, 1990.

Hickey, Gerald Cannon. *Accommodation in South Vietnam: The Key to Sociopolitical Solidarity*. Santa Monica, Calif.: Rand, 1967.

Huff, Lee W., and Richard G. Sharp. *The Advanced Research Projects Agency, 1958–1974*. Washington, D.C.: Richard J. Barber Associates, 1975.

Institute for Defense Analyses. JASON Division. *Air-Supported Anti-infiltration Barrier*. Study S-255, Aug. 1966.

———. *Project: Seesaw*. Study S-307, 1968.

Joint Thai-U.S. Military Research Development Center. *Thailand Quarterly Report, Oct. 1 to Dec. 31, 1963*. Advanced Research Projects Agency, 1963.

Jones, Norman H. *Support Capabilities for Limited War in Iran*. Santa Monica, Calif.: Rand, 1963.

Nierenberg, William A. "DCPG—the Genesis of the Concept." *IDA Journal of Defense Research, Series B, Tactical Warfare* (Fall 1969).

The 9/11 Commission Report: Final Report of the National Commission on Terrorist Attacks upon the United States: Official Government Edition. Washington, D.C.: Government Printing Office, 2004.

Office of the Assistant Secretary of Defense for Research and Engineering. *Report of the Ad Hoc Advisory Group on Psychology and the Social Sciences*, Dec. 19, 1957.

Orlansky, Jesse, and Jack Thorpe. *73 Easting: Lessons Learned from Desert Storm via Advanced Distributed Simulation Technology*. Alexandria, Va.: Institute for Defense Analyses, 1992.

OSD/ARPA. *Research Development Field Unit. Report of Task No. 13A Test of Armalite Rifle. AR-15*. By William P. Brooks Jr. Alexandria, Va.: Defense Documentation Center for Scientific and Technical Information, July 31, 1962.

Perry, Robert L. *History of Satellite Reconnaissance: The Perry Gambit & Hexagon Histories.* Vol. 1. Chantilly, Va: Center for the Study of National Reconnaissance, 2012.

Project Agile. *ARPA, Quarterly report: April 1–June 30 1963.* Battelle Memorial Institute, 1963.

Rand Vietnam Interview Series H: Villager's Impressions of Herbicide Operations. Santa Monica, Calif.: Rand, 1972.

Rand Vietnam Interview Series XN: Effects of Bombing of North Vietnam. Alexandria, Va.: Defense and Documentation Center for Scientific and Technical Information, 1972.

Reed, Sidney G., Richard H. Van Atta, and Seymour J. Deitchman. *DARPA Technical Accomplishments: An Historical Review of Selected DARPA Projects.* 3 vols. Alexandria, Va.: Institute for Defense Analyses, 1990.

Report of the Defense Science Board Task Force on the Investment Strategy for DARPA. Washington, D.C.: Office of the Undersecretary of Defense for Acquisition and Technology, 1999.

Ruffner, Kevin Conley. *Corona: America's First Satellite Program.* Washington, D.C.: History Staff, Center for the Study of Intelligence, Central Intelligence Agency, 1995.

SIMNET: Advanced Technology for the Mastery of Warfighting. Cambridge, Mass.: BBN Laboratories, 1985.

Tietzel, F. A., M. R. VanderLind, and J. H. Brown Jr. *Summary of ARPA-ASO, TTO Aerial Platform Programs, Vol. II: Remotely Piloted Helicopters.* Battelle Report. Alexandria, Va.: Defense Technical Information Center, July 1975.

U.S. Department of Agriculture Forest Service. *Operation Pink Rose: Final Report.* Advanced Research Projects Agency, May 1967.

U.S. Department of State. *Foreign Relations of the United States, Vietnam, 1961–1963.* Vol. 1. Washington, D.C.: Government Printing Office, 1988.

Van Atta, Richard H., and Michael J. Lippitz. *Transformation and Transition: DARPA's Role in Fostering an Emerging Revolution in Military Affairs.* 2 vols. Alexandria, Va.: Institute for Defense Analyses, 2003.

Warren, William F. *A Review of the Herbicide Program in South Vietnam.* FPO San Francisco: Commander in Chief, Scientific Advisory Group, 1968.

Watson, Robert J. *Into the Missile Age.* Washington, D.C.: Historical Office, Office of the Secretary of Defense, 1997.

Index

Page numbers beginning with 377 refer to end notes.

ILLUSTRATION CREDITS

Page 1
Top left: National Archives and Records Administration, St. Louis.
Top right: Getty Images
Center: Getty Images
Bottom: Getty Images

Page 2
Top: Lawrence Livermore National Laboratory
Center: National Archives and Records Administration, College Park
Bottom, left and right: National Archives and Records Administration, College Park

Page 3
Top: Karen Tweedy-Holmes
Center: Seymour Deitchman Family
Bottom: Stephen J. Lukasik

Page 4
Top, left and right: George H. Lawrence Family
Bottom, left and right: National Archives and Records Administration, College Park

Page 5
Top: Charles Bates
Center, left and right: National Archives and Records Administration, College Park
Bottom: DARPA

Page 6
Top: Seymour Deitchman Family
Center: Seymour Deitchman Family
Bottom: National Archives and Records Administration, College Park

Page 7
Top: DARPA
Center, left and right: National Archives and Records Administration, College Park
Bottom: National Archives and Records Administration, College Park

Page 8
Top: National Archives and Records Administration, College Park
Bottom: Allen Atkins

A NOTE ABOUT THE AUTHOR

Sharon Weinberger is the national security editor at *The Intercept* and the author of *Imaginary Weapons: A Journey Through the Pentagon's Scientific Underworld*. She was a 2015–2016 fellow at the Radcliffe Institute for Advanced Study at Harvard University. She has also held fellowships at the Woodrow Wilson International Center for Scholars, MIT's Knight Science Journalism program, the International Reporting Program at Johns Hopkins School of Advanced International Studies, and Northwestern University's Medill School of Journalism. She has written on military science and technology for *Nature, BBC, Discover, Slate, Wired,* and *The Washington Post,* among others.

A NOTE ABOUT THE TYPE

This book was set in Minion, a typeface produced by the Adobe Corporation specifically for the Macintosh personal computer, and released in 1990. Designed by Robert Slimbach, Minion combines the classic characteristics of old-style faces with the full complement of weights required for modern typesetting.

Composed by
North Market Street Graphics, Lancaster, Pennsylvania

Printed and bound by
Berryville Graphics, Berryville, Virginia

Designed by
Maggie Hinders